T0326489

Safety and Environmental Management

Safety and Environmental Management

Third Edition

Frank R. Spellman

BERNAN
Lanham • Boulder • New York • London

Published by Bernan Press
A wholly owned subsidiary of The Rowman & Littlefield Publishing Group, Inc.
4501 Forbes Boulevard, Suite 200, Lanham, Maryland 20706
www.rowman.com

Unit A, Whitacre Mews, 26–34 Stannary Street, London SE11 4AB, United Kingdom

Copyright © 2015 by Bernan Press

All rights reserved. No part of this book may be reproduced in any form or by any electronic or
mechanical means, including information storage and retrieval systems, without written permission
from the publisher, except by a reviewer who may quote passages in a review.

British Library Cataloguing in Publication Information Available

Library of Congress Cataloging-in-Publication Data

ISBN 13 : 978-1-59888-769-3
E-ISBN : 978-1-59888-770-9

♾™ The paper used in this publication meets the minimum requirements of American National
Standard for Information Sciences—Permanence of Paper for Printed Library Materials, ANSI/
NISO Z39.48-1992.

Printed in the United States of America

Contents

Preface   vii

1   Setting the Stage   1

2   Management Aspects   25

3   Safety and Environmental Management Terminology   37

4   Accident Investigation   51

5   Hazard Communication and Hazardous Waste   63

6   Emergency Response and Workplace Security   93

7   Fire & Hot Work Safety   105

8   Lockout/Tagout   127

9   Confined Space Safety   137

10   Personal Protective Equipment   179

11   Respiratory Protection   189

12   Hearing Safety   207

13   Electrical Safety   221

14   Ergonomics   231

15   Machine Guarding   237

16   Workplace Environmental Concerns   253

17   Safety and Health Training   339

Index   345

Preface

In the First and Second Editions of this text the point was made that everyone wants a safe and environmentally healthy workplace, but what each company and person is willing to do to achieve this important and worthwhile objective can vary a great deal. One thing is certain; it is impossible to comply with workplace safety and environmental regulations without management. Safety and environmental compliance go hand-in-hand with management.

Building on the success of the well-received earlier editions and on the suggestions provided by numerous users, *Safety and Environmental Management*, Third Edition, has been expanded and completely updated. We still cover the major elements required for a company to be successful in protecting its employees from physical and environmental hazards, but we have also incorporated updated and current information and regulatory requirements and included additional information on aspects that should be included in any organization's safety plan. In addition to Hazard Communication, Accident Investigating, Emergency Response, Fire Prevention, Ergonomics, Confined Space and Safety, and Lockout/Tagout, the additional elements include:

- Management Aspects
- Personal Protective Equipment (PPE)
- Respiratory Protection
- Hot Work Procedures
- Electrical
- Noise
- Thermal Stress
- First Aid/Bloodborne Pathogens

What is the difference between safety management and environmental management? A machine guard is a safety management issue, and airborne contaminants such as asbestos are environmental management issues. However, some safety management and environmental management aspects, such as those associated with paint spray and sand-blasting areas and hot work operations (i.e., welding, brazing, torch cutting, and grinding), are not so easy to classify. Some situations may be both a safety and environmental hazard.

This book will draw the following line between safety management and environmental management: safety management deals with managing acute hazards, whereas environmental management deals with chronic hazards. An acute effect is a sudden reaction to a severe condition (e.g., a machine pinch point severing a worker's fingers, hand, or arm, etc.); a chronic effect is a long-term deterioration due to a prolonged exposure to a milder adverse condition (e.g. breathing in hydrocarbon fumes that make the victim dizzy, sick, unconscious, or worse). Some hazards can be both acute and chronic. Noise is a good example because sudden exposure to high decibels or long-term exposure to 90–100 decibels can cause both short-term and long-term damage.

Degree of hazard is another point to be considered with regard to safety and environmental management; safety management is concerned with immediate hazards that are grave (i.e., hazards that cause fatalities on the job). The irony is that more fatalities result from environmental hazards than safety hazards, but the statistics do not reflect this difference because the environmental health fatalities are latent and are often never diagnosed. A good example is exposure to deadly asbestos. The worker may be exposed today but not become sick until 20–30 years from exposure time.

In this totally revised edition, we examine the pertinent standards of 29 CFR 1910, 1926 and CFR 40 environmental elements one by one with nontechnical, implementation-friendly explanations of the requirements and how to implement and comply with them. Most importantly, this book provides answers to a broad range of compliance questions, including who is obligated to observe the law, what the OSHA/EPA compliance obligations are, and how state OSHA standards compare to federal OSHA standards.

Finally, the reader is advised that this book is presented in conversational and engaging style, format, and tone. The critics say this is a Spellman trademark. This is to ensure the requirements of OSHA are clearly communicated to the reader.

Frank R. Spellman
Norfolk, VA

Chapter 1

Setting the Stage

It has been said that the attitude of the worker is what makes or breaks safety and environmental compliance in the workplace. Moreover, if you add enforcement, supervision, and awards for compliance, that is all you need to have a safe and environmentally sound work environment. Based on experience, I have found that these elements and others indeed contribute to making a workplace safe. But the mark is missed if the number one element is ignored and left out of the game plan to protect workers in the workplace. The number one element can be summed in one word:

Awareness! Awareness! Awareness!

Think about it. How can any worker be expected to be safe if he or she is not aware of the hazard, whether it is safety, environmental, or otherwise?

SAFETY AND ENVIRONMENTAL MANAGEMENT PRECEPTS

Whenever a person sees or hears the term "precepts" they almost immediately think of religion. "Am I about to get a lecture in religion" they might ask. Well, if you think that protecting the safety and environmental health of workers is a moral and ethical obligation, then, yes, you could say that complying for safety and health is sort of a religion.

Management of each workplace must decide at what level, along a broad spectrum, the safety and environmental health effort will be aimed, directed, and accomplished. Basically, this presents a continuum between personal freedom and individual responsibility.

Personal freedom? Yes, absolutely. When you mandate that anyone is to follow certain guideline, rules, regulations, etc., they may feel that their freedom to do their own thing is being infringed upon. And when it comes to guarding their safety and environmental health, they are right. But since the employer is like the captain of a ship, fully and unequivocally responsible for the health and welfare of his or her charges, they certainly expect those under them to abide by certain guidelines. In an industry, it is the safety manager, safety director, industrial hygienist (for our purposes this text will use the designation safety and health professional, recognizing the dual nature of the job), who sets the tone of the safety and environmental health program within the organization.

Back in the day, it was the safety and health professional who typically was a good neighbor Sam-type, a purveyor of public relations activities, such as posting motivational notes, labels and signs and compiling statistics. This role took a dramatic shift, however, in 1970 when the Occupational Safety and Health Act (OSH Act) was signed into law and created the Occupations Safety and Health Administration (OSHA), a federal agency whose regulations have had a large impact on the role of today's safety professional. It is interesting to note that the safety profession, especially the environmental health field, took a mega-leap forward when OSHA came into being.

Did You Know?

Safety and environmental health often are viewed largely as a simple matter of applying specific routines. In many cases the routines are repeated regularly despite obvious signs of their inadequacies. Greatly needed is an understanding that the sources of harm, which the safety and environmental health professional should be able to control, have basic origins although their consequences will differ in character and severity. This view furnishes the realization that hazards are not simply the agents most closely identified with injuries. Merely regulating them is not the sure way to limit their effect (Grimaldi & Simonds, 1989).

The purpose of this book is to provide tools and guidelines to safety and health professionals to help them execute their growing responsibilities and to deal with those hazards that are fixable. Dealing with upper management, workers, and applicable standards is one of the greatest challenges facing today's safety and health professional.

Did You Know?

For those Safety and Health Professionals who advocate for elimination of all workplace hazard, a word to the wise. Such a goal, although laudable, is unattainable and to reach for it is a poor plan because it ignores the need for 'fixing' the fixable. Only those hazards that are recognized, according to law, must be acted upon and corrected as soon as possible.

WHAT DOES A SAFETY AND ENVIRONMENTAL MANAGER DO?

When we ask new acquaintances standard questions (where they live, where they are from, etc.), sooner or later in the conversation, we'll ask the question, "What do you do for a living?"

If the answer is "I'm a professional," we might be impressed, but our question isn't completely answered. Professional? What type of professional? Many different professional disciplines or specialties are possible, from college professors, clergy, nurses, dentists, pharmacists, accountants, firefighters, police officers, botanists, and locomotive operators to "flight engineers" and "sanitation professionals."

So what kind of professional is our safety and environmental health professional? Unless he or she answers with a specific professional specialty, the natural tendency will be for the questioner to ask, "What kind of professional are you?" or "What type of professional service do you do?"

Our respondent might answer, "I am a safety and environmental health professional." Unless you are familiar with safety and environmental health, you might be taken aback by this statement. Safety and environmental health professional? What is that?

What a safety and environmental professional, or manager, is and does is what this text is all about. This text works to complete (and to thoroughly explain) the above paraphrased statement by Grimaldi and Simonds. For now, we will simply continue their statement by adding our own words:

To regulate the effect of safety and environmental hazards it is necessary to employ a means of removing or controlling the causes responsible for the presence of injurious agents. This is the essence of the practice of safety and environmental management, a true profession, for sure.

SAFETY AND ENVIRONMENTAL MANAGEMENT: A CLOSER LOOK

Simply put, the safety and environmental manager is a Jack or Jill of many engineering, scientific, and other professions. Does this mean that the safety and environmental manager is an expert in all engineering, scientific, and the other disciplines? Good question. The short answer is no. The long answer depends on your definition of "expert." If you define expert as does Webster (Expert—A person with a high degree of skill in or knowledge of a certain subject), you can answer yes, because trained and experienced safety and environmental health managers should possess a high degree of skill in or knowledge in a wide variety of disciplines. But this really doesn't answer our question, does it? To better answer the original question, "Is the safety and environmental manager an expert in all engineering, scientific and other disciplines," two words in this question need to be changed to render a more accurate answer. The words "expert" and "all" should be changed—"Expert" status denotes that the safety and environmental manager knows everything about safety and environmental health—a goal no one person could attain. To accomplish this amazing feat, the safety and environmental manager would have to be an "expert" in law, engineering, technical equipment, manufacturing processes, behavioral sciences, management, health sciences, finance, and insurance—as they relate to safety and environmental issues, to name just a few fields. While true that the safety and environmental manager is typically an "expert" in a particular area of safety and environmental health management, it is also true that he or she can't be expected to be an "expert" in everything. This point is amplified even further when we remember the following famous quotation "An expert is someone who has found all the mistakes in a very narrow field" (Niels Bohr, as quoted by Edward Teller). At this point and time, we are unfamiliar with anyone who has found all the mistakes in the practice of Safety and Environmental Management.

Let's take a look at what the safety and environmental manager really is. The safety and environmental manager is knowledgeable in many aspects of engineering, scientific, safety, environmental, and other related disciplines. While the safety and environmental manager is devoted to the application of engineering and scientific principles and methods to eliminate and control hazards, he or she must know a lot about many different engineering and scientific fields. The safety and environmental manager specializes in the recognition and control of hazards through the use of knowledge and skills related to engineering, science, and other disciplines.

In short, the safety and environmental manager must be a "generalist" with a firm grasp on how the many fields with safety and environmental health concerns connect with his or her area of responsibility and influence.

Did You Know?

Some safety and environmental managers are also personnel managers, and even more frequently they report to the personnel manager. Based on experience, both of these scenarios are not recommended. The safety and environmental manager job is too big for a collateral duty. The safety and environmental manager should report directly to the top manager in the company.

Why A Generalist?

Figure 1.1 illustrates why the safety and environmental manager should be a generalist—the importance of achieving an interdisciplinary education mixed with experience cannot be stressed enough. In addition to a "generalized" education honed smooth by years of on-the-job experience, the safety and environmental manager must know what role other organizational professionals, practitioners, and employees play in the organization to properly identify hazards, develop effective solutions to safety problems, and achieve safe operations and systems.

As a generalist, the safety and environmental manager functions to coordinate and facilitate the actions of other knowledgeable personnel in applying safety principles to particular problems.

Education and the Safety and Environmental Manager

What should a training package for a course of study in safety and environmental management include? Although professional education provides the minimum standard for professional credibility, having a degree certainly does not indicate that someone is smarter—or more effective professionally—than someone who lacks a degree. What it does mean is that the degree holder has accomplished a series of standardized

Figure 1.1   Safety and Environmental Management Interacts with other Entities/Functions within an Organization

tasks by which he or she may be measured. What a college education really does is provide the graduate with the tools he or she needs to do research, "to find answers." When you get right down to it, this is a worthy goal of any good advanced training course, to teach students how to problem-solve and to provide them with the tools that allow them to find pertinent answers independently.

So, why is a college education necessary? Can't an individual learn everything they need to know about safety and environmental management from on-the-job experience?

We can all learn from on-the-job experience. Absolutely nothing can replace real experience. But the question was: "Can't someone learn 'everything' they need to know about safety and environmental management from on-the-job experience?" A seasoned safety and environmental manager with many years of experience in the safety and environmental management profession would probably answer yes to this question. But what that safety and environmental manager is really saying is that he or she learned all he or she "needed to know" about safety and environmental management via on-the-job experience—on their way up through their profession—and often as the material they were learning was being discovered. Thirty or forty years ago, this statement had much validity. Today, however, in most cases, the demands of the safety and environmental management profession are vastly different than they were 30 years ago—thus training requirements are also much different.

Why?

The requirements for a more demanding, classroom-based, formal education, coupled with rigorous on-the-job training and years of experience are necessary simply because times have changed. With this change, the complexity of problems related to industrial safety has increased exponentially.

What kind of changes? Complexity of what problems? Anyone who pays attention to news programs on television or radio, or has scanned newspapers or surfed the Internet in recent years should have no problem recognizing the changes to which we are referring.

Let's take a look at a few of these changes and the corresponding industrial safety and environmental health management concerns they have generated.

Major Accidents and Disasters

Major advances (Herculean advances in some respects) in technology have been made in recent decades. These include advances in nuclear power, electronics, chemical processing, transportation, information management systems, manufacturing processes, and communication systems to name just a few of hundreds of growth industries and processes. Modern society's attempts to ensure "the good life" for us have at the same time generated many perils that are also responsible for many of the woes that beset us. Why? With technological advances come technological problems, many of which have a direct or indirect impact, not only on the health and safety of employees, but also on the public and the environment.

When a process or operation has been correctly engineered (properly designed and constructed), it will show clear and extensive evidence that the design included attention to safety and environmental health. Such a process or operation, under the watchful and experienced eye of a competent safety and environmental manager, can continue to operate in a way that reduces the chance of occupational injuries, illnesses, public exposure, and damage to the environment.

A safety and environmental manager, one well versed in the basics of civil, industrial, mechanical, electrical, chemical, and environmental engineering, takes information pertinent to safety and environmental management from each of these disciplines. Exactly what will such exposure afford the safety and environmental manager, our environment, and us? Let's take a look at some examples.

- *Civil engineering exposure*—Safety and environmental managers who have had exposure to this field understand the need for safe and sanitary handling, storage, treatment, and disposal of wastes. They have knowledge of the controls needed for air and water pollution. They understand the need for structural integrity of bridges, buildings, and other constructed facilities. They understand the planning required to build safe highways.
- *Industrial engineering exposure*—Industrial engineers try to fit tasks to people, rather than people to tasks. By doing this, they make work methods and environments physically more comfortable and safer. Safety and environmental managers need to understand the concepts involved with human factor engineering and ergonomics.

- *Mechanical engineering exposure*—The student of safety and environmental management soon discovers that the mechanical engineering field really got the ball rolling toward incorporating safety and health requirements for machines such as boilers, pumps, air compressors and elevators, and many other types of mechanized equipment and facilities. The safety and environmental manager must understand the operation and limitations of mechanized machines and ancillary processes.
- *Electrical engineering exposure*—Electrical engineers are concerned with the design of electrical safety devices such as interlocks, ground fault interrupters, and other items. Electrocution and fire caused by faulty electrical circuitry in the workplace are more common than you might think. The organizational safety and environmental manager must be cognizant of these potential hazards and know what needs to be done to prevent and correct such deficiencies.
- *Chemical engineering exposure*—Chemical engineers apply system safety techniques to process design through the design of less hazardous processes, that use less hazardous chemicals, chemicals that produce less waste, or chemicals that produce waste that can be easily reclaimed.
- *Environmental engineering exposure*—Like safety engineering, environmental engineering is a broad-based, interrelated discipline that incorporates the use of environmental science, engineering principles, and societal values. The safety engineer must have a good background in environmental science and environmental engineering practices, primarily because protecting the environment from pollution is often one of the safety and environmental manager's chief duties.

Preventing major accidents and disasters is a primary responsibility of the safety and environmental manager. Avoiding major conflagrations, explosions, or catastrophic failures to equipment such as boilers or airplanes, and preventing chemical releases and spills are just a few of the important responsibilities included in safety and environmental management.

Major accidents and disasters are terrible, obviously, to those directly affected. Serious injury, illnesses, or death resulting from such incidents take their toll; they have impact in ways we cannot imagine until we actually experience them ourselves. The point is, with proper, complete, careful planning and management, accidents and disasters can be prevented.

The potential for risks involved with many chemical processing operations is extreme. Bhopal, India, the location of the horrific and infamous gas leak in 1984 that killed thousands of people, is not unique or alone in the historical records detailing the horrors of major accidents and disasters (the Chernobyl and Three Mile Island Unit 2 nuclear power incidents are other famous examples). This text's intent is not to point the finger of blame at anyone for these incidents, but rather to point out that they do occur. More important for our concerns, these horrific events point to the need for safety and environmental management. What is hardest to bear in these tragic incidents (and in many others over the years) is that we recognize that they could have been prevented—and we fear that the same greed, carelessness, or disregard for human life that played a role in these incidents may someday affect us as well.

In one respect, the United States has been fortunate. Two of the world's most terrible industrial accidents—accidents that changed how nations handled the very concept of industrial safety and environmental management—had U.S. counter-parts. The difference? The accidents in the United States were on a far less serious scale. Only months after OSHA's post-Bhopal study on the U.S. chemical industry's chances of producing a similar incident (OSHA reported the possibility as very unlikely), a similar spill did occur in Institute, West Virginia where 100+ people became ill (there were no deaths). The result? The chemical manufacturing industry swiftly cleaned up its act. Three Mile Island terrified the American public into a more-than-healthy respect for the potential harm a nuclear reactor represents. Even with no deaths or injuries from Three Mile Island, nuclear power is no longer treated casually. The United States again reassessed its regulatory standards, learning the lessons in industrial safety Chernobyl had to teach.

Regulatory Influence

Over the years, several changes related to safety and environmental health in the workplace in the United States have come about because of regulations enacted by the U.S. Congress (most of which, at their conception at least, were based on the British example). The impetus for congressional action in this area was prompted by increased pressure on legislators by the public to force businesses to adopt safety and health measures and to provide hazard-free work places. The strong influence of governmental authority in regulating the safety and health of workers in the workplace cannot be overlooked. One of the most important pieces of legislation that directly affected the push for a safer and healthier workplace was the advent of workers' compensations laws. The main intention of the proponents of workers' compensation legislation was to advance occupational safety and health programs, which in turn sparked the need for highly trained safety and environmental managers to aid in the design, implementation, and management of them.

Did You Know?

Workers' compensation provided an initial structure to industrial safety. The first such laws were introduced in state legislatures in 1909, and now all states have comparable legislation. The purpose of workers' compensation legislation is to protect the worker by providing statutory compensation levels to be paid by the employer for various injuries that may be incurred by the worker. Typically, the employer carries insurance to pay for compensation claims. Insurance companies are valuable sources of technical advice to employers.

Legal Ramifications

Soon after laws are enacted the legal profession becomes involved. New regulation means eventual testing of the law through the court system—for safety and health reg-ulation, this means providing legal services for workers' complaints against employers for endangering their safety and health, and other related on-the-job concerns. This, of course, is what has happened, and it has had a profound effect in the workplace. In the old days, employers expected the employee to work long, hard hours without getting injured. If an employee was injured, the employer simply showed him the door and that was that. Workers now have a venue for action against dangerous conditions in the workplace. Times have changed.

Thus the goal of most employers quickly became to prevent such legal actions—because they are costly both in money and time. The primary way in which employers avoid expensive litigation is by ensuring compliance with applicable laws and regulations. They most often accomplish this by assigning the responsibility for ensuring workers' safety and health on the job to a designated safety and health official. Often this designated safety official is a fully qualified safety and environmental manager.

Recommended Training Program for Safety and Environmental Managers

For several years there was (and, to a lesser degree, still is) considerable confusion and disagreement about exactly what type of preliminary training, if any, a typical safety and environmental manager should have. Much of the confusion and disagreement stems from certain practices fostered by various industries throughout the industrial complex even to this day. Let's take a look at a few of these practices as they evolve based on organizational attitudes.

Sometimes a company considers safety and health within the organization as being the full responsibility of the employee. Management in these companies feel (obviously, they aren't thinking) that expensive safety and health programs, safety equipment, and safety and health managers to administer these programs are not necessary. "Safety and health is expensive—safety and health doesn't contribute to my bottom line—it takes money away from it." Have you heard this statement or a version of it before? Unfortunately, this attitude is much more common than you might think.

Even today in this age of regulation and enforcement (namely by the Occupational Safety and Health Administration—OSHA) the scenario described above is still practiced in various industries—at least until such an organization is cited by OSHA, or pays the expense brought about by litigation filed by workers and unions against the organization for violation of safe work practices.

When an organization is "hit in the head" by such actions, various (often predictable) reactions are to be expected. Typical reactions include paying the fine and continuing operations as before, assigning an employee to be "the safety and health person," or hiring a full-time professional safety and health manager.

Paying the Fine

When cited by OSHA, an organization may decide to pay any associated fines and then go back to business as usual. There actually exists a mind-set out there in the real work world (held by certain company owners and/or managers) that it is easier and cheaper to pay any fines levied by OSHA or other state and local authoritative bodies than to fully comply with their "costly" regulatory requirements. This attitude and practice is unfortunate, short-sighted, and illegal—but until recently, when Congress stiffened the penalties and dramatically increased the fines levied, many industries simply ignored OSHA and its requirements that provide a safe and healthy workplace for workers.

Assigning a Safety and Health Person

Another typical reaction employed by company owners and managers of companies cited by OSHA is to comply by assigning a worker from within the organization as "the safety and health person" for the organization. In some cases this practice has actually worked. For example, if the in-house employee (who probably has little or no safety and health management experience) is a highly motivated, active, and dedicated worker, he or she can successfully complete assigned safety and health management duties in a manner that both provides safety and health protection to the workers and may bring the company into compliance with the regulators.

However, let's take a look at what happens, more often than not, when a company chooses a worker for the safety and health manager position. In an industry, often the opinion is held that with the appointment of a "safety and health person," the company's responsibility for controlling hazards becomes the safety and health person's sole responsibility. Upper management and managers, all the way down the chain of command then "brush off" their responsibility for safety and health onto the safety and health person. Not only is this notion an example of convoluted thinking, but also it is decidedly wrong—the "ultimate" responsibility for the safety of workers cannot be delegated. Experience clearly shows that this practice ultimately destroys any real chance for safety achievement. Why? Let's take a closer look.

If a company follows the procedure described above and simply selects some worker to become the organization's "safety and health person," the success of the designated safety and health manager in accomplishing his or her safety duties is directly related to the amount of upper management support he or she receives. Often the safety and health person is simply pulled from the ranks of a workforce, assigned the safety and health person's duties, given no formal training in safety and health and zero support from upper management. Unless that person is exceptionally dedicated to safety and health goals and has the brains and guts to push for enforcement, he or she is doomed for failure—and thus so is the organization's safety and health program.

Another problem with pulling "someone" from the existing workforce to fill the company's safety and health position is readily apparent in the selection process itself. Who do you suppose is the most likely candidate to be chosen for the position? Let's take a look at three prospective candidates. Will it be Harry from the assembly line, who can outwork any three employees on any given day? Will it be Nancy from quality control with an eagle eye for defects, who saves the company big bucks? Or Larry, who is slower than slow, uncooperative, and basically just occupies the rarefied air in which he seems to hover?

You don't need to be a rocket scientist to determine that the likely candidate is often going to be Larry. The managerial mind-set might be: "Waste Harry or Nancy on this? Safety and health is too expensive to begin with without taking away my best workers. Larry's not working well where he is—he needs something different to keep him busy—to keep him out of the way. The safety and health job (that no one else really wants) is just perfect for Larry." Thus Larry fills the company safety and health person position, and you can judge for yourself the potential for Larry's ultimate success. Popular culture has given us a sterling example of such a safety and health officer—one instantly recognizable if you watch *The Simpsons*. Like Homer Simpson as safety and health officer of a nuclear power plant, Larry as the company safety and health person seems ridiculous and almost comical, doesn't he? That is until you think about the potential for risk he represents. Incompetence in the safety and health officer's position is ultimately not very funny, is it?

But what would have happened if the company had decided to "promote" Harry or Nancy to the safety and health person position? If either Harry or Nancy (for the sake of simplicity, let's stick with Harry) had been chosen as the company's safety person, one thing is certain, the choice was a good one in respect to attitude and work ethic. But the bigger question is: What exactly is expected of Harry? Or as Harry would certainly put it: What am I supposed to do? Where do I start?

"Where do I start?" Does this sound familiar? It should, because this is a natural question for the new "safety and health person" to ask. This question is complicated, of course, by other considerations—other questions. For example, Harry is going to want to know if he is assigned the safety and health person position full time, or as a collateral duty. If assigned the safety and health person's duties as a separate full-time job, then Harry probably has the initiative and drive needed to perform the safety and health person's duties in a credible manner, though the task will require Harry to learn as he goes which is difficult but not impossible. It's been done before).

Harry might learn, however, that he is expected to continue his full-time duties on the assembly line (employees who can do the work of three people are difficult to replace) and that the safety and health work is a collateral duty. Again, this is a typical practice in many industries, especially in companies with a workforce of fewer than twenty people or so.

If Harry is assigned the duties of company safety and health person, he is going to find out (since he is determined and conscientious) what several thousand other folks thrown into such positions in the past have already found out. That is, finding a more challenging or more mind-boggling collateral duty assignment than that of "the safety and health person job" may be impossible. This statement may seem strange to those managers who view safety and health as a duty that only requires "someone" (anyone) to keep track of accident statistics, to conduct safety and health meetings, and perhaps place an occasional safety and health notice or poster on the company bulletin boards. The fact is, in this age of highly technical safety and environmental health standards, government regulations, and insurance requirements, the safety and health person has much more to do and many more responsibilities than placing posters on bulletin boards. Our friend Harry will soon find out that safety is a full-time job.

Hiring a Professional Safety and Health Manager

Another safety and health person scenario in common practice today (typical for companies with 25 or more employees) is to hire a professional full-time safety and health manager as the company's safety and health person. This full-time position may be titled safety and health director, safety and environmental health manager, safety and health administrator, safety and health officer, safety and health official, safety and health coordinator, safety and health specialist, or some other variation. The title is not important; the job classification requirements or duties are.

Typically, a safety and environmental health professional is expected to organize, stimulate, and guide the company's safety and environmental health program, as well as stay up to date on safety and workplace health subjects and regulations and function as a resource person for all those involved in the work. Normally, in this capacity, the safety and environmental health professional, as a staff person, holds very few to no administrative powers over the operating components of the company. Is this a good practice? It depends. When you factor in the responsibility of company line officials who are primarily responsible for the operating components of the company, then yes, it is a good practice (in most cases).

Are you confused? If so, consider this. If the company safety and environmental health person is designated as "the person" solely (the keyword is "solely") responsible for the safety and health of workers involved in the operating components of the company, then instead of line management, which through chain of command holds the managerial power to control unsafe work practices and conditions, management often will relegate (or abdicate) this responsibility solely to the safety person. Obviously, this is not only an unhealthy situation but it could have titanic consequences. (Remember Bhopal, Chernobyl, or Three Mile Island? How could we forget?)

Just what responsibility—and enforcement power—does or should the safety and environmental health professional have? What can (and does) a safety and health officer do in those situations when he or she observes, for example, an imminently dangerous work situation or practice in progress? Should the safety and health professional just stand there and let the line manager in charge continue a dangerous and life-threatening operation? No, the safety and environmental health professional must take action.

What action? The potency of the safety and health professional lies in the capacity to use well-marshaled facts to persuade the line managers to act on behalf of the workers' safety. In other words, the safety and health professional who observes any unsafe act or unsafe condition should immediately point this out to the manager in charge.

What if the manager in charge is not available? What is the safety and environmental health professional supposed to do then? Good question—one that has been asked for decades (the answers received have not always been what you might expect). A company safety and environmental health professional should be given enough authority to stop work when required—when danger to the safety and health of workers, the public, or the environment is imminent.

Caution is advised, however. While the company safety and environmental health professional should have the authority to stop work because of unsafe conditions or unsafe work practices, this authority must not be abused. Interfering with line operations and stopping work as a show of power over nickel-and-dime trouble or for dreamed-up causes is not recommended.

Because of OSHA, insurance company requirements, and fear of litigation, common practice today is that larger companies hire a qualified safety professional—they don't promote from within. This brings us back to explaining exactly what qualifications the safety and environmental manager is typically expected to have today. Along with experience, the professional safety and environmental health manager is expected to be college graduate with a degree in safety and environmental health management, industrial hygiene, environmental health, engineering, business administration (with safety course completion), or some other affiliated field.

Certified Safety Professionals (CSPs)

The National Safety Council publishes every other year a list of colleges and universities that offer safety and environmental health training (the list grows in length each year). In addition, the Board of Certified Safety Professionals [an organization that sets the standards for registering safety engineers as Certified Safety Professionals (CSP)] has suggested a standard baccalaureate curriculum for the safety and environmental health professional. Note that the Board of Certified Safety Professionals does not infer that their suggested curriculum is the "only" academic program that can successfully prepare individuals who plan to enter the safety and health profession; instead, the content included in their proposed curriculum is minimal for meeting the educational requirements for becoming a CSP.

Let's take a look at some of the minimum college requirements.

- Calculus I & II
- Statistics
- Biology I & II
- Chemistry I & II
- Technical writing
- Composition
- Physics I & II
- Graphics
- Economics
- Psychology
- Public Speaking
- Applied Mechanics
- Materials & Processes
- Several Engineering electives
- Safety & Health I & II
- Industrial Hygiene
- Toxicology
- Computer Science
- Management
- Safety Engineering
- Fire Prevention & Protection
- Several other Safety Electives

Does completing the minimum educational requirements automatically qualify a candidate for certification as a safety professional? Not exactly, but the candidate is headed in the right direction. To qualify for and obtain certification as a certified safety professional (CSP), certain requirements must be completed. These requirements include:

- Graduation from an accredited college or university with a bachelor's degree in engineering or some other field such as occupational safety and health
- Completion of at least five years of professional safety experience.
- Achievement of a passing score on both the core examination (safety fundamentals) and the specialty examination (management aspects, construction, comprehensive, engineering, product safety, or system safety).

Did You Know?

A variety of resources have arisen to meet these needs of the safety and environmental Health Professional.

- Professional Certification
- Professional Societies
- National Safety Council
- Standards Institutes
- Trade Associations
- Government Agencies
  - Free consultation
  - The National Institute for Occupational Safety and Health (NIOSH) has a wealth of research data on the hazards of specific materials and processes.
  - NIOSH used these data to write criteria for recommended new standards.

If you are familiar with the requirements for registration and licensing for a professional engineer (PE), you will notice that the qualification requirements for CSP designation are very similar, especially regarding the two-tiered testing process. In achieving PE status, an engineer must first successfully complete a fundamentals examination, which leads to designation as an EIT (engineer in training). For CSP certification, candidates must also successfully complete a fundamentals (or core) examination, which leads to designation as an ASP (associate safety professional). This is not a recognized certification as such, but the designation does indicate a candidate's progression (as does the engineering EIT designation) toward certification.

The Safety Fundamentals Examination is administered semiannually by the Board of Certified Safety Professionals (BSCP) Savoy, Illinois. The examination is closed book and consists of approximately 280 multiple-choice questions.

The subject areas covered in the fundamentals examination are weighted as follows:

Basic and Applied Sciences	35%
Program Management and Evaluation	13%
Fire Prevention and Protection	13%
Equipment and Facilities	13%
Environmental Aspects	13%
System Safety and Product Safety	13%

The topics for the major areas given by the BSCP for the Safety Fundamentals Examination include:

1. **Basic and Applied Sciences**
   - **Mathematics**—includes algebra, trigonometry, calculus, statistics, and symbolic logic.
   - **Physics**—includes mechanics, heat, light, sound, electricity, magnetism, and radiation.
   - **Chemistry**—includes atomic structure, bonding, states of matter, chemical energetics and equilibrium, and chemical kinetics, addressing both inorganic and organic chemistry.
   - **Biological sciences**—includes heredity, diversity, reproduction, development, structure, and function of cells, organisms, and populations, with emphasis on human biology.
   - **Behavioral sciences**—includes individuals' differences, attitudes, learning, perception, and group behavior.
   - **Ergonomics**—includes the capabilities and limitations of humans and machines, simulation for design and training, principles of symbolic and pictorial displays, static and dynamic forces on the human frame, responses to environmental stress, and vigilance and fatigue.
   - **Engineering and technology**—includes applied mechanics, properties of materials, electrical circuits and machines, principles of engineering design, and computer science.
   - **Epidemiology**—includes the application of epidemiological techniques to the study and prevention of injuries and occupational illnesses.

2. **Program Management and Evaluation**
   - **Organizations, Planning, and Communication**—includes general principles of management applicable to the organization of safety function, safety program planning, and communication.
   - **Legal and Regulatory considerations**—includes federal, state, and local safety and health legislation, workers' compensation laws, regulatory agencies, contractual relationships, and public/product liability.
   - **Program Evaluation**—includes the identification, selection, and application of qualitative and quantitative measures to evaluate and document the need for specific control programs.
   - **Disaster and Contingency Planning**—includes the identification of types of emergencies, plan-of-action considerations, training, and equipment needs.
   - **Professional Conduct and Methods**—includes principles that define the rights and responsibilities of safety professionals in their relationships with each other and with other parties.

3. **Fire Prevention and Protection**
   - **Fire Prevention**—includes inspection procedures, hot work permits, flammable liquids handling and storage considerations, and other factors intended to control sources of ignition and fuel.
   - **Structural Design Standards**—includes emergency egress design, fire-resistant construction, and fire and smoke containment.

• **Detection and Control Systems and Procedures**—includes principles and design of fire detection and warning systems, fixed and portable extinguishing systems, fire and emergency drills, and fire brigade organization and training.

4. **Facilities and Equipment**
   • **Facilities and Equipment Design**—includes methods, standards, and principles of design for the control of injury-causing conditions associated with facilities and equipment in general.
   • **Mechanical Hazards**—includes identification and control of injury-causing conditions associated with facilities and equipment in general.
   • **Pressures**—includes identification and control of exposures associated with compressed gases, pressure vessels, and high-pressure systems.
   • **Electrical Hazards**—includes the nature of electrical shock, grounding, insulation, ground fault circuit interrupters (GFCIs), and inspections.
   • **Transportation**—includes identification and control of hazards associated with transportation of personnel and materials.
   • **Materials Handling**—includes identification and control of injury-causing conditions associated with both manual and mechanical materials handling, hoisting apparatus, conveyors, and powered industrial trucks.
   • **Illumination**—includes the assessment of the quality and quantity of illumination with respect to accident prevention.

5. **Environmental Aspects**
   • **Toxic Materials**—includes the identification and assessment of hazards associated with chemicals and other materials producing or capable of producing a harmful effect on biological mechanisms.
   • **Environmental Hazards**—includes the identification, assessment, and control of potentially harmful effluents.
   • **Noise**—includes the principles and methods of measurement, assessment, and control of noise exposures.
   • **Radiation**—includes the principles and methods of measurement, assessment, and control of radiation exposures, both ionizing and nonionizing.
   • **Thermal Hazards**—includes the identification and assessment of exposures associated with temperature extremes.
   • **Control Methods**—includes respiratory protection, protective clothing, ventilation, shielding, and process controls.

6. **System Safety and Product Safety**
   • **Product/System Design Considerations and Product Liability Administrative Aspects**—includes administrative and legal considerations associated with the prevention of product liability losses.
   • **Reliability, Quality Control and Systems Safety**—includes general concepts of product and system safety as they relate to reliability and quality control.

Upon successful completion of the safety fundamentals examination, the ASP designation is awarded. The CSP candidate must then decide which of the CSP specialty examinations he or she will sit for. Upon successful completion of the safety specialty examination, the CSP designation is awarded.

THE THREE ES PARADIGM

Throughout this chapter, we have stressed the importance of employing a qualified safety and environmental health manager to be responsible for the controlling, coordinating and managing of the company's safety and environmental health program. The keyword we have stressed throughout is "qualified." By now you should have a good feel for why "qualified" is important.

Creating a safe and healthy work environment is the primary goal of safety and environmental health managers and that safe work environment can be created, controlled, coordinated and managed by applying the three Es of safety: engineering, education, and enforcement.

We have pointed to the importance of engineering in safety, to the design of control systems to enhance occupational safety and health. Again, the requirements of OSHA, increased employer liability, and heightened worker awareness of the implications of unsafe and unhealthy work environments all contribute to this need. Experience has shown that the vast majority of engineers concerned with design and operation have not been adequately trained to recognize or control existing or potential safety and health hazards. Worse, the past tendency for safety considerations has been to call in the "engineers" only after equipment or process design and installation have been completed. This has been and is, obviously, a serious shortcoming. In this text we not only stress the need to "engineer out a hazard" but also to do so in the preliminary process, the design phase. We stress the proactive mode (developing preventive actions before accidents occur), not the reactive mode (deriving preventive actions from accidents) because it just makes good common sense to fix a problem before it becomes a problem.

Engineering includes virtually every preventive action necessary to correct a physical hazard (including such simple steps as maintaining good order and discipline within the workplace and maintaining housekeeping measures such as cleaning up spills, removing obstructions, etc.). Major preventive actions (i.e., "engineering out" a hazard) cover a broad range of possible steps, and require the application of physical sciences to control hazards in the workplace, including designing out hazards, modifying processes, and incorporating fail-safe devices. In other instances "engineering" may be devoted largely to more routine actions—substituting less hazardous materials, reducing the inventory of hazardous materials, and prescribing protective equipment (Grimaldi and Simonds, 1994).

Grimaldi and Simonds also point out that the scope and order of actions applied as engineering steps for the control of hazards usually include:

Step 1. Evaluate process or operation and identify its harmful agents (if any).
Step 2. Eliminate the harmful agents by redesign, or substitute a less harmful material, arrangement and so on.
Step 3. Shield or guard the hazard.
Step 4. Isolate the hazards.

Step 5. Use ventilation, wet processing, and other techniques to dilute the harmful effect.

Step 6. When steps 2 through 5 do not furnish the level of control needed, provide personal protective equipment (PPE).

Education (employee training), the second of the three Es of safety, is one of the most important preventive actions the safety and environmental health manager incorporates into the company safety and health program. The reason for this is not difficult to understand and appreciate. Simply put, employees can't be expected to perform their assigned functions unless they have been properly trained to perform them. Without proper training, the potential for injury or death is magnified many times. As an extreme example, if a manager tells an employee to operate a dangerous machine (say, a bulldozer or front-end loader) without ensuring that the employee is qualified (trained) on how to operate the machine, this is inviting trouble—disaster, in fact.

The safety and environmental manager is normally called upon to conduct safety training at all levels of organizational strata, but safety training usually begins with the supervisor. This makes sense since the supervisor is primarily responsible for ensuring that his or her workers are qualified (trained) to do their jobs.

Note that the safety and environmental health manager misses a golden opportunity to enhance the company's overall safety and environmental health profile whenever he or she does not employ the expertise of supervisors in the establishment and maintenance of the company's safety and health program. Experience has shown that no matter how hard the safety and environmental manager works to improve the company's safety and health program and no matter how conscientious he or she is, if the supervisors don't buy into the program, it is doomed to failure. Therefore, as company safety and environmental health manager you must employ the talent and expertise of supervisors. Without their support, you will have a rough row to hoe.

Exactly what training should the safety and environment manager provide to supervisors? Generally, the safety and health professional needs to train supervisors to recognize hazards and know the appropriate actions that need to be taken. More specifically, supervisor safety and health training should focus on four tenets:

- Develop safe working conditions and safe working practices.
- Conduct employee safety and environmental health training that stresses the real need for safety. This is best accomplished by personalizing the training (i.e., relate the training to workers' jobs, their functions within the company and show how it can and will benefit them).
- In personalizing safety and environmental health, what we are really talking about is employee participation. We have stated that it is important for the safety and environmental health manager to have the supervisors "buy in" to the company's safety and environmental health program—and this is true—but it should be stressed that worker safety and health is what safety and environmental management is all about. Thus, you must also have the workers "buy in" to your program. This is best accomplished by making the workers part of the company's safety and health program. Our experience has shown that any safety and health manager can learn a lot by just listening to workers—they have good ideas. Look at it this way, each and every

day the workers face the on-the-job hazards you are trying to eliminate or reduce. Their value to the safety and environmental health manager can't be over-stressed.
- Implement safety rules. While the safety engineer usually has a good "feel" for what an organization needs in the way of safety rules, safety rules often need to be tailored to the type of work performed. For example, if the company's primary work deals with material handling operations, the focus of safety rules will more than likely be on rules dealing with safe material handling operations.

Enforcement, the last of the three Es of safety, is more than just achieving compliance with company rules and procedures. This is especially the case today when company (and thus employee) practices must be in compliance with federal, state and local laws, and regulations and consensus standards.

When a company puts together a set of safety rules, safe work practices, and procedures, but fails to enforce them, the rules are not worth the paper they are written on. Enforcement normally implies punishment; however, the safety and environmental manager must not be regarded as the wielder of such. The safety and environmental manager makes a serious mistake whenever he or she becomes the company "enforcer" (often viewed by workers and supervisors as the company's Gestapo agent). Instead, the primary lesson the company safety and environmental manager must learn to be successful is to be an *advocate for* safe work practices in his/her company, not just a *regulator of* safe work practices.

A significant factor included in the enforcement element (one often overlooked) is the use of "praise" where a good safety effort emerges. The safety and environmental manager who does not recognize employees who complete their tasks in a correct, safe manner loses a golden opportunity to employ the positive to reinforce what is commonly viewed as the negative: enforcement. When an employee is singled out by the company safety and environmental manager and supervisors for outstanding performance of some function conducted safely—in full view and earshot of his co-workers—the effectiveness of the company's safety and environmental management program is often elevated to a higher level; one which is otherwise often difficult to attain.

THE SAFETY AND ENVIRONMENTAL MANAGER'S DUTIES

The company safety and environmental manager, as the organizer, stimulator, and guide of the safety and health program, performs a number of significant tasks. Probably the easiest way to illustrate what is typically required of a safety and environmental manager (and clearly reinforce our previous statement that a safety and environmental manager must be a Jack or Jill of all trades) is to take a look at an example job description for such a position.

Company Safety and Environmental Manager
<u>Examples of Duties</u>

- Conducts independent work activities based on broad directives.
- Manages Safety Division and Safety/Environmental Health activities for Company departments.

- Promotes awareness among Company employees of the benefits of safe operation.
- Plans, coordinates, and supervises safety activities; evaluates Company safety rules, policies, safe work practices, and training procedures; and recommends changes as appropriate.
- Coordinates implementation of safety rules, policies, safe work practices, and procedures.
- Provides liaison and coordination between Departmental Safety Officers/committees concerning safety activities; reviews and inspects operations and related maintenance procedures and recommends actions to ensure safety of Company employees.
- Investigates safety accidents and recommends policies/procedures and/or other actions to correct problems or prevent future accidents.
- Plans, coordinates, supervises, and reviews recording and reporting of safety data and information.
- Assists with the procurement of contractual services or materials and supplies for safety activities.
- Supervises safety and health training activities for all Company departments.
- Manages OSHA, SARA Title III, CERCLA, and other regulatory requirements.
- Directs development of trial or pilot safety and health programs.
- Reviews and critiques drawings, contract specifications and OSHA 200 logs of perspective outside contractors, ensuring compliance with 29 CFR 1910.119 Process Safety Management requirements, including pre-construction safety briefs for contractor supervisory personnel.
- Continuously improves knowledge of general and specific information pertaining to safety equipment, regulatory requirements, programs, and procedures.
- Collects, analyzes, and arranges information in such a manner as to fulfill management's decision-making needs.
- Assists various levels of management by originating and developing policy recommendations, operating plans, programs, measures, and controls that will permit the effective accident-free use of human and material resources.
- Manages Company's medical surveillance program, including CDL physical exam requirements, pulmonary function testing, audiometric testing, and environmental sampling/testing programs.
- Manages training program for HAZWOPER Compliance.
- Publishes annual Safety and Health Training/Medical Testing/Safety Inspection schedule.
- Investigates and mitigates employee reports of unsafe conditions.

THE "S" IN SAFETY

To this point in the chapter, we have basically rehashed the same information presented in the first and second editions of this text. Why? Because this information is still pertinent, current, and germane—with one exception. Consider the following:

September 11, 2001(9/11): Four commercial airplanes are hijacked on the East Coast. American Airlines Flight 11 out of Boston crashes into the North Tower of the

World trade Center with 92 people on board. United Airlines Flight 172, also out of Boston, crashes into the South Tower of the World Trade Center with 65 passengers on board. American Airlines Flight 77 out of Washington Dulles crashes into the Pentagon with 64 on board. Flight 93 out of Newark, New Jersey crashes into a rural, western Pennsylvania field with 44 people on board. An estimated additional 2,666 people on the ground die in the World Trade Center and another 125 are dead in the Pentagon. Nearly 3,000 people die in one day as a result of terrorist attacks (U.S. Department of State, 2002).

Are Safety and Environmental Managers responsible for preventing terrorism in the workplace? The short answer is yes. The compound answer is absolutely yes. Even though there is no written OSHA Standard mandating that a written workplace terrorism program is required, it should be included in any comprehensive safety or loss control program. For example, the workplace security plan could be attached to the emergency response plan. The point is, because of 9/11, companies are expanding their programs to incorporate terrorism and acts of subversion within the company. This makes sense when we consider that many safety and environmental managers are already responsible for addressing bomb threats and prevention of workplace violence. Adding security concerns to the wide range of responsibilities of safety and environmental managers is the logical next step.

The events of 9/11 changed many elements of our society. One of the major changes has to do with the workplace itself—the architecture, building plans, and company activities are now viewed in a different light. That is, we must factor in the "what if" scenario—what if our company comes under terrorist attack, from within or from outside the company?

Because of the need for increased security, the safety and environmental manager's job has expanded exponentially. Later we will discuss security concerns and prevention techniques in greater detail. For now it is important to point out that the "S" in safety and environmental management takes on a new connotation—security. We can say that 9/11 supersized the S in the practice of safety engineering.

ECONOMICS AND SAFETY AND HEALTH MANAGEMENT

It is often agreed that the safety and environmental manager is responsible for training, statistics, and industrial relations with respect to safety and health but often disputed is the safety and environmental manager's purchasing responsibilities. This does not make sense. With respect to the purchase of new equipment, the sooner the safety and environmental manager gets involved in the purchasing process, the sooner he can inject his opinion as to what is the safest equipment to purchase based upon a safety analysis.

Organizations often base safety and health decisions on dollars and cents. The safety and environmental health manager needs to learn quickly that he or she has to think in terms of the company's bottom line. The prevention of worker injuries and illnesses can be formulated as an economic objective; such a formulation is more meaningful to management than vague humanitarian aspirations. Based on experience, money matters and the rest is microminutae to upper management or owners in many organizations. Until an employee dies or is seriously injured, upper management generally does not get the point that safety and health programs that prevent injury or illness are not only a good practice but will also save money in the long run. Upper management must realize that accidents, injuries, and illnesses have undesirable costs that contribute nothing to the value of products manufactured by the organization or services performed.

If the safety and environmental manager is in an organization where the bottom line is more important than the safety and environmental health of its employees then the manager may want to look for another place of employment.

THOUGHT-PROVOKING QUESTIONS

- In what disciplines do safety and environmental managers need expertise?
- In what ways are safety and environmental managers "generalists"?
- What is the safety and environmental manager's role in industry?
- Why are both college degree study and hands-on experience important for safety and environmental managers today?
- Why is "prevention" more important than "reaction" in safety and environmental management?
- Aspects of civil, industrial, mechanical, electrical, chemical, and environmental engineering are critical to the safety and environmental management profession. Why? What do these disciplines provide for the safety and environmental manager?
- When new safety regulations are enacted, what is the general industry response?
- How have safety and health regulations affected the average workplace?
- Who is (or should be) responsible for safety in industrial facilities?
- What are the pros and cons of promoting from within versus hiring a safety and health manager?
- What problems are associated with assigning someone safety and health officer functions as collateral duty?
- What is a safety engineer's duty if he or she sees workers involved in unsafe work practices?
- Discuss the CSP certification process.
- List, define, and discuss the three Es of safety. Why and how is each element important?
- What are the six parts of the engineering steps for control of hazards?
- What four tenets should supervisor safety training focus on?
- What role does praise play in safety training?

REFERENCES AND RECOMMENDED READING

Allison, W.W. "Other Voices: Are We Doing Enough?" *Professional Safety*. February 1991: 31–32.

Banks, W.W., and F. Cerven. "Predictor Displays: The Application of Human Engineering in Process Control Systems." *Hazard Prevention*. Jan/Feb 1985: 26–32.

Benner, L., Jr. "Safety Training's Achilles Heel." *Hazard Prevention* 26, no. 1 (1990).

CoVan, J. *Safety Engineering.* New York: John Wiley & Sons, Inc., 1995.

Fawcett, H.K., and W.S. Wood, eds. *Safety and Accident Prevention in Chemical Operations.* New York: John Wiley & Sons, Inc., 1982.

Friend, M.A., and Kohn, J.P. *Fundamentals of Occupational Safety and Health*, 5th ed. Rockville, MD: Government Institutes, 2010.

Grimaldi, J.V., and R.H. Simonds. *Safety Management*, 5th ed. Homewood, IL: Irwin, 1994.

McCormick, E.J., ed. *Human Factors Engineering*, 3rd ed. New York: McGraw-Hill, 1970.

Speegle, M. *Safety, Health, and Environmental Concepts for the Process Industry.* Boston: Cengage Learning, 2012.

Spellman, F.R. *Safe Work Practices for Wastewater Treatment Plants.* Lancaster, PA: Technomic Publishing Company, 1996.

Spellman, F.R. *Surviving an OSHA Audit.* Lancaster, PA: Technomic Publishing Company, 1998.

U.S. Department of State. "September 11, 2001: Basic Facts." Accessed May 15, 2015. Washington, DC: U.S. Department of State, 2002. http://2001–2009.state.gov/coalition/cr/fs/12701.htm.

Waters, B. *Introduction to Environmental Management.* London: Routledge, 2013.

Chapter 2

Management Aspects

Leadership, patient observation, and problem solving are the keys that unlock the mystery of managing people.

INTRODUCTION

What is the difference between a newly degreed safety and environmental manager and a new management specialist? The main difference is quite obvious: the safety and environmental manager is a technical expert and the management specialist is a management specialist. Which of these two newly degreed specialists will make the best manager? This question is not so easy to answer. One might think that the management specialist has the advantage over the safety and environmental manager in managing people. But is this really the case?

To begin with, the individual person is the key factor. If the individual is frank, decisive, and assumes leadership readily, he or she might become a good manager. If the individual quickly perceives illogical and inefficient procedures and policies and develops and implements comprehensive systems to solve organizational problems, he or she might become a good manager. If the individual enjoys long-term planning and goal setting, he or she might become a good manager. If the individual is well informed, well read, and enjoys expanding his or her knowledge and passing it on to others, he or she might become a good manager. If the individual is forceful in presenting his or her ideas, he or she might become a good manager.

Along with the parameters listed above, each is qualified by the key phrase: "might become" a good manager. Simply, no individual, even a management specialist, comes with a guarantee for becoming an effective manager or having good management ability. One can meet all the parameters listed above (and others) and still lack that special innate skill (or set of skills) that is management ability (laced with a great deal of common sense). There is no template to rely on. We have witnessed many newly degreed managers step into the workplace for the first time thinking that because they have a college degree they are "instant" managers—this might be the case, but in reality they were managers in title only. Unfortunately, many of these well-intentioned individuals fall on their faces when trying to make and orchestrate even the most basic managerial decisions.

Management skill is a blessing, an asset, a cherished, intangible trait that is on just about everyone's mind but is nearly nebulous to grasp. There are those who say that you either have it or you don't. "It," of course, is the natural leadership ability to manage almost any situation and any worker. Others state that managers are born and not made. The truth lies somewhere in the middle. Someone with the desire to manage effectively might have the ability to do so. However, to manage and to manage effectively are opposite edges of a double-edged sword.

We don't have a template for the recent college grad that will assure management success—we do not think there is such a template. However, we are certain of a few things. To manage effectively, education is important; the desire to manage is very important; innate leadership qualities are even more important; and the actual ability to manage effectively is priceless.

THE RIGHT WAY

What is the right way to manage? This question has generated many different responses and resulted in so many different theories that it is difficult to keep up with them. From the "old" days when there was but one flagship way to manage: lead, follow, or get out of the way, to those more sophisticated mantras of participative, collective, and empowered management styles carrying the flag today. Determining the "right way to manage" has been an important quest.

The right way? Take your choice. The management style probably does not matter anyway if it includes some basic ingredients. The recipe for success is dependent on these ingredients—without them, the would-be manager would likely fail.

So, what are the "magic" ingredients? The ingredients that have worked for us are a combination of the old (proven factors) and the relatively new (behavioral approach). Actually, for us, success lies somewhere within the continuum between these two.

Let's look at some of the old (proven factors) first. These include the standard bearers of planning, organizing, controlling, and directing.

Planning simply advocates the old adages of a job well planned is a job well done and a job poorly planned is a job half-finished. Planning is essential in safety and environmental management. It is a proactive approach to safety and health, where waiting for accidents to occur is not an option. Instead, the planning factor allows for anticipating and planning how to deal with problems before they occur.

Organizing is all about fashioning a safety and environmental health program that will be accepted and followed by everyone in the organization. This is only possible, of course, if upper management fully supports the safety and environmental health program and the safety and environmental manager. Generally looked upon as a staffer and not a liner, the safety and environmental manager must not only know his or her safety and health program top to bottom but must also have full understanding of the entire organization and what makes it tick. If the safety and environmental manager has no idea where he or she fits into the organization, he or she needs to look for a new line of work in a different organization.

Controlling means just what it says; that is, a good safety and environmental health program with good support is doomed if it is not controlled (managed). Some form of metrics must be involved to continually measure and evaluate the performance of the program versus the number, type, and severity of on-the-job accidents, chemical spills, fires, incidents of workplace violence, and so on.

Directing is important if used correctly. Safety and environmental managers walk a thin line of authority. Safety and environmental managers who join an organization and begin directing workers to do this or that without working through and with the workers' supervisors are headed for a lot of heart burn and certain failure. About the only true directing the safety and environmental manager should do is that required to direct his/her own staff, if they have one. Being an advocate for safety is different from that of being an enforcer, and that is one line the safety and environmental manager never wants to cross.

The big four (planning, organizing, controlling, and directing) are old diehards that have served managers well in the past. However, to synergize these four important ingredients, the safety and environmental manager must be able to communicate. Reticence and safety and environmental management mix like gasoline and sand. This is not to say that the safety and environmental manager needs to be extroverted to the point of setting the standard of excessive demonstrative actions for the organization. On the contrary, the safety and environmental manager almost needs the skills of a used-car salesman and diplomat all rolled up into one effective delivery methodology. Because of the staff alignment in which most safety and environmental managers find themselves, the safety and environmental manager needs to bridge the gap between line and staff by communicating effectively. Based on experience, we have found that when a safety or health problem is apparent, it is best to ask the supervisor for his or her recommendations—we call this reverse empowerment. However, no matter what you call it, the ability to communicate is important. It is important to remember the goal is to protect workers from harm. Whatever communication technique works best should be employed on a consistent and straightforward basis.

The right way of safety and environmental management also includes ensuring compliance with federal, state and local safety and environmental health regulations. This can be a tricky proposition, however. The safety and environmental manager who walks into a company manager's office and throws a copy of some new OSHA, EPA, or DOT regulation on his or her desk and states that from now on we must do it (whatever it might be) the regulator's way or else, is using the wrong approach.

Few people like rules and regulations, and fewer like demands made upon them. When it is important to make sure the organization is in compliance with the regulators, the safety and environmental manager needs to put on his or her used-car salesperson's hat, because implementing any regulation in the workplace is a selling job. The managers and workers must buy into the proposition before they will comply with it.

Managing under the precepts of participative, collective, and empowered systems is not a pie in the sky management solution. However, like anything else, there are good parts of these concepts and bad parts. The key is to use the good and trash the bad.

Participative management requires the safety and environmental manager to:

1. have confidence in his or her subordinates
2. allow the subordinates the freedom to communicate
3. seek different ideas and different views
4. establish communication in all directions
5. ensure upward communication is accurate
6. establish decision-making as a team effort
7. ensure goal-setting is performed by group action
8. share control widely and work to make sure informal organized resistance does not exist.

This management style (approach) is doable and can be effective, but requires high maintenance; no element can be ignored.

We do not recommend the collective style of leadership. Why? We can sum this one up quite simply. What the organization needs to learn is that when everybody is accountable, nobody is accountable and to recognize that, where getting things done is concerned, the collective approach does not add to but detracts from individual decision making—the antithesis of participative management. Do you remember the old saying, "A camel is a horse that was designed by a committee?"

We discuss empowerment in detail later in our discussion of Total Quality Management (TQM). However, for now it is important to point out that empowerment is one of the latest buzzwords in a long litany of buzzwords currently being broadcast in various schools of management. Simply, empowerment of workers allows workers to do their jobs without over- or under-management (if you get off my back, you'll be amazed how productive I can be).

BEHAVIOR-BASED MODELS

Few topics invoke more dissent among safety and environmental professionals than the benefits to be derived from behavioral safety and health versus the traditional three Es of safety (engineering controls, education, and enforcement) strategy employed since the 1930s. There are compelling arguments on both sides. For example, the gurus of behavioral safety, namely E. Scott Geller and associates, argue that we should not fix the worker but instead fix the workplace. The other side counters with the only thing that makes safety work is fear (i.e., fear of losing a finger, arm, leg, life, or job). Which side is on the right side? The answer to this question is somewhat nebulous and, as mentioned, certainly arguable. Note that it is not our intention to argue for or against any particular safety and environmental management model. Instead, we simply discuss the tenets of the various models and leave final judgement on which model is best for a particular application to the reader.

What exactly is behavior-based safety? According to Mathis & McSween (1996), behavior-based safety is the infamous quick fix. They argue that "common wisdom in safety has it that the quick fix doesn't work." But then they ask the question: "Why?" We leave the answer to the experts. In the meantime, it is more important to consider what behavior safety is all about—what exactly is behavior-based safety?

Simply, behavior-based safety is all about using positive reinforcement to change unsafe individual behaviors. Opponents of this model argue that this is nothing more than the "Hawthorne Effect"—that environmental conditions impact work input and worker happiness.

The behavior-based safety model begins with a behavioral analysis to identify "at-risk" behaviors. These can be determined by using:

- Accident/incident/injury/near-miss reviews
- Use of incident/antecedent reports
- Employee interviews
- Job Hazard Analysis (JHA)
- Brainstorming

Did You Know?

Behavior-based Safety is the application of the science of behavior change to real-world problems (CCBS, 2015).

To get to the heart of the behavior-based model we must first begin at the beginning. Behavior-based safety is based loosely on the tenets of B.F. Skinner's basics of behavioral modification. Skinner, a famous psychological theorist, examined the way people act and why they react to things in the way they do. He and others did not think it was necessary to bother with the inner workings of the mind. Instead, they elected to focus on the observable behaviors of their subjects. From their body of work emerged a body of knowledge that explains behavior in terms of stimulus, response, and consequences. Behavior modification theory seeks to break all human behavior into these parts. The actual behavior modification methodology involves four techniques (Skinner, 1965). Two methods promote behaviors:

1. *Positive Reinforcement.* This technique rewards correct behavior with a positive consequence. When the desirable behavior occurs, something follows that is pleasurable. For teaching the correct behavior, this technique is unsurpassed. The pleasurable consequence follows the correct behavior and whoever receives the consequence knows exactly what behavior to repeat next time to receive the pleasure. Remember that the closer the reinforcement is to the behavior, the more power it has on future behaviors.

2. *Negative Reinforcement.* From the pain/pleasure principle, it is certain that behaviors that lead to pleasure will be repeated and behaviors that lead to pain will be avoided. Negative reinforcement relies on the second half of this principle, the tendency to avoid pain. As with positive reinforcement, behaviors that are negatively reinforced are more likely to be repeated when a similar circumstance is

encountered. However, in the case of negative reinforcement, the consequence is not pleasure but the removal of pain. Negative reinforcement relies on the removal of negative circumstances as the result of a behavior to reinforce the behavior. In other words, when some behavior leads to the removal of a negative consequence, that behavior is likely to be repeated when similar situations arise in the future. For example, if a person is cold (negative consequence) and turns on the heat (behavior), then they are not cold any more. Next time the circumstances are similar, they will likely behave in the same way.

Two methods discourage behaviors:

1. *Removal of positive consequences as result of undesirable behavior (Extinction).* The first technique used to reduce the likelihood that a behavior will be repeated is called *extinction.* Responses that are not reinforced are not likely to be repeated. (Ignoring worker misbehavior should extinguish that behavior.) By taking away the pleasurable consequence, the behavior eventually goes away, replaced by new behaviors that do lead to pleasurable consequences.
2. *Negative Consequences/Punishment.* Responses that bring painful or undesirable consequence will be suppressed, but many reappear if reinforcement contingencies change. (Penalizing late workers by withdrawing privileges should stop their lateness.) The behavioral modification technique of punishment is neither the end all, be all, nor the enigma of discipline. When used correctly it is no more out of place than rewards (which taken to extreme can be misused as well). With this justification aside, consider exactly how punishment fits into the model of behavioral modification.

The behavior-based safety system is considered the only proactive safety system. It involves direct intervention to change unsafe behavior, using scientific methodology to reinforce desired actions. Behavior-based safety involves four linked steps:

1. Identify safety-related behaviors that are critical to performance excellence
2. Gather data on work group conformance to safety excellence
3. Provide ongoing, two-way performance
4. Use accumulating behavior-based data to remove barriers to continuous improvement

Behavior-based safety focuses on human factors: Behavior = observable effect (no connotation of good or bad). It involves no mind reading or assumptions. The employee is simply a Black Box: This is the situation (input), this is the action (output).

One of the critical factors in behavior-based safety is observation. Observers apply the "ABC" principles of behavior-based safety observation (Krause, Hidley, and Hodson, 1990).

Antecedent: The issue, which precedes actions, but is the root cause.
Behavior: The action, which is the result of an antecedent.

Consequence: What happens, if anything, as a result of a behavior (positive or negative).

Observers learn to look at how the job is done and identify all task components on the checklist that are completed safely. Any step in the process completed unsafely— or any near miss—is marked as "at risk." The observer also provides feedback, emphasizing positives to reinforce safe performance and discussing at-risk performance. The observer looks for "footprints," or antecedents, to use in reaching agreement with the work team member on the need to change at-risk components.

So, is the behavior-based model the "right way" to manage safety and environmental health in the workplace? Judge for yourself; for us, the jury is still out. There is little doubt that the behavior-based model will work in some applications—the documented success stories seem unlimited. However, we also see some disadvantages of behavior-based safety. Specifically, this model requires "change" in the way things are done. We all know about change and its ramifications—people hate change. This model takes a lot of initial effort to train observers and make observations. Moreover, like any safety model, management must support it wholeheartedly. Finally, consultants are expensive.

On the other side of the coin, McSween sings the praises of behavior-based models: The behavior-based safety process is the only empirical approach to improving safety that has proven to be effective (McSween, 1998).

The bottom line: no matter what model one implements to manage an organizational safety program, the question becomes how does one achieve lasting results? The unfortunate reality is fairly simple—lasting change takes lasting effort. And experience has shown that this is not so easy.

BENCHMARKING

Benchmarking is another one of those relatively new buzzwords that has come to the forefront in management models. In regard to safety management, it can be defined as a process for rigorously measuring your safety program versus "best-in-class" programs and using the analysis to meet and exceed the best in class. Benchmarking versus best practices gives organizations a way to evaluate their safety programs—how effective and how cost effective they are. Benchmarking also shows companies both how well their programs stack up and how well those programs are implemented. Simply, benchmarking (1) is a new way of doing business; (2) is an objective-setting process; (3) forces an external view to ensure correctness of objective-setting; (4) forces internal alignment to achieve company safety goals; and (5) promotes teamwork by directing attention to those practices necessary to remain competitive. The benchmarking process is shown in Figure 2.1.

Start →Plan→Research→Observe→Analyze→Adapt

Figure 2.1 The Benchmarking Process

Our focus in this text is on the use of the benchmarking tool to improve safety and environmental management. Benchmarking versus best practices gives the safety and environmental manager a way to evaluate his or her operations overall.

What Benchmarking Can Reveal

- How effective the organization or process is
- How cost effective the organization or process is
- How well their operations stack up, and how well those operations are implemented

Potential Results of Benchmarking

- Benchmarking is an objective-setting process
- Benchmarking is a new way of doing business
- Benchmarking forces an external view to ensure correctness of objective setting
- Benchmarking forces internal alignment to achieve plant goals
- Benchmarking promotes teamwork by directing attention to those practices necessary to remain competitive
- Benchmarking may indicate direction of required change rather than specific metrics
  - Costs must be reduced
  - Customer satisfaction must be increased
  - Return on assets must be increased
  - Improved maintenance
  - Improved operational practices
  - Best practices translated into operational units of measure

Targets

- Consideration of available resources converts benchmark findings to targets
- A target represents what can realistically be accomplished in a given time frame
- Can show progress toward benchmark practices and metrics
- Quantification of precise targets should be based on achieving a benchmark

Did You Know?

Benchmarking can be performance based, process based, or strategic based and can compare financial or operational performance measures, methods or practices, or strategic choices.

Benchmarking: The Process

When forming a benchmarking team, the goal should be to provide a benchmark that evaluates and compares privatized and re-engineered organizational operations to your particular operation in order to be more efficient, remain competitive, and make continual improvements. It is important to point out that benchmarking is more than simply setting a performance reference or comparison; it is a way to facilitate learning for continual improvements. The key to the learning process is looking outside one's own plant to other plants that have discovered better ways of achieving improved performance.

Benchmarking Steps

As shown in Table 2.1, the benchmarking process consists of five steps.

Table 2.1 Benchmarking Steps

Step 1	Planning	Managers must select a process (or processes) to be benchmarked. A benchmarking team should be formed. The process of benchmarking must be thoroughly understood and documented. The performance measure for the process should be established (i.e., cost, time, and quality).
Step 2	Research	Information on the "best-in-class" performer must be determined through research. The information can be derived from the industry's network, industry experts, industry and trade associations, publications, public information, and other award-winning operations.
Step 3	Observation	The observation step is a study of the benchmarking subject's performance level, processes, and practices that have achieved those levels and other enabling factors.
Step 4	Analysis	In this phase, comparisons in performance levels among facilities are determined. The root causes for the performance gaps are studied. To make accurate and appropriate comparisons, the comparison data must be sorted, controlled for quality, and normalized.
Step 5	Adaptation	This phase is putting what is learned throughout the benchmarking process into action. The findings of the benchmarking study must be communicated to gain acceptance, functional goals must be established, and a plan must be developed. Progress should be monitored and, as required, corrections in the process made.

Did You Know?

Benchmarking should be interactive. It should also recalibrate performance measures and improve the process itself.

THE TOTAL QUALITY MANAGEMENT (TQM) PARADIGM

What is Total Quality Management (TQM)? TQM is not a traditional part of American culture. In the old days if an organization had an abundance of resources, waste was affordable. At present, with increasing competition, and an economy growing at a slower rate, doing it right the first time is more important. Simply, "TQM is an integrated system of principles, methods, and best practices that provide a framework for organizations to strive for excellence in everything they do" (*Total Quality Engineering Inc.*, 2003).

No two organizations have the same TQM implementation, but certain characteristics or elements are uniform. Consider the following:

1. TQM incorporates dynamic people concepts
2. TQM uses strategic planning with objectives and measurements
3. TQM uses benchmarking and program evaluation techniques
4. TQM focuses on continuous improvement
5. TQM revolves around four key concepts: Customers, Waste, Time, and Excellence

To determine and define "Customers," answer the following questions:

1. Who are they?
2. What do they need and want?
3. What do they expect?
4. Are they internal or external customers?

To determine and define "Waste," consider the following:

1. Waste is not only what is thrown away
2. Waste is anything that is inefficient, wastes resources, or generates waste. Includes product rework, excess inventory, disposal costs, excess labor, and injuries
3. Waste drives up cost

To determine and define "Time," consider the following:

1. Cycle time: from the start of an activity to its completion
2. Time is money
3. Wasted time affects both product and operation cost

To determine and define "Excellence," consider the following:

1. Excellence determined by customers (in TQM process) in comparison to competition
2. Continual pursuit
3. Outcome of total quality process

TQM incorporates dynamic people concepts, such as participation, empowerment, and ownership.

What does all this have to do with safety and environmental management? Accidents and incidents can be viewed as defective safety products. TQM is a significant management paradigm for non-line, special functions like safety and environmental health where high participation and holistic management can be taken advantage. Remember that safety and environmental management do not generate income, so they must be cost-effective.

THOUGHT-PROVOKING QUESTIONS

• What personal qualities indicate the possibility of good management ability?

- Discuss "the right way" to manage.
- Discuss "planning," "organizing," "controlling," and "directing."
- What's the fifth essential element that synergizes planning, controlling, organizing and directing? Why?
- How can you ensure compliance in your facility as safety manager, through persuasion, not muscle?
- Discuss participative management.
- What are some of the drawbacks to collective management?
- Relate "The camel is a horse that was designed by a committee" to collective management.
- Discuss behavior-based management versus the three Es of safety.
- How can using positive reinforcement lead to safe workplace behaviors?
- What happens when you take away the chocolate chip cookies? (i.e., the reward)
- What are the ABCs of behavior-based observation? Discuss.
- Define benchmarking in light of safety and environmental management.
- How can benchmarking be used to improve safety management? Discuss seven different ways.
- Name and discuss four different potential results of benchmarking.
- Describe the benchmarking process and the steps.
- What is total quality management?
- Discuss how TQM uses the four key concepts.

REFERENCES AND RECOMMENDED READING

Cambridge Center for Behavioral Studies. "Introduction to Behavioral Safety." Accessed February 28, 2015. http://www.behavior.org/resource.php?id=330.

Cooper, Dominic. *Behavioral Safety: A Framework for Success.* Franklin, IN: BSMS (2009).

Code of Federal Regulations. Occupational Safety and Health Standards, title 29, sec. 1910.

Krause, T.R., J.H. Hidley, and S.J. Hodson. *The Behavior-Based Safety Process: Managing Involvement for an Injury-Free Culture.* New York: Van Nostrand Reinhold (1990).

Mathis, T., and T. McSween. "Behavior-Based Safety: The Infamous Quick Fix." *Industrial Safety & Hygiene News.* February 1996.

McSween, T. "Culture: A behavioral perspective." In Proceedings of Light Up Safety in the New Millennium: A Behavioral Safety Symposium, 43–49. Des Plaines, IL: America Society of Safety Engineers, 1998.

Sarkus, D.J. *The Safety Coach: Unleash the 7 C's for World Class Safety Performance.* Donora, PA: Championship Publishing, 2001.

Skinner, B.F. *Science and Human Behavior.* New York: The Free Press, 1965.

Sulzer-Azeroff, B., and W.E. Iischeid. "Assessing the Quality of Behavioral Safety Initiatives." *Professional Safety* 44, no. 4 (1999): 31–36.

Total Quality Management. Poway, CA: Total Quality Engineering Inc, 2003.

Chapter 3

Safety and Environmental Management Terminology

Every profession has its own language for communication. Safety and environmental management is no different. To work even at the edge of safety and environmental management, you must acquire a fundamental vocabulary of the components that make up the process of administering safety and environmental management.

Administering safety and environmental management?

Absolutely. Remember, the safety and environmental manager is a practitioner of safety and environmental management, a specialized field.

Anytime we look to a definition for meaning, we are wise to remember the words of that great philosopher who provides meanings in terms we can all understand. "90% of baseball is mental, the other half is physical." (Yogi Berra)

INTRODUCTION

In this chapter, we define the "terms" or "tools" (concepts and ideas) used by safety and environmental managers in applying their skills to make our technological world safer. We present these concepts early in the text rather than later (as is traditionally done in an end of book glossary), so you can become familiar with the terms early, before the text approaches the issues those terms describe. The practicing safety and environmental manager or student of safety and environmental management should know these concepts. Without them it is difficult (if not impossible) to practice safety and environmental management. Several other chapters contain vocabulary specific to those more specialized fields.

Safety and environmental management has extensive and unique terminology, most with well-defined meanings, but a few terms [especially *safety, accident, injuries*, and *engineering* (as used in the safety context)] often are not only poorly defined, but are defined from different and conflicting points of view. For our purpose, we present the definitions of key terms, highlighting and explaining those poorly defined terms (showing different views from different sides) where necessary.

We do not define every safety and environmental management term—only those terms and concepts necessary to understand the technical jargon presented in this text. For those practicing safety and environmental managers and students of safety and environmental management who want a complete, up-to-date, and accurate "dictionary" of terms used in the safety profession, we recommend the American Society of Safety Engineers (ASSE) text, *The Dictionary of Terms Used in The Safety Profession*, Fourth Edition (2001) This concise, informative, and valuable safety asset can be obtained from the ASSE at www.asse.org/dictionary-of-terms-used-in-the-safety-profession-fourth-edition/

TERMINOLOGY

Abatement period: The amount of time given to an employer to correct a hazardous condition that has been cited.

Absorption: The taking up of one substance by another, such as a liquid by a solid or a gas by a liquid.

Accident: This term is often misunderstood and is often mistakenly used interchangeably with *injury*. The meanings of the two terms are different, of course. Let's look at the confusion caused by the different definitions supplied to the term *accident*. The dictionary defines an accident as "a happening or event that is not expected, foreseen or intended." Defined another way: "an accident is an event or condition occurring by chance or arising from an unknown or remote cause." The legal definition is: "an unexpected happening causing loss or injury which is not due to any fault or misconduct on the part of the person injured, yet entitles some kind of legal relief."

Are you confused? Stand by.

Now you should have a better feel for what an accident really is; however, another definition, perhaps one more applicable to our needs is provided by safety experts—the authors of the *ASSE Dictionary of Safety Terms*, 2001. Let's see how they define *accident*.

"An accident is an unplanned and sometimes injurious or damaging event which interrupts the normal progress of an activity and is invariably preceded by an unsafe act or unsafe condition thereof. An accident may be seen as resulting from a failure to identify a hazard or from some inadequacy in an existing system of hazard controls. Based on applications in casualty insurance, an event that is definite in point of time and place but unexpected as to either its occurrence or its results" (p. 1).

In this text we use the ASSE's definition of *accident*.

Accident analysis (see accident investigation): A comprehensive, detailed review of the data and information compiled from an accident investigation. An accident analysis should be used to determine causal factors only, not to point the finger of blame at any one. Once the causal factors have been determined, corrective measures should be prescribed to prevent recurrence.

Accident prevention: The act of preventing a happening that may cause loss or injury to a person.

Accommodation: The ability of the eye to become adjusted after viewing the visual display terminal (VTD) so as to be able to focus on other objects, particularly objects at a distance.

ACGIH: American Conference of Governmental Industrial Hygienists is an organization of professional personnel in governmental agencies or educational institutes engaged in occupational safety and health programs. ACGIH establishes recommended occupational exposure limits for chemical substances and physical agents.

Acoustics: In general, the experimental and theoretical science of sound and its transmission; in particular, that branch of the science that has to do with the phenomena of sound in a particular space such as a room or theater. Safety engineering is concerned with the technical control of sound, and involves architecture and construction, studying control of vibration, soundproofing, and the elimination of noise—to engineer out the noise hazard.

Acid: Any chemical that undergoes dissociation in water with the formation of hydrogen ions. Acids have a sour taste and may cause severe skin burns. Acids turn litmus paper red and have pH values of 0 to 6.

Action level: Term used by OSHA and NIOSH (National Institute for Occupational Safety and Health—a federal agency that conducts research on safety and health concerns) and is defined in the Code of Federal Regulations (CFR), Title 40, Protection of Environment. Under OSHA, action level is the level of toxicant, which requires medical surveillance, usually 50% of the PEL (Personal Exposure Level). Note that OSHA also uses the action level in other ways besides setting the level of "toxicant." For example, in its hearing conservation standard, 29 CFR 1910.95, OSHA defines the action level as an eight-hour time-weighted average (TWA) of 85 decibels measured on the A-scale, slow response, or equivalently, a dose of 50% (see Chapter 27 for more information on hearing conservation). Under CFR 40 §763.121, action level means an airborne concentration of asbestos of 0.1 fiber per cubic centimeter (f/cc) of air calculated as an 8-hour time-weighted average.

Acute: Health effects, which show up a short length of time after exposure. An acute exposure runs a comparatively short course and its effects are easier to reverse than those of a chronic exposure.

Acute toxicity: The discernible adverse effects induced in an organism within a short period of time (days) of exposure to an agent.

Adenocarcinoma: A tumor with glandular (secreting) elements.

Adenosis: Any disease of a gland.

Adhesion: A union of two surfaces that are normally separate.

Adsorption: The taking up of a gas or liquid at the surface of another substance, usually a solid (e.g., activated charcoal adsorbs gases).

Aerosols: Liquid or solid particles that are so small. They can remain suspended in air long enough to be transported over a distance.

Air contamination: The result of introducing foreign substances into the air so as to make the air contaminated.

Airline respirator: A respirator that is connected to a compressed, breathable air source by a hose of small inside diameter. The air is delivered continuously or intermittently in a sufficient volume to meet the wearer's breathing requirements.

Air pollution: Contamination of the atmosphere (indoor or outdoor) caused by the discharge (accidental or deliberate) of a wide range of toxic airborne substances.

Air-purifying respirator: A respirator that uses chemicals to remove specific gases and vapors from the air or that uses a mechanical filter to remove particulate matter. An air-purifying respirator must only be used when there is sufficient oxygen to sustain life and the air contaminant level is below concentration limits of the device.

Air sampling: Safety engineers are interested in knowing what contaminants workers are exposed to, and the contaminant concentrations. Determining the quantities and types of atmospheric contaminants is accomplished by measuring and evaluating a representative sample of air. The types of air contaminants that occur in the workplace depend upon the raw materials used and the processes employed. Air contaminants can be divided into two broad groups, depending upon physical characteristics: (1) gases and vapors and (2) particulates.

Allergens: An antigenic substance capable of producing immediate hypersensitivity.

Ambient: Descriptive of any condition of the environment surrounding a given point. For example, ambient air means that portion of the atmosphere, external to buildings, to which the general public has access. Ambient sound is the sound generated by the environment.

Asphyxiation: Suffocation from lack of oxygen. A substance (e.g., carbon monoxide), that combines with hemoglobin to reduce the blood's capacity to transport oxygen produces chemical asphyxiation. Simple asphyxiation is the result of exposure to a substance (such as methane) that displaces oxygen.

Atmosphere: In physics, a unit of pressure whereby 1 atmosphere (atm) equals 14.7 pounds per square inch (psi).

Attenuation: The reduction of the intensity at a designated first location as compared with intensity at a second location, which is farther from the source (reducing the level of noise by increasing distance from the source is a good example).

Audible range: The frequency range over which normal hearing occurs—approximately 20 Hz through 20,000 Hz. Above the range of 20,000 Hz, the term ultrasonic is used. Below 20 Hz, the term subsonic is used.

Audiogram: A record of hearing loss or hearing level measured at several different frequencies—usually 500 to 6000 Hz. The audiogram may be presented graphically or numerically. Hearing level is shown as a function of frequency.

Audiometric testing: Objective measuring of a person's hearing sensitivity. By recording the response to a measured signal, a person's level of hearing sensitivity can be expressed in decibels, as related to an audiometric zero, or no-sound base.

Authorized person (see competent or qualified person): A person designated or assigned by an employer or supervisor to perform a specific type of duty or duties, to use specified equipment, and/or to be present in a given location at specified times (for example, an authorized or qualified person is used in confined space entry).

Auto ignition temperature: The lowest temperature at which a vapor-producing substance or a flammable gas will ignite even without the presence of a spark or flame.

Baghouse: Term commonly used for the housing containing bag filters for recovery of fumes from arsenic, lead, sulfa, etc. Many different trade meanings, however.

Base: A substance that (1) liberates hydroxide (OH) ions when dissolved in water, (2) receives hydrogen ions from a strong acid to form a weaker acid, and (3) neutralizes an acid. Bases react with acids to form salts and water. Bases have a pH greater than 7 and turn litmus paper blue.

Baseline data: Data collected prior to a project for later use in describing conditions before the project began. Also commonly used to describe the first audiogram given (within six months) to a worker after he or she has been exposed to the action level (85 dBA—to establish his or her baseline for comparison to subsequent audiograms.

Bel: A unit equal to 10 decibels (see decibel).

Benchmarking: A process for rigorously measuring company performance versus "best-in-class" companies and using analysis to meet and exceed the best in class.

Behavior-based management models: A management theory, based on the work of B.F. Skinner, that explains behavior in terms of stimulus, response, and consequences.

Bioaerosols: Mold spores, pollen, viruses, bacteria, insect parts, animal dander, etc. that are small enough to be suspended in the air.

Biohazard (biological hazard): Organisms or products of organisms that present a risk to humans.

Boiler code: ANSI/ASME Pressure Vessel Code whereby a set of standards prescribing requirements for the design, construction, testing, and installation of boilers and unfired pressure vessels.

Boyle's Law: The product of a given pressure and volume is constant with a constant temperature.

Carcinogen: A cancer-producing agent.

Carpal Tunnel Syndrome: An injury to the median nerve inside the wrist.

Catalyst: A substance that alters the speed of, or makes possible, a chemical or biochemical reaction but remains unchanged at the end of the reaction.

Catastrophe: A loss of extraordinary large dimensions in terms of injury, death, damage, and destruction.

Casual factor (accident cause): A person, thing, or condition that contributes significantly to an accident or to a project outcome.

Charles's Law: Law stating that the volume of a given mass of gas at constant pressure is directly proportional to its absolute temperature (temperature in kelvin).

Chemical change: Change that occurs when two or more substances (reactants) interact with each other, resulting in the production of different substances (products) with different chemical compositions. A simple example of chemical change is the burning of carbon in oxygen to produce carbon dioxide.

Chemical hazards: Include mist, vapors, gases, dusts, and fumes.

Chemical spill: An accidental dumping, leakage, or splashing of a harmful or potentially harmful substance.

Chronic: Persistent, prolonged, and repeated. Chronic exposure occurs when repeated exposure to or contact with a toxic substance occurs over a period of time, the effects of which become evident only after multiple exposures.

Coefficient of friction: A numerical correlation of the resistance of one surface against another surface.

Combustible gas indicator: An instrument that samples air and indicates whether an explosive mixture is present and the percentage of the lower explosive limit (LEL) of the air-gas mixture that has been reached.

Combustible liquid: Liquids having a flash point at or above (100°F) 37.8°C.

Combustion: Burning, defined in chemical terms as the rapid combination of a substance with oxygen, accompanied by the evolution of heat and usually light.

Competent person: As defined by OSHA, one who is capable of recognizing and evaluating employee exposure to hazardous substances or to unsafe conditions, and who is capable of specifying protective and precautionary measures to be taken to ensure the safety of employees as required by particular OSHA regulations under the conditions to which such regulations apply.

Confined space: A vessel, compartment, or any area having limited access and (usually) no alternate escape route, having severely limited natural ventilation or an atmosphere containing less than 19.5 percent oxygen, and having the capability of accumulating a toxic, flammable, or explosive atmosphere, or of being flooded (engulfing a victim).

Containment: In fire terminology, restricting the spread of fire. For chemicals, restricting chemicals to an area that is diked or walled off to protect personnel and the environment.

Contingency plan: (commonly called the emergency response plan) Under CFR 40 § 260.10), a document that sets forth an organized, planned, and coordinated course of action to be followed in the event of an emergency that could threaten human health or the environment.

Convection: The transfer of heat from one location to another by way of a moving medium.

Corrosive material: Any material that dissolves metals or other materials, or that burns the skin.

Cumulative injury: A term used to describe any physical or psychological disability that results from the combined effects of related injuries or illnesses in the workplace.

Cumulative trauma disorder: A disorder caused by the highly repetitive motion required of one or more parts of a worker's body, which in some cases can result in moderate to total disability.

Dalton's Law of Partial Pressures: In a mixture of theoretically ideal gases, the pressure exerted by the mixture is the sum of the pressures exerted by each component gas of the mixture.

Decibel (dB): A unit of measure used originally to compare sound intensities and subsequently electrical or electronic power outputs; now also used to compare voltages. In hearing conservation it is a logarithmic unit used to express the magnitude of a change in level of sound intensity.

Decontamination: The process of reducing or eliminating the presence of harmful substances such as infectious agents to reduce the likelihood of disease transmission from those substances.

Density: A measure of the compactness of a substance; it is equal to its mass per unit volume and is measured in kilograms per cubic meter/lb per cubic foot (D = mass/volume).

Dermatitis: Inflammation or irritation of the skin from any cause. Industrial dermatitis is an occupational skin disease.

Design load: The weight, which can be safely supported by a floor, equipment, or structure, as defined by its design characteristics.

Dike: An embankment or ridge of either natural or man made materials used to prevent the movement of liquids, sludges, solids, or other materials.

Dilute: Adding material to a chemical by the user or manufacturer to reduce the concentration of active ingredient in the mixture.

Dose: An exposure level. Exposure is expressed as weight or volume of test substance per volume of air (mg/l), or as parts per million (ppm).

Dosimeter: Provides a time-weighted average dose over a period of time such as one complete work shift.

Dusts: Various types of solid particles that are produced when a given type of organic or inorganic material is scraped, sawed, ground, drilled, heated, crushed, or otherwise deformed.

Electrical grounding: Precautionary measures designed into an electrical installation to eliminate dangerous voltages in and around the installation, and to operate protective devices in case of current leakage from energized conductors to their enclosures.

Emergency plan: See contingency plan.

Emergency response: The response made by firefighters, police, health-care personnel, and/or other emergency services upon notification of a fire, chemical spill, explosion, or other incident in which human life and/or property may be in jeopardy.

Energized ("live"): Having voltage applied to the conductors of an electrical circuit such conductors and to surfaces which a person might touch; having voltage between such surfaces and other surfaces which might complete a circuit and allow current to flow.

Energy: The capacity for doing work. Potential energy (PE) is energy deriving from position; thus a stretched spring has elastic PE, and an object raised to a height above the earth's surface, or the water in an elevated reservoir, has gravitational PE. A lump of coal and a tank of oil, together with oxygen needed for their combustion, have chemical energy. Other sorts of energy include electrical and nuclear energy, light, and sound. Moving bodies possess kinetic energy (KE). Energy can be converted from one form to another, but the total quantity stays the same (in accordance with the conservation of energy principle). For example, as an orange falls, it loses gravitational PE, but gains KE.

Engineering: The application of scientific principles to the design and construction of structures, machines, apparatus, manufacturing processes, and power generation and utilization, for the purpose of satisfying human needs. Safety engineering is concerned with control of environment and humankind's interface with it, especially safety interaction with machines, hazardous materials, and radiation.

Engineering controls: Methods of controlling employee exposures by modifying the source and reducing the quantity of contaminants released into the workplace environment.

Epidemiological Theory: The models used for studying and determining epidemiological relationships can also be used to study causal relationships between environmental factors and accidents or diseases.

Ergonomics: A multidisciplinary activity dealing with interactions between man and his total working environment, plus stresses related to such environmental elements as atmosphere, heat, light, and sound, as well as all tools and equipment of the workplace.

Etiology: The study or knowledge of the causes of disease.

Exposure: Contact with a chemical, biological, or physical hazard.

Exposure ceiling: The concentration level of a given substance that should not be exceeded at any point during an exposure period.

Fall-arresting system: A system consisting of a body harness, a lanyard or lifeline, and an arresting mechanism with built-in shock absorber, designed for use by workers performing tasks in locations from which falls would be injurious or fatal, or where other kinds of protection are not practical.

Fire: A chemical reaction between oxygen and a combustible fuel.

Flammable liquid: Any liquid having a flash point below (100°F) 37.8°C.

Flammable solid: A nonexplosive solid liable to cause fire through friction, absorption of moisture, spontaneous chemical change, or heat retained from a manufacturing process, or that can be ignited readily and when ignited, burns so vigorously and persistently so as to create a serious hazard.

Flash point: The lowest temperature at which a liquid gives off enough vapor to form an ignitable moisture with air, and produce a flame when a source of ignition is present. Two tests are used: open cup and closed cup.

Foot-candle: A unit of illumination. The illumination at a point on a surface which is one foot from, and perpendicular to, a uniform point source of one candle.

Fume: Airborne particulate matter formed by the evaporation of solid materials, for example metal fume emitted during welding. Usually less than one micron in diameter.

Gas: A state of matter in which the material has very low density and viscosity, can expand and contract greatly in response to changes in temperature and pressure, easily diffuses into other gases, and readily and uniformly distributes itself throughout any container.

Grounded system: A system of conductors in which at least one conductor or point is intentionally grounded, either solidly or through a current-limiting (current transformer) device.

Ground-fault circuit interrupter (GFCI): A sensitive device intended for shock protection, which functions to de-energize an electrical circuit or portion thereof within a fraction of a second, in case of leakage to ground of current sufficient to be dangerous to persons but less than that required to operate the overcurrent protective device of the circuit.

Hazard: The potential for an activity, condition, circumstance, or changing conditions or circumstances to produce harmful effects. Also an unsafe condition.

Hazard analysis: A systematic process for identifying hazards and recommending corrective action.

Hazard assessment: A qualitative evaluation of potential hazards in the interrelationships between and among the elements of a system, upon the basis of which the occurrence probability of each identified hazard is rated.

Hazard Communication Standard (HazCom): An OSHA workplace standard found in 29 CFR 1910.1200 that requires all employers to become aware of the chemical hazards in their workplace and relay that information to their employees. In addition, a contractor conducting work at a client's site must provide chemical information to the client regarding the chemicals that are brought onto the work site.

Hazard and operability (HAZOP) analysis: A systematic method in which process hazards and potential operating problems are identified, using a series of guide words to investigate process deviations.

Hazard identification: The pinpointing of material, system, process, and plant characteristics that can produce undesirable consequences through the occurrence of an accident.

Hazard control: A means of reducing the risk from exposure to a hazard.

Hazardous material: Any material possessing a relatively high potential for harmful effects upon persons.

Hazardous substance: Any substance that has the potential for causing injury by reason of its being explosive, flammable, toxic, corrosive, oxidizing, irritating, or otherwise harmful to personnel.

Hazardous waste: A solid, liquid, or gaseous waste that may cause or significantly contribute to serious illness or death, or that poses a substantial threat to human health or the environment when the waste is improperly managed.

Hearing conservation: The prevention of, or minimizing of, noise-induced deafness through the use of hearing protection devices, the control of noise through engineering controls, annual audiometric tests, and employee training.

Heat cramps: A type of heat stress that occurs as a result of salt and potassium depletion.

Heat exhaustion: A condition usually caused by loss of body water from exposure to excess heat. Symptoms include headache, tiredness, nausea, and sometimes fainting.

Heatstroke: A serious disorder resulting from exposure to excess heat. It results from sweat suppression and increased storage of body heat, characterized by high fever, collapse, and sometimes convulsions or coma.

Homeland security: Federal cabinet level department created to protect America as a result of 9/11. The new Department of Homeland Security (DHS) has three primary missions: Prevent terrorist attacks within the United States, reduce America's vulnerability to terrorism, and minimize the damage from potential attacks and natural disasters.

Hot work: Work involving electric or gas welding, cutting, brazing, or similar flame or spark-producing operations.

Human factor engineering (used in the United States) or **Ergonomics** (used in Europe): For practical purposes, the terms are synonymous, and focus on human beings and their interactions with products, equipment, facilities, procedures, and environments used in work and everyday living. The emphasis is on human beings (as opposed to engineering, where the emphasis is more strictly on technical engineering considerations) and how the design of things influences people. Human factors, then, seek to change the things people use and the environments in which they use these things to better match the capabilities, limitations, and needs of people (Sanders & McCormick, 1993).

Ignition temperature: The temperature at which a given fuel can burst into flame.

Illumination: The amount of light flux a surface receives per unit area. May be expressed in lumens per square foot or in foot-candles.

Impulse noise: A noise characterized by rapid rise time, high peak value, and rapid decay.

Incident: An undesired event that, under slightly different circumstances, could have resulted in personal harm or property damage; any undesired loss of resources.

Indoor air quality (IAQ): refers to the effect, good or bad, of the contents of the air inside a structure, on its occupants. Usually, temperature (too hot and cold), humidity (too dry or too damp), and air velocity (draftiness or motionless) are considered "comfort" rather than indoor air quality issues. Unless they are extreme, they may make someone unhappy, but they won't make a person ill. Nevertheless, most IAQ professionals will take these factors into account in investigating air quality situations.

Industrial hygiene: The American Industrial Hygiene Association (AIHA) defines industrial hygiene as "that science and art devoted to the anticipation, recognition, evaluation, and control of those environmental factors or stresses—arising in the workplace—which may cause sickness, impaired health and well-being, or significant discomfort and inefficiency among workers or among citizens of the community."

Ingestion: Entry through the mouth.

Injury: A wound or other specific damage.

Interlock: A device that interacts with another device or mechanism to govern succeeding operations. For example: an interlock on an elevator door prevents the car from moving unless the door is properly closed.

Ionizing radiation: Radiation that becomes electrically charged or changed into ions.

Irritant: A substance that produces an irritating effect when it contacts the skin, eyes, nose, or respiratory system.

Job hazard analysis: (also called job safety analysis) The breaking down into its component parts of any method or procedure, to determine the hazards connected therewith and the requirements for performing it safely.

Kinetic energy: The energy resulting from a moving object.

Laboratory Safety Standard: A specific hazard communication program for laboratories, found in 29 CFR 1910.1450. These regulations are essentially a blend of hazard communication and emergency response requirements for laboratories. The cornerstone of the Lab Safety Standard is the requirement for a written Chemical Hygiene Plan.

Lockout/tagout procedure: An OSHA procedure found in 29 CFR 1910.147. A tag or lock is used to tag out or log out a device, so that no one can inadvertently actuate the circuit, system, or equipment that is temporarily out of service.

Log and summary of occupational injuries and illnesses (OSHA-200 Log): A cumulative record that employers (generally of more than 10 employees) are required to maintain, showing essential facts of all reportable occupational injuries and illnesses.

Loss: The degradation of a system or component. Loss is best understood when related to dollars lost. Examples include death or injury to a worker, destruction or impairment of facilities or machines, destruction or spoiling of raw materials, and creation of delay. In the insurance business, loss connotes dollar loss, and we have seen underwriters who write it as LO$$ to make that point.

Lower explosive limit (LEL): The minimum concentration of a flammable gas in air required for ignition in the presence of an ignition source. Listed as a percent by volume in air.

Material safety data sheet (MSDS): Chemical information sheets provided by the chemical manufacturer that include information such as: chemical and physical characteristics; long- and short-term health hazards; spill control procedures; personal protective equipment (PPE) to be used when handling the chemical; reactivity with other chemicals; incompatibility with other chemicals; and manufacturer's name, address, and phone number. Employee access to and understanding of MSDS are important parts of the HazCom Program.

Medical monitoring: The initial medical examination of a worker, followed by periodic examinations. The purpose of medical monitoring is to assess workers' health, determine their fitness to wear personal protective equipment, and maintain records of their health.

Metabolic heat: Produced within a body as a result of activity that burns energy.

Mists: Tiny liquid droplets suspended in air.

Molds: are the most typical form of fungus found on earth, comprising approximately 25% of the earth's biomass (McNeel and Kreutzer, 1996).

Monitoring: Periodic or continuous surveillance or testing to determine the level of compliance with statutory requirements and/or pollutant levels, in various media, or in humans, animals or other living things.

Mycotoxins: Some molds are capable of producing *mycotoxins*, natural organic compounds that are capable of initiating a toxic response in vertebrates (McNeel and Kreutzer, 1996).

Nonionizing radiation: That radiation on the electromagnet spectrum that has a frequency of 10^{15} or less and a wavelength in meters of 3×10^{-7}.

Occupational Safety and Health Act (OSH Act): A federal law passed in 1970 to assure, so far as possible, every working man and woman in the nation safe and healthful working conditions. To achieve this goal, the Act authorizes several functions, such as encouraging safety and health programs in the workplace and encouraging labor-management cooperation in health and safety issues.

OSHA Form 300: Log and Summary of Occupational Injuries and Illnesses.

Oxidation: When a substance either gains oxygen, or loses hydrogen or electrons in a chemical reaction. One of the chemical treatment methods.

Oxidizer: Also known as an oxidizing agent, a substance that oxidizes another substance. Oxidizers are a category of hazardous materials that may assist in the production of fire by readily yielding oxygen.

Oxygen-deficient atmospheres: The legal definition of an atmosphere where the oxygen concentration is less than 19.5% by volume of air.

Particulate matter: Minute, separate particle of liquid or solid material released directly into the air such as diesel soot and combustion products resulting from the burning of wood.

Performance standards: A form of OSHA regulation standards that lists the ultimate goal of compliance but does not explain exactly how compliance is to be accomplished. Compliance is usually based on accomplishing the act or process in the safest manner possible, based on experience (past performance).

Permissible exposure limit (PEL): The time-weighted average concentration of an airborne contaminant that a healthy worker may be exposed to 8 hours per day or 40 hours per week without suffering any adverse health effects. Established by legal means and enforceable by OSHA.

Personal protective equipment (PPE): Any material or device worn to protect a worker from exposure to or contact with any harmful substance or force.

Preliminary assessment: A quick analysis to determine how serious the situation is, and to identify all potentially responsible parties. The preliminary assessment uses readily available information, for instance, forms, records, aerial photographs, and personnel interviews.

Pressure: The force exerted against an opposing fluid or thrust distributed over a surface.

Radiant heat: The result of electromagnetic nonionizing energy that is transmitted through space without the movement of matter within that space.

Radiation: Consists of energetic nuclear particles and includes alpha rays, beta rays, gamma rays, e-rays, neutrons, high-speed electrons, and high-speed protons.

Reactive: A substance that reacts violently by catching on fire, exploding, or giving off fumes when exposed to water, air, or low heat.

Reactivity hazard: The ability of a material to release energy when in contact with water. Also, the tendency of a material, when in its pure state or as a commercially produced product, to vigorously polymerize, decompose, condense, or otherwise self-react and undergo violent chemical change.

Reportable quantity (RQ): The minimum amount of a hazardous material that, if spilled while in transport, must be reported immediately to the National Response Center. Minimum reportable quantities range from 1 pound to 5,000 pounds per day.

Resource Conservation and Recovery Act (RCRA): A federal law enacted in 1976 to deal with municipal and hazardous waste problems and to encourage resource recovery and recycling.

Risk: The combination of the expected frequency (event/year) and consequence (effects/event) of a single accident or a group of accidents; the result of a loss-probability occurrence and the acceptability of that loss.

Risk assessment: A process that uses scientific principles to determine the level of risk that actually exists in a contaminated area.

Risk characterization: The final step in the risk assessment process, it involves determining a numerical risk factor. This step ensures that exposed populations are not at significant risk.

Risk management: The professional assessment of all loss potentials in an organization's structure and operations, leading to the establishment and administration of a comprehensive loss control program.

Safety: A general term denoting an acceptable level of risk of, relative freedom from, and low probability of harm.

Safety factor: Based on experimental data, the amount added (e.g., 1000-fold) to ensure worker health and safety.

Safety standard: A set of criteria specifically designed to define a safe product, practice, mechanism, arrangement, process, or environment, produced by a body representative of all concerned interests and based upon currently available scientific and empirical knowledge concerning the subject or scope of the standard.

Secondary containment: A method using two containment systems so that if the first is breached, the second will contain all of the fluid in the first. For USTs, secondary containment consists of either a double-walled tank or a liner system.

Security assessment: A security assessment is an intensified security test in scope and effort, the purpose of which is to obtain an advanced and very accurate idea of how well the organization has implemented security mechanisms and, to some degree, policy.

Sensitizers: Chemicals that in very low dose trigger an allergic response.

Short-term exposure limit (STEL): The time-weighted average concentration to which workers can be exposed continuously for a short period of time (typically 15 minutes) without suffering irritation, chronic or irreversible tissue damage, or impairment for self-rescue.

Silica: Crystalline silica (SiO_2) is a major component of the earth's crust and is responsible for causing silicosis.

Specific gravity: The ratio of the densities of a substance to water.

Threshold limit value (TLV): The same concept as PEL except that TLVs do not have the force of governmental regulations behind them, but are based on recommended limits established and promoted by the American Conference of Governmental Industrial Hygienists.

Time-weighted average (TWA): A mathematical average of exposure concentration over a specific time. [(exposure in ppm x time in hours) 1/4 time in hrs. = time-weighted average in ppm]

Total quality management (TQM): A way of managing a company that revolves around a total and willing commitment to quality of all personnel at all levels to quality.

Toxicity: The relative property of a chemical agent with reference to a harmful effect on some biologic mechanism and the condition under which this effect occurs. The quality of being poisonous.

Toxicology: The study of poisons, which are substances that can cause harmful effects to living things.

Unsafe condition: Any physical state that deviates from that which is acceptable, normal, or correct in terms of past production or potential future production of personal injury and/or damage to property; any physical state that results in a reduction in the degree of safety normally present.

Upper explosive limit (UEL): The maximum concentration of a flammable gas in air required for ignition in the presence of an ignition source.

Vulnerability assessment: A vulnerability assessment is very regulated, controlled, cooperative, and documented evaluation of an organization's security posture from outside-in and inside-out, for the purpose of defining or greatly enhancing security policy.

Workers' compensation: A system of insurance required by state law and financed by employers, which provides payments to employees and their families for

occupational illnesses, injuries, or fatalities incurred while at work and resulting in loss of wage income, usually regardless of the employer's or employee's
negligence.

Zero energy state: The state of equipment in which every power source that can
produce movement of a part of the equipment, or the release of energy, has been
rendered inactive.

THOUGHT-PROVOKING QUESTIONS

- Define "accident" from a safety point of view.
- Define "injuries" from a safety point of view.
- Define "engineering" from a safety point of view.

REFERENCES AND RECOMMENDED READING

American Society of Safety Engineers. *Dictionary of Terms Used in the Safety Profession*, 4th
    ed. Des Plaines, IL: ASSE, 2001.
Bird, F.E., and G.L. Germain. *Damage Control.* New York: American Management Associa-
    tion, 1966.
Boyce, A. *Introduction to Environmental Technology.* New York: Van Nostrand Reinhold,
    1997.
Center for Chemical Process Safety. *Guidelines for Hazard Evaluation Procedures*, 3rd ed.
    Hoboken, NJ: John Wiley & Sons, 2008.
Code of Federal Regulations. Occupational Safety and Health Standards, title 29, sec. 1910.
Code of Federal Regulations. Protection of Environment, title 40.
Della-Giustina, D. *Safety and Environmental Management.* Lanham, MD: Government Insti-
    tutes Press, 2007.
Fletcher, J.A. *The Industrial Environment—Total Loss Control.* Ontario, Canada: National
    Profile Limited, 1972.
Fundamentals of Industrial Hygiene, 6th ed. Edited by B.A. Plog. Chicago: National Safety
    Council, 2012.
Haddon, W., Jr., E.A. Suchman, and D. Klein. *Accident Research.* New York: Harper & Row,
    1964.
McNeel, S., and R. Kreutzer. "Fungi & Indoor Air Quality." *Health & Environment Digest* 10,
    no. 2 (1996): 9–12.
Rose, C.F. "Antigens." In *ACGIH Bioaerosols Assessment and Control*, edited by Janet
    Macher, 25.1-25.11. Cincinnati: American Conference of Governmental Industrial Hygien-
    ists, 1999.
Sanders, M.S., and E.J. McCormick. *Human Factors in Engineering and Design*, 7th ed. New
    York: McGraw-Hill, 1993.

Chapter 4

Accident Investigation

In the past, blame for accidents on workers or users has been management's device for transferring responsibility for errors. Any blame must be scrutinized more carefully, especially in light of the accident causers or contributors pointed out in [Figure 4.1].

—*Willie Hammer*

INTRODUCTION

In his well-known work, *Occupational Safety Management and Engineering*, Willie Hammer (a highly respected expert in the field of safety) made an important point that all safety and environmental health professionals should "save" into their human micro-computer memories (i.e., into their brains): "Although accident investigation does have its uses, it must be considered the hardest way to learn about hazards and accident prevention is through accidents" (p. 273). We agree with Mr. Hammer.

OSHA (2003) points out that "thousands of accidents occur throughout the United States every day. The failure of people, equipment, and surroundings to behave or react as expected causes most of the accidents. Accident investigations determine why these failures occur. By using the information gained through an investigation, a similar or perhaps disastrous accident may be prevented. Conduct accident investigations with accident prevention in mind. Investigations are NOT to place blame."

Although at the present time no set-in-concrete method, technique, protocol, model, paradigm, or scheme on how to conduct an accident investigation is accepted totally by the majority of practicing safety professionals, some routine steps should be taken in accomplishing an accident investigation. But before accident investigation should come accident prevention.

This text focuses on accident prevention. Anyone can talk about what should or should not have been done after an accident occurs. Those who investigated (and many still are) the tragedies at Bhopal and the Space Shuttle *Challenger* had the huge advantage of hindsight (which equals 20–20 vision, of course). However, finding the cause of an accident, after the fact, is not—or should not be—the safety and environmental manager's primary mission. The safety and environmental manager must take a proactive prevention approach versus the typical (too often relied on) reactive approach. Does this make sense? We hope so.

The next logical question might be could we absolutely prevent the occurrence of all accidents? Of course not. When human beings are involved, accidents will occur. In fact, one view asserts that accidents are usually caused by a couple of factors, with humans as the major culprits. According to Heinrich (1950):

> The occurrence of an injury invariably results from a completed sequence of factors, the last one of these being the injury itself. The accident which caused the injury is in turn invariably caused or permitted directly by the unsafe act of a person and/or a mechanical or physical hazard.

Is Heinrich correct? Unfortunately, our experience has shown that he is—to a point. Accidents are sometimes more complicated, of course, the result of a sequence of events or happenings beyond the control of one individual. One thing is certain, however, people are at the center of all accidents (other than those precipitated by Mother Nature: hurricanes, tornadoes, earthquakes, typhoons, lightning, floods, meteor impacts, and so forth). To a point? How can we say that people are responsible for mechanical or physical hazards? How can we say that some of the terrifying and destructive acts of Mother Nature are her responsibility only to a point? Good questions. The answers may not seem clear, but they are as clear as dew on grass.

Let's take one point at a time. How is it that a person can be blamed for an accident caused by mechanical or physical hazards? Quite easily, actually. Remember, some person or persons designed, built, installed (and often ignored) mechanical and physical hazards in the first place. So, when we say that people are responsible for mechanical and/or physical hazards, that is exactly what we mean.

Acts of nature? You might also ask, "Since when have people had the power to generate hurricanes, typhoons, floods, and other such natural disasters?" Never, that we know of. But—hang on—think about it. Let's look at hurricanes and typhoons first, then floods.

When human beings decide to build cities, buildings, businesses, and homes in a traditionally hurricane- or typhoon-prone area (like directly on the oceanfront on the east coast of the United States, in part of hurricane alley, or the lowland areas of India, have people then put themselves at risk for the damage, destruction, and death that follows? If people had built on property in areas not usually subject to hurricane or typhoon damage, would the chances of their being affected by either one decrease significantly? Probably.

Another classic example of people's disregard of the whims of Mother Nature involves floods. Consider the flooding that occurred in the midwestern United States during 1993, 1995, and 1997. Why were so many people affected by the massive flooding events that occurred along those major midwestern rivers? Many rivers exhibited "century" water levels—several years in a row. Several answers are possible, but one thing is absolutely certain, human beings were the main players. People settled (built those cities, those businesses, and those homes) in the flood plains of those river systems. People made alterations, here and there, to the rivers' natural courses. Bottom line: People forgot the golden axiom: Don't mess with Mother Nature.

"The accident which caused the injury is in turn invariably caused or permitted directly by the unsafe act of a person and/or a mechanical or physical hazard." In this text, based on our own experience, we support this view.

Did You Know?

What causes accidents? This question should be the primary focus of every accident prevention program (Della-Giustina, 2007).

WHAT IS AN ACCIDENT?

Earlier we defined an *accident* (based on the ASSE definition) as:

> An unplanned and sometimes injurious or damaging event, which interrupts the normal progress of an activity and is invariably preceded by an unsafe act or unsafe condition thereof. An accident may be seen as resulting from a failure to identify a hazard or from some inadequacy in an existing system of hazard controls. Based on applications in casualty insurance, an event that is definite in point of time and place but unexpected as to either its occurrence or its results.

A basic dictionary definition of accident might define *accident* as "an unexpected, unwanted event." In other dictionaries, you might find accident defined as "an event occurring by chance or from unknown causes." The first dictionary definition is rather obvious—how else could you define accident in its most basic terms? The second definition is contradictory, in the way we use the word "accident" in this text. Accidents have no "unknown causes" (see Figure 4.1). All accidents (no matter how they are defined) have causes (Slote, 1987). In accident investigation, the investigator's job is to find those causes.

WHAT IS AN ACCIDENT INVESTIGATION?

Simply put, accident investigation is a vital part of any effective safety and health program. The key word we emphasize is "effective." Anyone can devise a safety and environmental health program that he or she might assume fits the bill as an effective "safety program"—but that does not necessarily mean that the program is really effective. One thing is certain: If company policy does not insist upon investigation of all accidents (and even so-called near misses), then the company's safety and environmental policy is definitely not effective.

Why is it so important to investigate accidents? The simple answer is we must investigate accidents to reduce the likelihood of their being repeated. Note that we are talking about all types of accidents, near misses and other minor occurrences included. If not corrected, those "minor mishaps" and "near misses" may result in much more

Figure 4.1   Causes of Accidents

serious accidents the next time around. We have all heard "If we do not learn from the past (our mistakes), we are doomed to repeat it (them)." Why would any logical person want to take a chance on repeating a near-miss or minor accident that could generate quite a different result next time around?

To reduce the likelihood of accidents being repeated, their causes must be identified, so that remedial actions can be taken to ensure they do not happen again. The accident investigation process also helps investigators and safety and environmental managers compile facts to be used as legal or liability evidence in the event of claims or lawsuits for losses or injuries. The new safety and environmental management professional soon discovers that if one of his or her company's workers is seriously injured (or worse) while on the job, the likelihood of litigation emanating from the worker or the family is quite good. For this reason (and others), many insurance companies require complete compilation of such facts.

Most texts make the point that whenever an accident investigation is conducted, the purpose of the investigation is not to place blame or find fault. We disagree with this view. When an accident occurs, someone is certainly at fault. To determine corrective action to be taken requires that blame for an accident be fixed. This is not to say that the safety and environmental manager who investigates an accident must be a Gestapo agent. Heavy-handed tactics are not required and definitely not recommended. Instead, the safety and environmental manager should remember the words of Sergeant Joe Friday (of television's "Dragnet" fame) who said: "Just the facts, ma'am." A good accident or near-miss investigation is impersonal, and completely fact-oriented.

The safety and environmental manager is not a disciplinarian. As Willie Hammer (1989) points out: "Whether disciplinary action should be taken if blame is fixed on an individual is beyond the duties of safety [professionals]. They are obligated to find out who was or may have been responsible for the accident; whether it was a worker, supervisor, manager, or other party" (p. 273). We share Mr. Hammer's view on this topic; if the facts collected point the finger of blame at some individual, this should be reported. Higher authority should handle the disposition of the incident. Recommendations for punishment or disciplinary action, and/or personal opinions are beyond the purview of safety and environmental management and the safety and environmental manager.

THE ACCIDENT INVESTIGATION PROCESS

We recommend a proactive approach to accident prevention. We are certain that preventing accidents is much better than to have to react to them and their consequences. We also understand that accidents happen. Even the best-managed safety program will not erase the potential for some accidents to occur.

The accident investigation process begins, of course, with an accidental occurrence. The accident may be nothing more than a slight injury to a worker, or minor damage to company or non-company property. On the other hand, the accident may be major—or even catastrophic. The accident investigation process should begin as soon after the occurrence as possible, because the investigator must gather facts from fresh evidence. Looking for evidence at the scene of an accident that has been tampered with or cleaned up makes the investigator's job much more difficult. Another problem with waiting too long to look at the accident is that witnesses who have just witnessed an accident (with the incident still fresh in their minds) are much more likely to relate what they saw more accurately to the investigator.

You must be aware of another problem—the reluctance or refusal of workers, witnesses, and/or supervisors to report accidents. Frequently workers have reasons they may not want to go on record about an accident. Company policy must insist that all accidents be reported.

All accidents? It depends. For example, many companies require that any personal injury resulting in the need for more than minor first aid at the work site, or that results in lost work time must be reported immediately (or as soon as possible) to the work center supervisor. This is important for two reasons: (1) to ensure that an injured worker receives proper medical attention and (2) to ensure that the incident is properly recorded in the OSHA 300 log.

Does this mean that when a worker receives a minor bruise or scratch while working at the job site, he or she should not report the accident? Again, it depends on company policy and the type of work being done. For example, in a wastewater treatment plant, a minor scratch received while repairing a treatment process machine where the worker is exposed to raw wastewater may be more than just a minor occurrence. A worker who is scratched, then exposed to the waste stream may not recognize the injury as a major event then (and thus will not report it), but three days later, when any pathogens contained in the wastewater have entered the body and had a chance to "do their work," a simple scratch may turn into a serious event, possibly life-threatening, if not properly treated.

After an accident is reported, the work center supervisor normally accomplishes the first investigation. In fact, many companies use an accident investigation form called a "Supervisor's" or "Employer's First Report of Accident." This is a good practice, because the supervisor not only possesses the most knowledge about his or her employees, but also knows the equipment and work practices in his or her area. In companies employing a full-time safety and environmental manager, company policy may or may not require that the supervisor wait until the safety and environmental manager arrives to conduct the investigation as a team. Experience shows us that allowing the supervisor to investigate first is better. The supervisor is usually right on the scene or nearby when the incident occurs, and waiting for the safety and environmental manager to show up on scene may require too long a delay. Remember that the idea is to investigate as soon as possible. The exception might be, of course, incidents involving a fatality or multiple fatalities. In such cases, the safety engineer should be called immediately (along with OSHA) to investigate. But for minor occurrences, the supervisor should have the first look, the safety engineer should read his or her report, and if deemed pertinent, should perform a follow-up investigation. This follow-up investigation is not intended to check out the supervisor's ability to conduct investigations but rather to determine causal factors and mitigation procedures. The safety and environmental manager is the company's safety and health expert, and his or her job is to recommend steps (remedial actions) to ensure that like incidents do not occur again. Experience has also demonstrated that an immediate first look (conducted by the supervisor) works well to gather the majority of facts—then the follow-up investigation (conducted by the safety engineer) works well to "fine tune" the investigation to the point where items overlooked by one set of eyes might be picked up by another set.

Whatever type of form and/or protocol the supervisor uses to make his or her first report of accident, the supervisor's main mission must be to determine the what, where, when, how, and possibly the why. To answer these questions, the supervisor must gather facts. Normal practice involves using five methods to gather accident information:

1. Interviewing the victim,
2. Interviewing accident witnesses,
3. Investigating of the accident scene, including taking photographs when possible,
4. Re-enacting of the accident, and
5. Reconstructing the accident.

Findings must be recorded on the Supervisor's or Employer's First Report of Accident. A typical accident report form is shown in Figure 4.2.

Let's take a look at the form shown in Figure 4.2 and examine how it might be used in conjunction with the Supervisor's five-step fact gathering process.

Items 1–6 are straightforward enough, but keep in mind that this document might be needed by the insurance vendor, and/or could be used as a legal document in a

SUPERVISOR'S FIRST REPORT OF ACCIDENT

1. **Name:**

2. **Social Security Number:**

3. **Department:**

4. **Time of Incident:**

5. **Date of Incident:**

6. **Job Classification:**

7. **Briefly describe the accident/illness (include estimated lost work days), exact accident location, time of occurrence, equipment involved, etc.**

8. **Victim's remarks.**

9. **Unsafe acts, conditions causing the accident/illness.**

10. **Corrective actions required (note specific actions taken).**

11. **Supervisor's Signature:**

12. **Date:**

Figure 4.2   Sample "Supervisor's First Report of Accident" Form

hearing or court of law. Thus, it is important the items 1–6 be filled in accurately and completely.

Item 7 is also straightforward, but be sure to carefully describe the accident/illness as it actually occurred. Sometimes when a worker gets injured, a week, a month, or a year later he or she claims that the injury received was a different, much more serious injury (which in some cases might be true, but in most cases is not true). The supervisor should list estimated lost time, because if lost time exceeds more than one workday, then this incident must be entered in the OSHA 300 log. The exact accident location, time of occurrence, and equipment involved are important.

Item 8 is the victim's remarks section. This section is intended for the victim to provide his or her description and analysis of the accident. After the victim writes out his or her remarks and the supervisor reads them, the supervisor should then ask questions of the victim. This is directly related, of course, to the supervisor's first method of fact gathering—interviewing the accident victim. Interviewing the worker that had the accident will probably be the most important means of getting the facts on the accident that occurred. However, this procedure takes care on the part of the supervisor; pointing the finger of blame at any worker while trying to obtain a factual accounting of what really occurred is pointless and futile.

During the victim interviewing process, the supervisor should take notes, but should wait until the accident has been completely investigated before completing the accident investigation form. Looking and listening is more important than writing out a report at this stage of the investigation.

Human nature dictates that if victims feel that they have done something wrong, something unsafe that caused the incident, they may be hesitant about giving all the details—especially the details that would demonstrate that they are at fault. The supervisor should be aware of this tendency and guard against the supply of false information.

In many such investigations, the accident facts often do not appear clear or reasonable. Often this is because the worker is overwrought or stressed because of the accident. If you judge that this might be the case, you might want to schedule a second interview later, after the victim has had the opportunity to calm down.

The investigator should also seek the victim's opinion on how he or she would prevent such an incident from occurring again. Asking the accident victim his or her opinion on measures that should or could be taken to prevent a recurrence of the accident is always a good idea.

Did You Know?

We have found that when victims are asked their opinion on how to prevent future recurrences of the same incident, the answers received sometimes reveal an underlying problem that might not have been foreseen or thought of, but which must definitely be dealt with. Invariably the victim will give his or her opinion on how to prevent a reoccurrence, and then reveal that they had made this same recommendation earlier (in some cases, several times) to their immediate supervisor, or to their divisional safety and health representative, and nothing had been done to correct the deficiency. Obviously, this is a matter that must be looked at closely. Investigating the aftermath of an accident is difficult enough without finding out that the company's system of reporting unsafe conditions is inadequate—a serious matter, to say the least.

A final word on item 8. If the victim is unable to complete this section because of injury or worse, all other sections should be completed and this section is left blank until the victim is able to provide the information. If the victim will never be able to contribute to this section because of disability or death, the supervisor should state the situation in section 8 and move on with the rest of the investigation.

Item 9 is to be completed by the supervisor, and should list all the factors that may have contributed to the accident/illness event. The supervisor investigates specifically to determine if any unsafe acts or conditions were involved with causing the accident or illness. Findings are usually not based solely on victim and witness interviews, but also on close scrutiny of the accident scene itself. The investigator may want to take photographs, videotapes, tape recordings, and/or transcripts for later study. If photography is used in accident investigation, poorly done photographs can be worthless, and a waste of time.

Often an investigator can ascertain pertinent causal factors by conducting a re-enactment of the accident. This re-enactment might include a walk-through or dry run of the events leading up to the accident.

Another method that is often employed (especially in the aftermath of plane crashes and other large-scale disasters) to determine conditions that led to the accident occurring is the reconstruction of the accident. (NOTE: the safety engineer, a structural engineer, or some other expert specialty usually conducts this procedure. However, when reconstruction is rather straightforward and does not require the expertise of engineers and/or others, the supervisor may be able to handle minor or uncomplicated reconstructions on his or her own.)

Item 10 should list the specific actions that need to be completed and the estimated time frame for completion. This is a critically important step. Ending an investigation with a piece of paper—only a written report—is not sufficient. The report should be routed to top managers, public officials, insurance personnel, or others who can do something about the recommendations provided in the investigation report. Remember that the primary mission of the investigation is to determine cause and to initiate steps to ensure that such an accident does not occur again. The supervisor's recommendations should lead to follow-up investigations and to engineering changes, procedural changes, changes in policy, rules or regulations, and/ or improved training. The idea is to learn from such incidents, from mistakes, and not to repeat them. The importance of the training element cannot be over-stressed.

After proper completion of this entire process, a cause (or causes) for the accident will be found, and measures and remedial actions taken to reduce or prevent the likelihood of a recurrence.

A word of caution is advised, however. In completing the form shown in Figure 4.2 the supervisor must avoid certain pitfalls, which include making statements like "this incident was caused by carelessness by an accident-prone employee" (we feel that this is a convenient way to categorize someone, but is totally unfounded) or "caused by a dumb worker mistake—not paying attention when he/she should have been." Such statements should not be included in the report and do absolutely nothing to further the goal of the investigation, which is to provide "Just the facts, ma'am."

Again, that the supervisor does not assign blame during the investigation process is very important. While, as we stated earlier, blame can always be placed on the cause of an accident, it may seem contradictory to state now that blame should not be assigned during the investigation process. However, if you think about it, blame cannot be assigned (if it need be at all) until all the facts are in—and they have been completely and carefully analyzed.

Item 11 and **12** are self-explanatory. What should the supervisor do with his or her completed First Report of Accident form? This depends, of course, on organizational requirements. Often organizations require that a copy of the First Report be forwarded to several individuals or divisions: higher level supervisor, safety engineer, human resource manager, insurance vendor, and so forth.

The point is the completed investigation report is worthless if it is not forwarded to those who need to review it—those who have the authority to ensure that remedial actions are taken to prevent recurrence.

ACCIDENT INVESTIGATIONS AND THE SAFETY PROFESSIONAL

We stated earlier that the supervisor should conduct the initial accident investigation, and our experience has supported our view. However, each organization is different. Some organizations insist that the company safety and environmental professional either conduct the investigation, accompany and assist the supervisor doing the investigation, or be a member of a team put together to perform such investigatory functions.

We stand by our recommendation that the safety professional conduct his or her investigation after the supervisor has completed his or her First Report of Accident. This is the case, of course, if the organization is large enough to have a full-time safety professional, who should be considered the organization expert on accident investigations (Ferry, 1990).

Supervisors tasked to perform accident investigations must have a certain level of training to be able to conduct them properly. Often the organizational safety and environmental professional provides the training. To do this correctly, obviously the safety and environmental professional has to have a higher degree of skill than that needed by the other members.

How does the safety and environmental manager learn the skills needed to properly perform accident investigations? The most common way in which this is accomplished is through on-the-job experience. Very few college curriculums spend much time (if any) teaching this skill. However, several NIOSH-approved training centers provide this training.

Remember the end goal of accident investigations: "All of the facts and findings in the world will not do any good if no action is taken on them" (Ferry, 1990).

ACCIDENT INVESTIGATIONS: THE REALITY

To have an effective organizational safety and environmental program, accident investigations are an important part of the process, the paradigm that cannot be overlooked. We have said this before. We also stated that the ideal approach to use in safety and environmental management is to examine and investigate all undesired events. The key words, of course, are "all undesired events." But is this realistic? Maybe. Maybe not. It depends.

Safety and environmental practitioners in the real world have found this goal hard to achieve. The real world, with all its limitations, gets in the way. "So," Slote (1987) points out, "the reader may very well ask. What really is to be investigated and how far should the investigation be carried? The answers vary accordingly to the beliefs of management within any one organization" (p. 119). In many organizations, the most difficult task the safety and environmental professional has to perform is to convince upper management that accidents are *not* the cost of doing business. Instead, the safety and environmental manager must convince them that accidents are preventable. A good accident investigation procedure can find the causal factors and ensure that a similar occurrence does not occur again. When you get right down to it, isn't the point that accidents are preventable the real "backbone" of any good safety management program? We think it is.

THOUGHT-PROVOKING QUESTIONS

- What are the uses of accident investigations?
- What is their chief drawback?
- What causes accidents?
- Define accident.
- Why investigate noninjury accidents or near misses?
- How should the safety and environmental manager handle blame?
- Discuss accident investigation and lawsuits.
- What considerations should a safety and environmental manager use in handling witnesses?
- What organizational support should the company offer? Why is it important?
- Why is the "Supervisor's First Report of Accident" important?
- What five methods should be used to gather accident information?
- Why are victim interviews so important?
- What should you look out for in victim interviews?
- Why would photographs be important?
- What are the action steps, and why are they important?
- When and who is responsible for stating blame for causing an accident?
- Who should see the investigation report?
- Who should investigate an accident first?
- Where can you learn investigative skills?
- Discuss the dilemma of the newly hired safety and environmental manager with programs not fully in place, in light of a serious or fatal accident.
- What is the purpose of the recommended administrative controls?
- What are the current National Safety Council's statistics for injuries and fatalities?
- Why do companies often seem to consider safety unimportant until a fatality occurs?
- How do you strike a balance between what to investigate and what not to investigate? How do you decide how far to carry an investigation?

- How can you use the accident investigation process as a tool for involving upper management in safety and health?
- What is the difference between a traditionalist approach and a systems approach to safety and health issues?
- Some safety and environmental management experts feel that an accident victim's supervisor should never be the person to investigate the accident. What do you think?

REFERENCES AND RECOMMENDED READING

Bird, F.E., Jr. *Management Guide to Loss Control.* Atlanta, GA: Institute Press, 1984.

Della-Giustina, D. *Safety and Environmental Management.* Lanham, MD: Government Institutes Press, 2007.

Ferry, T.S. *Modern Accident Investigation and Analysis.* New York: John Wiley & Sons, 1988.

Ferry, T.S. *Safety and Health Management Planning.* New York: Van Nostrand Reinhold, 1990.

Fragala, G. "A Modern Approach to Injury Reporting." *Professional Safety Magazine.* January 1983.

Hammer, W. *Occupational Safety Management and Engineering*, 4th ed. Englewood Cliffs, NJ: Prentice Hall, 1989.

Hammer, W. "Missile Base Disaster." *Heating, Piping, and Air Conditioning.* December 1968.

Head, G.L. "Selling Safety: Using Management's Language." *Professional Safety Magazine.* April 1984.

Heinrich, H.W. *Industrial Accident Prevention*, 3rd ed. New York: McGraw-Hill, 1950.

Kingsley, H., and L. Benner, Jr. *Investigating Accidents with STEP.* Des Plaines, IL: American Society of Safety Engineers, 1987.

Manuele, F.A., Accident Investigation and Analysis, an Evaluative Review, *Professional Safety Magazine*, American Society of Safety Engineers, October, 1982.

Manuele, F.A. "How Do You Know Your Hazard Control Program is Effective?" *Professional Safety Magazine.* June 1981.

National Safety Council. "Celebrate National Safety Month." *Safety Focus Newsletter.* May–June 1998.

National Safety Council. *Accident Prevention Manual for Industrial Operations*, 10th ed. Chicago: 2009.

National Safety Council. *Accident Facts.* Chicago: 2009.

Occupational Safety & Health Administration. *Accident Investigation.* Washington, D.C.: U.S. Department of Labor, 2003. Accessed May 15, 2015. https://www.osha.gov/dcsp/products/topics/incidentinvestigation/index.html.

Slote, L. *Handbook of Occupational Safety and Health.* New York: John Wiley & Sons, 1987.

Spellman, F.R., and N.E. Whiting. *Safety Engineering: Principles and Practices*, 2nd ed. Lanham, MD: Government Institutes Press, 2005.

Wood, R.H. "Accident Photography." In *Readings in Accident Investigation*, edited by Ted S. Ferry. Springfield, IL: Charles C. Thomas, 1984.

Chapter 5

Hazard Communication and Hazardous Waste

A FAILURE TO COMMUNICATE[1]

The Bhopal, India Incident, the ensuing chemical spill, and the resulting tragic deaths of thousands and injuries to many more are well known. However, not all of the repercussions from this incident are as well known. After Bhopal arose a worldwide outcry. "How could such an incident occur? Why wasn't something done to protect the inhabitants? Weren't there safety measures taken or in place to prevent such a disaster from occurring?" Lots of questions, few answers. The major problem was later discovered to be a failure to communicate. That is, the workers, residents, and visitors had no idea of how dangerous a chemical spill could be. Many found out the hard way.

The Bhopal Incident, among other industrial disasters, was the motivating force behind the genesis of OSHA's Hazard Communication Standard (29 CFR 1910.1200). This standard commonly known as the Worker's Right-to-Know Standard; that is, HazCom (as it is commonly known) requires employers to inform their workers and others of the hazardous chemicals they may be exposed to.

There is no all-inclusive list of chemicals covered by the HazCom Standard; however, the regulation refers to "any chemical which is a physical or health hazard." Those specifically deemed hazardous include:

- Chemicals regulated by OSHA in 29 CFR Part 1910, Subpart Z, Toxic and Hazardous Substances
- Chemicals included in the American Conference of Governmental Industrial Hygienists' (ACGIH) latest edition of Threshold Limit Values (TLVs) for *Chemical Substances and Physical Agents in the Work Environment*
- Chemicals found to be suspected or confirmed carcinogens by the National Toxicology Program in the *Registry of Toxic Effects of Chemicals Substances* published by NIOSH or appearing in the latest edition of the *Annual Report on Carcinogens,*

---

[1] Much of information in this chapter is based on F.R. Spellman (2015). *Occupational Safety & Health Simplified for the Industrial Workplace*. Lanham, MD: Bernan Press.

or by the International Agency for Research on Cancer in the latest editions of its IARC *Monographs*

Because OSHA's Hazard Communication is a dynamic (living) standard, it has been easily amended and adjusted to comply with ongoing worldwide changes to make employer and worker chemical safety compliance requirements more pertinent and applicable. In light of this ongoing desire for currency and applicability, Federal OSHA published a revised Hazard Communication Standard (HazCom) on March 26, 2012, to align with the United Nations' Globally Harmonized System of Classification and Labeling of Chemicals. It affects how chemical hazards are classified, the elements incorporated into a label, and the format of the safety data sheet (SDS). In addition, terminology and several definitions have changed, including the definition of a hazardous chemical.

Under its "HazCom" or "Right-to-Know Law," OSHA requires employers who use or produce chemicals on the worksite to inform all employees of the hazards that might be involved with those chemicals. HazCom says that employees have the right to know what chemicals they are handling or could be exposed to. HazCom's intent is to make the workplace safer. Under the HazCom Standard, the employer is required to fully evaluate all chemicals on the worksite for possible physical and health hazards. All information relating to these hazards must be made available to the employee 24 hours each day. The standard is written in a performance manner, meaning that the specifics are left to the employer to develop.

HazCom also requires the *employer* to ensure proper labeling of each chemical, including chemicals that might be produced by a *process* (process hazards). For example, in the wastewater industry, deadly methane gas is generated in the waste stream. Another common wastewater hazard is the generation of hydrogen sulfide (which produces the characteristic rotten-egg odor) during degradation of organic substances in the waste stream which can kill quickly. OSHA's HazCom requires the employer to label methane and hydrogen sulfide hazards so that workers are warned and safety precautions are followed.

Labels must be designed to be clearly understood by all workers. Employers are required to provide both training and written materials to make workers aware of what materials they are working with and what hazards they might be exposed to. Employers are also required to make Safety Data Sheets (SDS) available to all employees. An SDS is a fact sheet for a chemical posing a physical or health hazard at work. SDS must be in English and contain the following information:

- Identity of the chemical (label name)
- Physical hazards
- Control measures
- Health hazards
- Whether it is a carcinogen
- Emergency and first aid procedures
- Date of preparation of latest revision
- Name, address, and telephone number of manufacturer, importer, or other responsible party

Blank spaces are not permitted on an SDS. If relevant information in any one of the categories is unavailable at the time of preparation, the SDS must indicate no information was available. Your facility must have an SDS for each hazardous chemical it uses. Copies must be made available to other companies working on your worksite (outside contractors, for example), and they must do the same for you. The facility Hazard Communication Program must be in writing and, along with SDSs, made available to all workers 24 hours each day/each shift.

BETTER COMMUNICATION FOR WORKER SAFETY AND HEALTH

In an effort to provide better worker protection from hazardous chemicals and to help American businesses compete in a global economy, OSHA has revised its Hazardous Communication (HazCom) standard to align with the United Nations' Globally Harmonized System of Classification and Labeling of Chemicals—referred to as GHS—incorporating the quality, consistency, and clarity of hazard information that workers receive by providing harmonized criteria for classifying and labeling hazardous chemicals and for preparing safety data sheets for these chemicals.

The GHS system is a new approach that has been developed through international negotiations and embodies the knowledge gained in the field of chemical hazard communication since the HazCom standard was first introduced in 1983. Simply, HazCom with GHS means better communication of chemical hazards for workers on the job.

BENEFITS OF HAZCOM WITH GHS[2]

Practicing safety and environmental professionals are familiar with OSHA's original 1983 Hazard Communication Standard. Many are now becoming familiar with the phase-in of the new combined HazCom and GHS standard. The first thing they learn is that the Globally Harmonized System (GHS) is an international approach to hazard communication, providing agreed criteria for classification of chemical hazards and a standardized approach to label elements and safety data sheets. The GHS was negotiated in a multi-year process by hazard communication experts from many different countries, international organizations, and stakeholder groups. It is based on major existing systems around the world, including OSHA's Hazard Communication Standard and the chemical classification and labeling systems of other US agencies.

The result of this negotiation process is the United Nations' document entitled "Globally Harmonized System of Classification and Labeling of Chemicals," commonly referred to as The Purple Book. This document provides harmonized classification criteria for health, physical, and environmental hazards of chemicals. It also includes standardized label elements that are assigned to these hazard classes and

---

[2] Based on information from OSHA's (2014). *Modification of the Hazardous Communication Standard (HCS) to Conform with the United Nations' (UN) Globally Harmonized System of Classification and Labeling of Chemicals (GHS).* Accessed 01/16/15 @ https://www.osha.gov/dsg/hazcom/hazcom-faq.html.

categories, and provide the appropriate signal words, pictograms, and hazard and precautionary statements to convey the hazards to users. A standardized order of information for safety data sheets is also provided. These recommendations can be used by regulatory authorities such as OSHA to establish mandatory requirements of hazard communication but do not constitute a model regulation.

OSHA's motive to modify the Hazard Communication Standard (HCS) was to improve safety and health of workers through more effective communications on chemical hazards. Since it was first promulgated in 1983, the HCS has provided employers and employees extensive information about the chemicals in their workplaces. The original standard is performance oriented, allowing chemical manufacturers and importers to convey information on labels and material data sheets in whatever format they choose. While the available information has been helpful in improving employee safety and health, a more standardized approach to classifying the hazards and conveying the information will be more effective and provide further improvements in American workplaces. The GHS provides such a standardized approach, including detailed criteria for determining what hazardous effects a chemical poses, as well as standardized label elements assigned by hazard class and category. This will enhance both employer and worker comprehension of the hazards, which will help to ensure appropriate handling and safe use of workplace chemicals. In addition, the safety data sheet requirements establish an order of information that is standardized. The harmonized format of the safety data sheets will enable employers, workers, health professionals, and emergency responders to access the information more efficiently and effectively, thus increasing their utility.

Adoption of the GHS in the United States and around the world will also help to improve information received from other countries—because the United States is both a major importer and exporter of chemicals, American workers often see labels and safety data sheets from other countries. The diverse and sometimes conflicting national and international requirements can create confusion among those who seek to use hazard information effectively. For example, labels and safety data sheets may include symbols and hazard statements that are unfamiliar to readers or not well understood. Containers may be labeled with such a large volume of information (overkill) that important statements are not easily recognized. Given the differences in hazard classification criteria, labels may also be incorrect when used in other countries. If countries around the world adopt the GHS, these problems will be minimized, and chemicals crossing borders will have consistent information, thus improving communication globally.

Phase-In Period for The Hazard Communication Standard

Table 5.1 summarizes the phase-in dates required under the revised Hazard Communication Standard (HCS).

During the phase-in period, employers would be required to be in compliance with either the existing HCS or the revised HCS, or both. OSHA recognizes that hazard communications programs will go through a period of time where labels and SDSs under both standards will be present in the workplace. This will be considered

Table 5.1 Phase-In Period for Hazard Communication Revisions

Effective Completion Date	Requirement(s)	Who
December 1, 2013	Train employees on the new label elements and safety data sheet (SDS)	Employers
June 1, 2015*	Compliance with all modified provisions of this final rule	Chemical manufacturers, importers, distributors, and employers
December 1, 2015	The distributor shall not ship containers labeled by the chemical manufacturer or importer unless it is a GHS label	
June 1, 2016	Update alternative workplace labeling and hazard	Employers
	Communication program as necessary, and provide additional employee training for newly identified physical or health hazards	

* This date coincides with the EU implementation date for classification of mixtures.

acceptable, and employers are not required to maintain two sets of labels and SDSs for compliance purposes.

It is important to point out that prior to OSHA's effective compliance date for full implementation of the revised HCS, employee training must be conducted. This is the case because American workplaces will receive SDS and new labeling before the full compliance date is to be met. Thus, employees will need to be trained early to enable them to recognize and understand the new label elements (i.e., pictograms, hazard statements, precautionary statements, and signal words) and the SDS format.

MAJOR CHANGES TO THE HAZARD COMMUNICATION STANDARD

There are three major areas of change in the modified HCS: in hazard classification, labels, and safety data sheets.

- **Hazard Classification:** The definitions of hazard have been changed to provide specific criteria for classification of health and physical hazards, as well as classification of mixtures. These specific criteria will help to ensure that evaluations of hazardous effects are consistent across manufacturers and that labels and safety data sheets are more accurate as a result.
- **Labels:** Chemical manufacturers and importers will be required to provide a label that includes a harmonized signal word, pictogram, and hazard statement for each hazard class and category. Precautionary statements must be provided.
- **Safety Data Sheets (SDSs):** Will now have a 16-section format.

Note: The GHS does not include harmonized training provisions but recognizes that training is essential to an effective hazard communication approach. The revised Hazard Communication Standard (HCS) requires that workers be re-trained within two years of the publication of the final result to facilitate recognition and understanding of the new labels and safety data sheets.

Hazard Classification

Not all HCS provisions are changed in the revised HCS. The revised HCS is simply a modification to the existing standard, designed to make it universal and worker-friendly. The parts of the standard that did not relate to the GHS (such as the basic framework, scope, and exemptions) remained largely unchanged. There have been some modifications in terminology in order to align the revised HCS with language used in the GHS. For example, the term "hazard determination" has been changed to "hazard classification" and "material safety data sheet" was changed to "safety data sheet."

Under both the current Hazard Communication Standard (HCS) and the revised HCS, an evaluation of chemical hazards must be performed considering the available scientific evidence concerning such hazards. Under the current HCS, the hazard determination provisions have definitions of hazard and the evaluator determines whether or not the data on a chemical meet those definitions. It is a performance-oriented approach that provides parameters for the evaluation, but not specific, detailed criteria. The hazard classification approach in the revised HCS is quite different. The revised HCS has specific criteria for each health and physical hazard, along with detailed instructions for hazard evaluation and determinations as to whether mixtures or substances are covered. It also establishes both hazard classes and hazard categories—for most of the effects, the classes are divided into categories that reflect the relative severity of the effect. The current HCS does not include categories for most of the health hazards covered, so this new approach provides additional information that can be related to the appropriate response to address the hazard. OSHA has included the general provisions for hazard classification in paragraph (d) of the revised rule and added extensive appendices that address the criteria for each health or physical effect.

Label Changes Under The Revised HCS

Under the current HCS, the label preparer must provide the identity of the chemical and the appropriate hazard warnings. This may be done in a variety of ways, and the method to convey the information is left to the preparer. Under the revised HCS, once the hazard classification is completed, the standard specifies what information is to be provided for each hazard class and category. Labels will require the following elements:

- *Pictogram*: a symbol plus other graphic elements, such as a border, background pattern, or color that is intended to convey specific information about the hazards of a chemical. Each pictogram consists of a different symbol on a white background within a red square set on a point (i.e., a red diamond) (see Figure 5.1). There are nine pictograms under the GHS. However, only eight pictograms are required under the HCS. Note that the environment pictogram shown in Figure 5.1 is not mandatory; however, the other eight are mandatory.
- *Signal Words*: words used to indicate the relative level of severity of hazard and alert the reader to a potential hazard on the label. The signal words used are

"danger" and "warning" (see Figure 5.2). "Danger" is used for the more severe hazards, while "warning" is used for less severe hazards.

- *Hazard Statement*: a statement assigned to a hazard class and category that describes the nature of the hazard(s) of a chemical, including, where appropriate, the degree of hazard.
- *Precautionary Statement*: a phrase that describes recommended measures to be taken to minimize or prevent adverse effects resulting from exposure to a hazardous chemical or improper storage or handling of a hazardous chemical.

In the revised HCS, OSHA is lifting the stay on enforcement regarding the provision to update labels when new information on hazards becomes available. Chemical manufacturers, importers, distributors, or employers who become newly aware of any significant information regarding the hazards of a chemical shall revise the labels for the chemical within six months of becoming aware of the new information. If the chemical is not currently produced or imported, the chemical manufacturer, importer, distributor, or employer shall add the information to the label before the chemical is shipped or introduced into the workplace again.

The current standard provides employers with flexibility regarding the type of system to be used in their workplaces, and OSHA has retained that flexibility in the

HAZCOM STANDARD PICTOGRAMS

Health Hazard	Flame	Exclamation Mark
• Carcinogen • Mutagenicity • Reproductive Toxicity • Respiratory Sensitizer • Target Organ Toxicity • Aspiration Toxicity	• Flammables • Pyrophorics • Self-Heating • Emits Flammable Gas • Self-Reactives • Organic Peroxides	• Irritant (skin and eye) • Skin Sensitizer • Acute Toxicity (harmful) • Narcotic Effects • Respiratory Tract Irritant • Hazardous to Ozone Layer (Non-Mandatory)
Gas Cylinder	**Corrosion**	**Exploding Bomb**
• Gases Under Pressure	• Skin Corrosion/ Burns • Eye Damage • Corrosive to Metals	• Explosives • Self-Reactives • Organic Peroxides
Flame Over Circle	**Environment** (Non-Mandatory)	**Skull and Crossbones**
• Oxidizers	• Aquatic Toxicity	• Acute Toxicity (fatal or toxic)

Figure 5.1   Globally Harmonized Labels

Figure 5.2   Sample Signal Word Labels

Did You Know?

Under the revised HCS, pictograms must have red borders. OSHA believes that the use of the red frame will increase recognition and comprehensibility. Therefore, the red frame is required regardless of whether the shipment is domestic or international. Moreover, the revised HCS requires that all red borders printed on the label have a symbol printed inside it. If OSHA were to allow blank red borders, workers may be confused about what they mean and concerned that some information is missing. OSHA has determined that prohibiting the use of blank red borders on labels is necessary to provide the maximum recognition and impact of warning labels and to ensure that users do not get desensitized to the warnings placed on labels.

revised Hazard Communication Standard (HCS). Employers may choose to label workplace containers either with the same label that would be on shipped container for the chemical under the revised rule, or with label alternatives that meet the requirements for the standard. Alternative labeling systems such as the National Fire Protection Association (NFPA) 704 Hazard Rating and the Hazardous Material Identification System (HMIS) are permitted for workplace containers. However, the information supplied on these labels must be consistent with the revised HCS, and have no conflicting hazard warnings or pictograms.

SDS Changes Under the Revised HCS

The information required on the (Material) Safety Data Sheet (SDS) will remain essentially the same as in the current standard (HazCom 1994). HazCom 1994 indicates what information has to be included on an SDS but does not specify a format for presentation or order of information. The revised Hazard Communication Standard (HCS 2012) requires that the information on the SDS be presented in a specified sequence. The revised SDS should contain 16 headings (Table 5.2).

Table 5.2   Minimum Information for an SDS

1. Identification of the substance or mixture and of the supplier	• GHS Product Identifier • Other means of identification • Recommended use of the chemical and restrictions on use • Supplier's details (including name, address, phone number, etc.) • Emergency phone number
2. Hazards Identification	• GHS classification of the substance/mixture and any national or regional information • GHS label elements, including precautionary statements. (Hazard symbols may be provided as a graphical reproduction of the symbols in black and white or the name of the symbol, e.g., flame, skull and crossbones) • Other hazards that do not result in classification (e.g., dust explosion hazard) or are not covered by GHS
3. Composition/ Information on Ingredients	**Substance** • Chemical identity • Common name, synonyms, etc. • CAS number, EC number, etc. • Impurities and stabilizing additives that are themselves classified and which contribute to the classification of the substance **Mixture** • The chemical identity and concentration or concentration ranges of all ingredients that are hazardous within the meaning of the GHS and are present above their cutoff levels.
4. First aid measures	• Description of necessary measures, included according to the different routes of exposure that is, inhalation, skin and eye contact, and ingestion • Most important symptoms/effects, acute and delayed • Indication of immediate medical attention and special treatment needed, if necessary
5. Firefighting measures	• Suitable (and unsuitable) extinguishing media • Specific hazards arising from the chemical (e.g., nature of any hazardous combustion products) • Special protective equipment and precautions for firefighters
6. Accidental release measures	• Personal precautions, protective equipment, and emergency procedures • Environmental precautions • Methods and materials for containment and clean up
7. Handling and storage	• Precautions for safe handling • Conditions for safe storage, including any incompatibilities
8. Exposure controls/ personal protection	• Control parameters, for example, occupational exposure limit values or biological limit values • Appropriate engineering controls • Individual protection measures, such as personal protective equipment
9. Physical and chemical properties	• Appearance (physical state, color, etc.) • Odor • Odor threshold • pH • Melting point/freezing point • Initial boiling point and boiling range • Flash point • Evaporation rate • Flammability (solid, gas) • Upper/lower flammability or explosive limits • Vapor pressure

(Continued)

Table 5.2   Minimum Information for an SDS (*Continued*)

9. Physical and chemical properties (*Continued*)	• Vapor density • Relative density • Solubility (ies) • Partition coefficient: n-octanol/water • Autoignition temperature • Decomposition temperature
10. Stability and reactivity	• Chemical stability • Possibility of hazardous reactions • Conditions to avoid (e.g., static discharge, shock, or vibration) • Incompatible materials • Hazardous composition products
11. Toxicological information	Concise but complete and comprehensible description of the various toxicological (health) effects and available data used to identify those effects, including: • Information on the likely routes of exposure (inhalation, ingestion, skin and eye contact) • Symptoms related to the physical, chemical, and toxicological characteristics • Delayed and immediate effects and also chronic effects from short- and long-term exposure
12. Ecological information	• Ecotoxicity (aquatic and terrestrial, where available) • Persistence and degradability • Bioaccumulative potential • Mobility in soil • Other adverse effects
13. Disposal considerations	• Description of waste residues and information on their safe handling and methods of disposal including the disposal of any contaminated packaging
14. Transportation information	• UN number • Transport Hazard class(es) • Packing group, if applicable • Marine pollutant (Yes/No) • Special precautions that a user needs to be aware of or needs to comply with in connection with transport or conveyance either within or outside the premises
15. Regulatory information	• Safety, health, and environmental regulations specific for the product in question
16. Other information including information on preparation and revision of SDS	

HAZCOM AND THE OCCUPATIONAL SAFETY AND HEALTH PROFESSIONAL

The occupational safety and health professional must take a personal interest in ensuring that the facility is in full compliance with the Hazard Communication Standard for three major reasons: (1) it is the law; (2) it is consistently the number one cause of citations issued by OSHA for noncompliance; and (3) compliance with the standard goes a long way toward protecting workers.

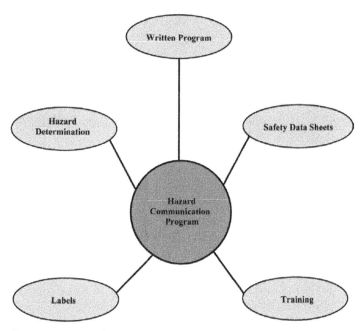

Figure 5.3   Elements Required for a Hazard Communication Program.

Figure 5.3 shows the elements of the Hazard Communication Standard, which the safety and health professional must ensure are part of the facility's HazCom Program: hazard determination, written hazard communication program, labels and other forms of warning, safety data sheets, and employee training. Another required element is not shown in Figure 5.3: **trade secrets**. For the sake of brevity in describing HazCom requirements, and because we are presenting a HazCom Program that is typically required in the non-chemical manufacturing type workplace, we don't discuss the **trade secrets** element beyond the brief mention that it allows the chemical manufacturer, importer, or employer to withhold the specific chemical identity, including the chemical name and other specific identification of a hazardous chemical from the material data sheet under certain conditions. For more information on the trade secrets, review 29 CFR 1910.1200 (Hazard Communication Standard).

Hazard determination primarily affects the chemical manufacturer and importers, not the facility employer and safety official, unless they choose not to rely on the evaluation performed by the chemicals manufacturer or importer to satisfy this requirement.

The chemical manufacturers and importers must supply Safety Data Sheets (SDS) to the purchaser. Purchasers (employers) are required to have an SDS in the workplace for each hazardous chemical they use.

Note: We cannot over-emphasize the need for the safety and environmental health professional to ensure that an SDS is present on the worksite for every chemical on the worksite. It is absolutely essential that you ensure that all workers know where these SDS are located.

The employer must provide employee training on the hazard communication program. Training on the hazardous chemicals in their work areas must be provided to employees upon their initial assignment. Whenever a new physical or health hazard is introduced into the workplace (one that training has not previously been accomplished on), the employer must provide the training. Specifically, employee training shall include:

1. methods and observation that may be used to detect the presence or release of a hazardous chemical in the work area;
2. the physical and health hazards of the chemicals in the work area;
3. the measures employees can take to protect themselves from these hazards, including specific procedures the employer has implemented to protect employees from exposure to hazardous chemicals, such as appropriate work practices, emergency procedures, and personal protective equipment to be used;
4. the details of the hazard communication program developed by the employer, including an explanation of the labeling system and the material safety data sheet, and how employees can obtain and use the appropriate hazard information.

Note: As with all OSHA-required training, you must not only ensure that the training is conducted, you must also ensure that it has been properly documented.

Labels and other forms of warning are elements of HazCom that the safety and environmental manager must pay particular attention to. Specifically, the chemical manufacturer, importer, or distributor must ensure that each container of hazardous chemicals leaving the workplace is labeled, tagged, or marked with the following information:

- Identity of the hazardous chemical(s);
- Appropriate hazard warnings;
- Name and address of the chemical manufacturer, importer, or other responsible party.

The employer's (thus, the safety and environmental health person's) responsibilities include: signs, placards, process sheets, batch tickets, operating procedures, or other such written materials in lieu of affixing labels to individual stationary process containers, as long as the alternative method identifies the containers to which it applies and conveys the information required on the label. The written materials must be readily accessible to the employees in their work areas throughout each shift.

The employer must not remove or deface existing labels on incoming containers of hazardous chemicals, unless the container is immediately marked with the required information.

The safety and environmental health practitioner must ensure that labels or warnings in his or her workplace are legible, in English, and are prominently displayed on the container or readily available in the work area throughout each work shift. Employers with employees who speak other languages may need to add the information in their language to the material presented, as long as the information is also presented in English.

If existing labels already convey the required information, the employee need not affix new labels.

If the employer becomes newly aware of any significant information regarding the hazards of a chemical, the employer must revise the labels for the chemical within three months of becoming aware of the new information. Labels on containers of hazardous chemicals shipped after that time shall contain the new information.

Note: Hazard warnings or labels is an area in which the facility safety and environmental manager, supervisors, and employees must be constantly vigilant to ensure that they are in place and legible.

The employer is required to develop a *written Hazard Communication program*. This particular requirement is often cited as the most common noncompliance violation found in industry today. The written HazCom program must be present, maintained, and readily available to all workers and visitors in each workplace. The written program must contain a section for labels and other warning devices and for safety data sheets, and employee information must be provided and training conducted. The written program must include a list of hazardous chemicals known to be present, using an identity that is referenced on the appropriate safety data sheet, the methods the employer uses to inform employees of the hazards of nonroutine tasks, and the hazards associated with chemicals contained in unlabeled pipes in their work areas.

DEFINITIONS OF HAZCOM TERMS

The Hazard Communication Program defines various terms as follows: (These terms either appear in a company's Hazard Communication Program or are definitions appropriate to SDS).

Chemical: Any element, compound, or mixture of elements and/or compounds.
Chemical name: The scientific designation of a chemical in accordance with the nomenclature system developed by the International Union of Pure and Applied Chemistry (IUPAC) or the Chemical Abstracts Service (CAS) Rules of Nomenclature, or a name which will clearly identify the chemical for the purpose of conducting a hazard evaluation.
Combustible liquid: Any liquid having a flashpoint at or above 100° F (37.8° C), but below 200° F (93.3° C).
Common name: Any designation or identification, such as code name, code number, trade name, brand name, or generic name used to identify a chemical other than its chemical name.
Compressed gas: A gas or mixture of gases in a container having an absolute pressure exceeding 40 psi at 70° F (21.1° C); or a gas or mixture of gases in a container having an absolute pressure exceeding 104 psi at 130° F (54.4° C) regardless of the pressure at 70° F (21.1 ° C); or a liquid having a vapor pressure exceeding 10 psi at 100° F (37.8° C), as determined by ASTM D-323-72.

Container: Any bag, barrel, bottle, box, can, cylinder, drum, reaction vessel, storage tank, or the like that contains a hazardous chemical.

Explosive: A chemical that causes a sudden, almost instantaneous release of pressure, gas, and heat when subjected to sudden shock, pressure, or high temperature.

Exposure: The actual or potential subjection of an employee to a hazardous chemical through any route of entry, in the course of employment.

Flammable aerosol: An aerosol that, when tested by the method described in 16 CFR 1500.45, yields a flame projection exceeding 18 inches at full valve opening, or a flashback (flame extending back to the valve) at any degree of valve opening.

Flammable gas: A gas that at ambient temperature and pressure forms a flammable mixture with air at a concentration of 13 percent by volume or less, or a gas that at ambient temperature and pressure forms a range of flammable mixtures with air wider than 12 percent by volume, regardless of the lower limit.

Flammable liquid: A liquid having flashpoint 100° F (37.8° C).

Flammable solid: A solid, other than a blasting agent or explosive as defined in 29 CFR 1910.109 (a), that is likely to cause fire through friction, absorption of moisture, spontaneous chemical change or retained heat from manufacturing or processing, or which can be ignited, and that when ignited, burns so vigorously and persistently so as to create a serious hazard. A chemical shall be considered to be a flammable solid if, when tested by the method described in 16 CFR 1500.44, it ignites and burns with a self-sustained flame at a rate greater than one-tenth of an inch per second along its major axis.

Flashpoint: The minimum temperature at which a liquid gives off a vapor in sufficient concentration to ignite.

Hazard warning: Any words, pictures, symbols, or combination thereof appearing on a label or other appropriate forms of warning that convey the hazards of the chemical(s) in the container.

Hazardous chemical: Any chemical that is a health or physical hazard.

Hazardous chemical inventory list: An inventory list of all hazardous chemicals used at the site, and containing the date of each chemical's SDS insertion.

Health hazard: A chemical for which there is statistically significant evidence based on at least one study conducted in accordance with established scientific principles that acute or chronic health effects may occur in exposed employees.

Immediate use: The use under the control of the person who transfers the hazardous chemical from a labeled container, and only within the work shift in which it is transferred.

Label: Any written, printed, or graphic material displayed on or affixed to containers or hazardous chemicals.

Safety data sheet (SDS): The written or printed material concerning a hazardous chemical, developed in accordance with 29 CFR 1910.

Mixture: Any combination of two or more chemicals if the combination is not, in whole or in part, the result of a chemical reaction.

NFPA Hazardous Chemical Label: A color-code labeling system developed by the National Fire Protection Association (NFPA) which rates the severity of the health hazard, fire hazard, reactivity hazard, and special hazard of the chemical.

Organic peroxide: An organic compound that contains the bivalent 0-0 structure and that may be considered to be a structural derivative of hydrogen peroxide, where one or both of the hydrogen atoms has been replaced by an organic radical.

Oxidizer: A chemical (other than a blasting agent or explosive as defined in 29 CFR 1910.198 (a)) that initiates or promotes combustion in other materials thereby causing fire either of itself or through the release of oxygen or other gases.

Physical hazard: A chemical for which there is scientifically valid evidence that it is a combustible liquid, a compressed gas explosive, flammable, an organic peroxide, an oxidizer, pyrophoric, unstable (reactive), or water reactive.

Portable container: A storage vessel that is mobile, such as a drum, side-mounted tank, tank truck, or vehicle fuel tank.

Primary route of entry: The primary means (such as inhalation, ingestion, skin contact, etc.) whereby an employee is subjected to a hazardous chemical.

"Right-To-Know" work station: Provides employees with a central information work station where they can have access to site SDS, Hazardous Chemical Inventory List, and the company's written Hazard Communication Program.

"Right-To-Know" station binder: A Station Binder located in the "Right-To-Know" work station that contains the company's Hazard Communication Program, the Hazardous Chemicals Inventory List and corresponding SDS, and the Hazard Communication Program Review and Signature Form.

Pyrophoric: A chemical that will ignite spontaneously in air at a temperature of 130° F (54.4° C) or below.

Signal word: A word used to indicate the relative level of severity of hazard and alert the reader to a potential hazard on the label. The signal words used in this section are "danger" and "warning." "Danger" is used for the more severe hazards, while "Warning" is used for the less severe.

Stationary container: A permanently mounted chemical storage tank.

Unstable (reactive chemical): A chemical that in its pure state or as produced or transported will vigorously polymerize, decompose, condense or will become self-reactive under conditions of shock, pressure, or temperature.

Water reactive (chemical): A chemical that reacts with water to release a gas that is either flammable or presents a health hazard.

Work center: Any convenient or logical grouping of designated unit processes or related maintenance actions.

HAZCOM AUDIT ITEMS

If your facility has a written HazCom program similar to the one above, you are well along the road on your trek toward compliance. If your HazCom program is audited by OSHA, the goal, of course, is for any auditor who might visit your facility is to be able to readily "see" that you're in compliance. Often an auditor will not even review your written HazCom program if he or she can plainly see you are in compliance.

Let's take a look at some of the HazCom items OSHA will be looking at. You must be able to answer "yes" to each of the following items, if site-applicable.

- Are all chemical containers marked with contents' name and hazards?
- Are the storage cabinets used to hold flammable liquids labeled "Flammable—Keep Fire Away"?
- For a fixed extinguishing system, is a sign posted warning of the hazards presented by the extinguishing medium?
- Are all aboveground storage tanks properly labeled?
- If you store hazardous materials (including gasoline) in aboveground storage tanks, are tanks or other containers holding hazardous materials appropriately labeled with chemical name and hazard warning?
- Are all chemicals used in spray painting operations correctly labeled?
- If you store chemicals, are all containers properly labeled with chemical name and hazard warning?

Along with checking these items, the OSHA auditor will make notes on the chemicals he or she finds in the workplace. During the walk-around, the auditor is likely to seek out any flammable materials storage lockers you have in your workplace. The auditor will list many of the items stored in the lockers. Later, when the walk-around is completed, the auditor will ask you to provide a copy of the SDS for each chemical in his or her notes.

To avoid a citation, you must not fail this major test. If the auditor, for example, noticed during the walk-around that employees were using some type of solvent or cleaning agent in the performance of their work, he or she will want to see a copy of the SDS for that particular chemical. If you can't produce a copy, you are in violation and will be cited. Be careful on this item—it is one of the most commonly cited offenses. Obviously, the only solution to this problem is to ensure that your facility has an SDS for each chemical used, stored, or produced, and that your chemical inventory list is current and accurate. Save yourself a big hassle—ensure SDSs are available to employees for each chemical used on site.

Keep in mind that the OSHA auditor will look at each work center within your company and that each different work center will present its own specialized requirements. If your company has an environmental laboratory, for example, the auditor will spend considerable time in the lab, ensuring you are in compliance with OSHA's laboratory standards, and that you have a written Chemical Hygiene Plan.

HAZARDOUS WASTE HANDLING

The most alarming of all man's assaults upon the environment is the contamination of air, earth, rivers, and sea with dangerous and even lethal materials. This pollution is for the most part irrecoverable; the chain of evil it initiates not only in the world that must support life but also in living tissues is for the most part irreversible. In this now universal contamination of the environment, chemicals are the sinister, and little–recognized partners of radiation in changing the very nature of the world—the very nature of life (Rachel Carson, 1962).

In 1990, R. B. Smith reported that the United States Environmental Protection Agency (EPA) estimated that the United States generates 570 million tons of hazardous waste annually. This waste includes toxic, biologic, and radioactive waste. But the broader human interaction, the safety and health considerations we have with wastes, most concerns the safety and environmental manager. These may overlap and directly interface with the classic environmental spans of control alluded to by Rachel Carson.

Rachel Carson combined the insight and sensitivity of a poet with the realism and observations of science in her classic and highly influential book *Silent Spring*, in ways no environment writer had before. To us, today, with the impact of that critically important, visionary work proven true, that Rachel Carson was ostracized, vilified, laughed at, and lambasted (particularly by chemical manufacturers) for that work strikes us as puzzling.

While those with vested interest in the book's failure worked to disparage it, Rachel Carson was not disregarded by those who understood. Examined with an unbiased eye, her message was clear: the chemicals we use commonly, in quantity, if not properly handled, treated, and disposed, not only pose a short-term threat to human life, but they also pose a long-term threat to the environment as a whole. Her plea is also clear: to end the poisoning of earth. With the clarity of vision provided by 20/20 hindsight, we now see (and have known for many years) that Rachel Carson was well ahead of her time. The concerns that *Silent Spring* addressed in 1962, while based on limited data, have since been confirmed.

In this section we discuss the hazards of handling hazardous materials—especially hazardous wastes, all of which should be the focus of the safety engineer. We illustrate the nature of hazardous waste, the problem, and the possible consequences.

AMERICA: A THROWAWAY SOCIETY

America as a whole has lost the habit of earlier generations to "use it up, wear it out, make it do, or do without." A new American characteristic, one that might be further described as habit, trend, custom, practice or tendency is to discard those objects we no longer want, whether or not they still have useful life. We have become a "throw-away society."

While many of us conscientiously recycle our bottles, cans, newspapers, and plastic containers, we often simply discard other, larger items we have no more use for, simply because throwing them away is easier than finding an avenue to recycle or reuse them. When an item loses its value to us because it is broken, shabby, no longer fashionable, or no longer needed for whatever reason, discarding it should not be an insurmountable problem. But it is—especially whenever the item we throw away is a hazardous substance, one that is persistent, nonbiodegradable, and poisonous.

What is the magnitude of the problem with hazardous substance and waste disposal? Let's take a look at a few facts.

- Hazardous substances—including industrial chemicals, toxic waste, pesticides, and nuclear waste—are entering the marketplace, the workplace, and the environment in unprecedented amounts.

- The United States produces almost 300 million metric tons of hazardous waste each year—with a present population of 320 million this amounts to more than one ton for every person in the country.
- Through pollution of the air, the soil and water supplies, hazardous wastes pose both short- and long-term threats to human health and environmental quality.

WHAT IS A HAZARDOUS SUBSTANCE?

Hazardous wastes can be informally defined as a subset of all solid and liquid wastes that are disposed of on land rather than being shunted directly into the air or water, and which have the potential to adversely affect human health and the environment. We often believe that hazardous wastes result mainly from industrial activities, but households also play a role in the generation and improper disposal of substances that might be considered hazardous wastes. Hazardous wastes (via Bhopal and other disastrous episodes) have been given much attention, but surprisingly little is known of their nature and of the actual scope of the problem. In this section, we examine definitions of hazardous materials, substances, wastes, etc., and attempt to bring hazardous wastes into perspective both as a major environmental and as a safety and health concern.

Unfortunately, defining a hazardous substance is largely a matter of choice between the definitions offered by the various regulatory agencies and pieces of environmental legislation, each defining it somewhat differently. Many of the different terms are used interchangeably. Even experienced professional Certified Hazardous Materials Managers have been known to interchange these terms, even though they are generated by different official sources and have somewhat different meanings, dependent upon the nature of the problem being addressed. To understand the scope of the dilemma in defining a hazardous substance let's take a look at the terms that are in common use today, used interchangeably, and often thought to mean the same thing.

Hazardous Material

A hazardous material is a substance (gas, liquid, or solid) capable of causing harm to people, property, and the environment. The United States Department of Transportation (DOT) uses the term hazardous materials to cover nine categories identified by the United Nations Hazard Class Number System, including:

- Explosives
- Gases (compressed, liquefied, dissolved)
- Flammable Liquids
- Flammable Solids
- Oxidizers
- Poisonous Materials
- Radioactive Materials
- Corrosive Materials
- Miscellaneous Materials

Hazardous Substances

The term hazardous substance is used by the EPA for chemicals that, if released into the environment above a certain amount, must be reported and, depending on the threat to the environment, for which federal involvement in handling the incident can be authorized. The EPA lists hazardous substances in its 40 CFR Part 302, Table 302.4.

The Occupational Safety and Health Administration (OSHA) uses the term hazardous substance in 29 CFR 1910.120 (which resulted from Title I of SARA and covers emergency response) differently than does the EPA. Hazardous substances (as defined by OSHA) cover every chemical regulated by both DOT and the EPA.

Extremely Hazardous Substances

Extremely hazardous substance is a term used by the EPA for chemicals that must be reported to the appropriate authorities if released above the threshold reporting quantity (RQ). The list of extremely hazardous substances is identified in Title III of the Superfund Amendments and Reauthorization Act (SARA) of 1986 (40 CFR Part 355). Each substance has a threshold reporting quantity.

Toxic Chemicals

The EPA uses the term toxic chemical for chemicals whose total emissions or releases must be reported annually by owners and operators of facilities that manufacture, process, or otherwise use listed toxic chemicals. The list of toxic chemicals is identified in Title III of SARA.

Hazardous Wastes

The EPA uses the term hazardous wastes for chemicals regulated under the Resource, Conservation and Recovery Act (RCRA-40 CFR Part 261.33). Hazardous wastes in transportation are regulated by DOT (49 CFR Parts 170-179).

For our purposes in this text, we define a hazardous waste as any hazardous substance that has been spilled or released into the environment. For example, chlorine gas is a hazardous material. When chlorine is released to the environment, it becomes a hazardous waste. Similarly, when asbestos is in place and undisturbed, it is a hazardous material. When it is broken, breached, or thrown away, it becomes a hazardous waste.

Hazardous Chemicals

OSHA uses the term hazardous chemical to denote any chemical that poses a risk to employees if they are exposed to it in the workplace. Hazardous chemicals cover a broader group of chemicals than the other chemical lists.

RCRA'S DEFINITION OF A HAZARDOUS SUBSTANCE

For the purposes of this text, to form the strongest foundation for understanding the main topic of this chapter (hazardous waste handling) and because RCRA's definition for a hazardous substance can also be used to describe a hazardous waste, we use RCRA's definition.

RCRA defines a substance as hazardous if it possesses any of the following four characteristics: reactivity, ignitability, corrosiveness, or toxicity. Briefly,

- **Ignitability** refers to the characteristic of being able to sustain combustion and includes the category of flammability (ability to start fires when heated to temperatures below 140°F or less than 60°C).
- **Corrosive** substances (or wastes) may destroy containers, contaminate soils and groundwater, or react with other materials to cause toxic gas emissions. Corrosive materials provide a specific hazard to human tissue and aquatic life where the pH levels are extreme.
- **Reactive** substances may be unstable or have tendency to react, explode, or generate pressure during handling. Pressure-sensitive or water-reactive materials are included in this category.
- **Toxicity** is a function of the effect of hazardous materials (or wastes) that may come into contact with water or air and be leached into the groundwater or dispersed in the environment.

The toxic effects that may occur to humans, fish, or wildlife are our principal concerns here. Toxicity (until 1990) was tested using a standardized laboratory test, called the extraction procedure (EP Toxicity Test). The EP Toxicity Test was replaced in 1990 by the Toxicity Characteristics Leaching Procedure (TCLP), because the EP test failed to adequately simulate the flow of toxic contaminants to drinking water. The TCLP test is designed to identify wastes likely to leach hazardous concentrations of

Table 5.3   Maximum Concentration of Contaminants for TCLP Toxicity Test

Contaminant	Regulatory Level (mg/L)	Contaminant	Regulatory Level (mg/L)
Arsenic	5	Lead	5
Barium	100	Lindane	0.4
Benzene	0.5	Mercury	0.2
Cadmium	1	Methoxychlor	10
Carbon tetrachloride	0.5	Methyl ethyl ketone	200
Chlordane	0.03	Nitrobenzene	2
Chlorobenzene	100	Pentachlorophenol	100
Chloroform	6	Pyridine	5
Chromium	5	Selenium	1
Cresol	200	Silver	5
2,4-D	10	Tetrachloroethylene	0.7
1,4-Dichlorobenzene	7.5	Toxaphene	0.5
1,5-Dichloroethane	0.5	Trichloroethylene	0.5
2.4-Dinitrololuene	0.13	2,4,5-Trchlorophenol	400
Endrin	0.02	2,4,6-Trchlorophenol	2
Heptachlor	0.008	2.4,5-TP (Silvex)	1
Hexachlorobenezene	0.13	Vinyl chloride	0.2
Hexachloroethane	3		

Source: USEPA (1990), 40 CFR 261.24.

particular toxic constituents into the surrounding soils of groundwater as a result of improper management.

TCLP extracts constituents from the tested waste in a manner designed to simulate the leaching actions that occur in landfills. The extract is then analyzed to determine if it possesses any of the toxic constituents listed in Table 5.3. If the concentrations of the toxic constituents exceed the levels listed in the table, the waste is classified as hazardous.

WHAT IS A HAZARDOUS WASTE?

A general rule of thumb states that any hazardous substance that is spilled or released into the environment is no longer classified as a hazardous substance but as a hazardous waste. The EPA uses the same definition for hazardous waste as it does for hazardous substance. The four characteristics described in the previous section (reactivity, ignitability, corrosivity, or toxicity) can also be used to identify hazardous substances as well as hazardous wastes.

Note that the EPA lists substances that it considers hazardous waste. These lists take precedence over any other method used to identify and classify a substance as hazardous (i.e., if a substance is listed in one of the EPA's lists described below, it is a hazardous substance, no matter what).

EPA Lists of Hazardous Wastes

EPA-listed hazardous wastes are organized into three categories: Nonspecific source wastes, specific source wastes, and commercial chemical products. All listed wastes are presumed to be hazardous, regardless of their concentrations. EPA developed these lists by examining different types of wastes and chemical products to determine whether they met any of the following criteria:

- Exhibit one or more of the four characterizations of a hazardous waste.
- Meet the statutory definition of hazardous waste.
- Are acutely toxic or acutely hazardous.
- Are otherwise toxic.

These lists are described briefly, as follows:

- **Nonspecific source wastes** are generic wastes commonly produced by manufacturing and industrial processes. Examples from this list include spent halogenated solvents used in degreasing and wastewater treatment sludge from electroplating processes, as well as dioxin wastes, most of which are "acutely hazardous" wastes because of the danger they present to human health and the environment.
- **Specific source wastes** are from specially identified industries such as wood preserving, petroleum refining, and organic chemical manufacturing. These wastes typically include sludges, still bottoms, wastewaters, spent catalysts, and residues, such as wastewater treatment sludge from pigment production.

- **Commercial chemical products** (also called "P" or "U" list wastes because their code numbers begin with these letters) include specific commercial chemical products or manufacturing chemical intermediates. This list includes chemicals such as chloroform and creosote, acids such as sulfuric and hydrochloric, and pesticides such as DDT and kepone (40 CFR 261.31,32 & 33).

Note that the EPA ruled that any waste mixture containing a listed hazardous waste is also considered a hazardous waste and must be managed accordingly. This applies regardless of what percentage of the waste mixture is composed of listed hazardous wastes. Wastes derived from hazardous wastes (residues from the treatment, storage, and disposal of a listed hazardous waste) are considered hazardous waste as well (EPA, 1990).

WHERE DO HAZARDOUS WASTES COME FROM?

Hazardous wastes are derived from several waste generators. Most of these waste generators are in the manufacturing and industrial sectors and include chemical manufacturers, the printing industry, vehicle maintenance shops, leather products manufacturers, the construction industry, metal manufacturing, and others. These industrial waste generators produce a wide variety of wastes, including strong acids and bases, spent solvents, heavy metal solutions, ignitable wastes, cyanide wastes, and many more.

WHY ARE WE CONCERNED ABOUT HAZARDOUS WASTES?

From the safety and environmental manager's perspective, any hazardous waste release that could alter the environment and/or impact the health and safety of employees in any way is a major concern. The specifics of the safety engineer's concern lie in acute and chronic toxicity to organisms, bioconcentration, biomagnification, genetic change potential, etiology, pathways, change in climate and/or habitat, extinction, persistence, esthetics such as visual impact, and most importantly, the impact on the health and safety of employees.

Remember, we have stated consistently that when a hazardous substance or hazardous material is spilled or released into the environment, it becomes a hazardous waste. This is important because specific regulatory legislation has been put in place regarding hazardous wastes, responding to hazardous waste leak/spill contingencies, and for proper handling, storage, transportation, and treatment of hazardous wastes, the goal being, of course, protecting the environment, and ultimately, protecting the health and safety of our employees and the surrounding community.

Why are we so concerned about hazardous substances and hazardous wastes? This question is relatively easy to answer based on experience, publicity, and actual hazardous materials incidents, which have resulted in tragic consequences to the environment and to human life.

HAZARDOUS WASTE LEGISLATION

Humans are strange in many ways. We may know that a disaster is possible, is likely, could happen, is predictable. But do we act before someone dies? Not often enough. We often ignore the human element. We forget the victim's demise. We simply do not want to think about it, because if we think about it, we must come face-to-face with our own mortality. The safety and environmental manager, though, must think about it—constantly, and before such travesties occur—to prevent them from ever occurring.

Because of Bhopal and other similar (but less catastrophic) chemical spill events, the United States Congress (pushed by public concern) developed and passed certain environmental laws and regulations to regulate hazardous substances/wastes in the United States. This section focuses on the two regulatory acts most crucial to the current management programs for hazardous wastes. The first (mentioned several times throughout the text) is the Resource Conservation and Recovery Act (RCRA). Specifically, RCRA provides guidelines for prudent management of new and future hazardous substances/wastes. The second act (more briefly mentioned) is the Comprehensive Environmental Response, Compensation, and Liability Act (CERCLA), otherwise known as Superfund, which deals primarily with mistakes of the past: inactive and abandoned hazardous waste sites.

Resource Conservation and Recovery Act

The Resource Conservation and Recovery Act (RCRA) is the United States' single most important law dealing with the management of hazardous waste. RCRA and its amendment Hazardous and Solid Waste Act (HSWA-1984) deal with the ongoing management of solid wastes throughout the country, with emphasis on hazardous waste. Keyed to the waste side of hazardous materials, rather than broader issues dealt with in other acts, RCRA is primarily concerned with land disposal of hazardous wastes. The goal is to protect groundwater supplies by creating a "cradle-to-grave" management system with three key elements: a tracking system, a permitting system, and control of disposal.

1. **Tracking system**—a manifest document accompanies any waste that is transported from one location to another.
2. **Permitting system**—helps assure safe operation of facilities that treat, store, or dispose of hazardous wastes.
3. **Disposal control system**—controls and restrictions governing the disposal of hazardous wastes onto, or into, the land (Masters, 2007).

RCRA regulates five specific areas for the management of hazardous waste (with the focus on treatment, storage, and disposal). These are:

1. Identifying what constitutes a hazardous waste and providing classification of each.
2. Publishing requirements for generators to identify themselves, which includes notification of hazardous waste activities and standards of operation for generators.

3. Adopting standards for transporters of hazardous wastes.
4. Adopting standards for treatment, storage, and disposal facilities.
5. Providing for enforcement of standards through a permitting program and legal penalties for noncompliance (Griffin, 2009).

Arguably, RCRA is our single most important law dealing with the management of hazardous waste. It certainly is the most comprehensive piece of legislation that EPA has promulgated to date.

CERCLA

The mission of the Comprehensive Environmental Response, Compensation, and Liabilities Act of 1980 (CERCLA) is to clean up hazardous waste disposal mistakes of the past, and to cope with emergencies of the present. More often referred to as the Superfund Law, as a result of its key provisions a large trust fund (about $1.6 billion) was created. Later, in 1986, when the law was revised, this fund was increased to almost $9 billion. The revised law is designated as the **Superfund Amendments and Reauthorization Act of 1986 (SARA)**. The key requirements under CERCLA are listed in the following. Briefly,

1. CERCLA authorizes the EPA to deal with both short-term (emergency situations triggered by a spill or release of hazardous substances), as well as long-term problems involving abandoned or uncontrolled hazardous waste sites for which more permanent solutions are required.
2. CERCLA has set up a remedial scheme for analyzing the impact of contamination on sites under a hazard ranking system. From this hazard ranking system, a list of prioritized disposal and contaminated sites is compiled. This list becomes the National Priorities List (NPL) when promulgated. The NPL identifies the worst sites in the nation, based on such factors as the quantities and toxicity of wastes involved, the exposure pathways, the number of people potentially exposed, and the importance and vulnerability of the underlying groundwater.
3. CERCLA also forces those parties who are responsible for hazardous waste problems to pay the entire cost of cleanup.
4. Title III of SARA requires federal, state, and local governments and industry to work together in developing emergency response plans and reporting on hazardous chemicals. This requirement is commonly known as the Community Right-To-Know Act, which allows the public to obtain information about the presence of hazardous chemicals in their communities and releases of these chemicals into the environment.

OSHA

Moretz (1989) points out that OSHA's hazardous waste standard specifically addresses the safety of the estimated 1.75 million workers who deal with hazardous waste: hazardous waste workers in all situations including treatment, storage, handling, and disposal; firefighters; police officers; ambulance personnel; and hazardous materials response team personnel.

Occupational Health magazine summarizes the requirements of this standard:

- Each hazardous waste site employer must develop a safety and health program designed to identify, evaluate, and control safety and health hazards and provide for emergency response.
- There must be preliminary evaluation of the site's characteristics prior to entry by a trained person to identify potential site hazards and to aid in the selection of appropriate employee protection methods.
- The employer must implement a site control program to prevent contamination of employees. At a minimum, the program must identify a site map, site work zones, site communications, safe work practices, and the location of the nearest medical assistance. Also required in particularly hazardous situations is the use of the two-person rule (buddy system) so that employees can keep watch on one another and provide quick aid if needed.
- Employees must be trained before they are allowed to engage in hazardous waste operations or emergency response that could expose them to safety and health hazards.
- The employer must provide medical surveillance at least annually and at the end of employment for all employees exposed to any particular hazardous substance at or above established exposure levels and/or those who wear approved respirators for thirty days or more on site.
- Engineering controls, work practices, and PPE, or a combination of these methods, must be implemented to reduce exposure below established exposure levels for the hazardous substances involved.
- There must be periodic air monitoring to identify and quantify levels of hazardous substances and to ensure that proper protective equipment is being used.
- The employer must set up an information program with the names of key personnel and their alternates responsible for site safety and health and the requirements of the standard.
- The employer must implement a decontamination procedure before any employee or equipment leaves an area of potential hazardous exposure; establish operating procedures to minimize exposure through contact with exposed equipment, other employees, or used clothing; and provide showers and changing rooms where needed.
- There must be an emergency response plan to handle possible on-site emergencies prior to beginning hazardous waste operations. Such plans must address personnel roles; lines of authority, training, and communications; emergency recognition and prevention; safe places of refuge; site security; evacuation routes and procedures; emergency medical treatment; and emergency alerting.
- There must be an off-site emergency response plan to better coordinate emergency action by local services and to implement appropriate control actions.

HAZARDOUS WASTE SAFETY PROGRAM

For the purposes of this text, hazardous waste handling includes work activities that include the collection, storage, treatment, disposal, and cleanup of hazardous waste materials. We also focus on standard industrial wastes and their handling. Industrial wastes include:

1. Acids
2. Abrasives
3. Bases
4. Animal products/by-products
5. Biologic substances
6. Carcinogenic substances
7. Explosives
8. Solvents
9. Salts
10. Pesticides
11. Oils
12. Combustible materials
13. Metals
14. Reactive materials
15. Organic materials

As a safety and environmental manager, the fact that you must perform a comprehensive system safety analysis should be evident. This will allow you to recognize, evaluate, and control a wide variety of hazards and associated risks and to provide this information to all employees affected by exposure via a written Hazardous Waste Safety Program.

The comprehensive site characterization and safety analysis is required to identify specific site hazards and to determine the appropriate safety and health control measures needed to protect employees from the identified hazards.

As mentioned, the safety and environmental manager must ensure that appropriate site control measures are implemented to control employee exposure to hazardous substances before clean-up work begins. As a minimum, site control should include:

- a site map;
- site work zones;
- the use of a "buddy system";
- site communications including alerting means for emergencies;
- the standard operating procedures or safe work practices; and
- identification of the nearest medical assistance.

All employees working on site who have the potential for exposure to hazardous substances, health hazards, or safety hazards must receive appropriate training and obtain certification before they can be allowed to engage in hazardous waste operations that could expose them to hazardous substances and safety or health hazards.

Employees engaged in hazardous waste operations must be included in a company medical surveillance program.

Engineering controls, work practices, and personal protective equipment for employee protection must be implemented to protect employees from exposure to hazardous substances and safety and health hazards.

Monitoring must be performed to assure proper selection of engineering controls, work practices (such as confined space entry) and personal protective equipment (PPE) so that employees are not exposed to levels that exceed permissible exposure limits (or published exposure levels if there are no permissible exposure limits) for hazardous substances.

Decontamination procedures for all phases of decontamination must be developed and implemented.

An emergency response plan must be developed and implemented by all employers who engage in hazardous waste operations. The plan must be written and available for inspection by employees and appropriate regulatory agencies.

THOUGHT-PROVOKING QUESTIONS

- What does "Right-To-Know" mean? What does it entail for the safety and environmental manager?
- HazCom is a performance standard. What does this mean for the safety and environmental manager?
- What responsibilities do employers hold for informing and training employees?
- What is an SDS, and how are SDSs used?
- What are the three chief reasons a safety and environmental manager should be involved with implementing the HazCom standards?
- What should employee training include?
- What importance does labeling hold for HazCom? Why?
- Why is a written HazCom program important?
- How are responsibilities for HazCom distributed?
- What's a "Right-To-Know station," and what should it contain?
- What's important about the chemical inventory list?
- What are the regulations for warnings?
- How are outside contractors affected by HazCom regulations and safety programs?
- What would an auditor look for to determine your HazCom program's effectiveness?
- What problems are associated with hazardous waste, environmentally and societally? What specifically concerns the safety and environmental managers?
- What are the principle sources and generators of hazardous waste?
- Define and discuss hazardous substance, hazardous material, hazardous waste, extremely hazardous substance, toxic chemicals, and hazardous chemicals.
- When do hazardous materials become hazardous waste?
- What are the three hazardous waste categories? How does each category affect the safety and environmental manager?
- What effect has hazardous waste legislation had on industry?
- Discuss RCRA, CERCLA, and the EPA's distinction between these terms. How do their regulations affect industry and the environment?

- What are RCRA's four characteristics of hazardous substances? Why are they important?
- What are the nine categories used in the United Nations Hazard Class Number System?
- What are RCRA's regulation areas and CERCLA's requirements? How can they affect your responsibilities as safety and environmental manager?
- What is bioremediation?
- What is a hazardous waste? Explain.
- What's the difference between hazardous waste, hazardous substance, and toxic chemical?
- To be a hazardous waste, does a substance have to be listed by EPA first? Explain.
- Identify and discuss the safety and environmental manager's chief concerns over hazardous waste.

REFERENCES AND RECOMMENDED READING

Blackman, W.C. *Basic Hazardous Waste Management*, 3rd ed. Boca Raton, FL: Lewis Publishers, 2001.

Carson, R. *Silent Spring*. Boston: Houghton Mifflin Company, 1962.

Code of Federal Regulations. Other Regulations Relating to Transportation, title 49, sec. 170–179.

Code of Federal Regulations. The Control of Hazardous Energy (Lockout/Tagout), title 29, sec 1910.147.

Code of Federal Regulations. Toxicity characteristic, title 40, sec 261.24.

Code of Federal Regulations. Lists of Hazardous Wastes, title 40, sec 261.31-33.

Code of Federal Regulations. Purpose, scope and applicability, title 40, sec 264.1.

Code of Federal Regulations. Content of contingency plan, title 40, sec 264.52(b).

Code of Federal Regulations. Designation of hazardous substances, title 40, sec 302.4.

Coleman, R.J., and K.H. Williams. *Hazardous Materials Dictionary*. Lancaster, PA: Technomic Publishing Company, 1988.

"General Requirements of OSHA's Final Hazardous Waste Standard." *Occupational Hazards.* November 1989: 12.

Griffin, R.D. *Principles of Hazardous Materials Management*, 2nd ed. Chelsea, MI: Lewis Publishers, 2009.

Kharbanda, O.P., and E.A. Stallworthy. *Waste Management*, UK: Gower Publishing, 1990.

Knowles, P.C., ed. *Fundamentals of Environmental Science and Technology*. Rockville, MD: Government Institutes, Inc., 1992.

Lindgren, G.F. *Managing Industrial Hazardous Waste: A Practical Handbook*. Chelsea, MI: Lewis Publishers, 1989.

Masters, G.M. *Introduction to Environmental Engineering and Science*, 3rd ed. New York: Prentice–Hall, 2007.

Moretz, S. "Industry Prepares for OSHA's Final Hazardous Waste Role." *Occupational Hazards* November 1989: 39–42.

OSHA's Hazard Communication Standard. Rockville, MD: Government Institutes, 1996.

Portney, P.R., ed. Public Policies for Environmental Protection. Washington, DC: Resources for the Future, 1993.

RCRA, Public Law 98–616, Hazardous and Solid Wastes Act, amendments PL 94–580 (42 USC 6901), 1984.

SARA (CERCLA), Public Law 99–499, Superfund Amendments & Reauthorization Act (1986), Amended 142 USC 9601, 1980.

Smith, R.B. "Manufacturing Companies of All Sizes Benefit from Waste–Reduction Policy." *Occupational Health & Safety* January 1990.

Spellman, F.R. *Environmental Science & Technology: Concepts & Applications.* Rockville, MD: Government Institutes, 2006.

Spellman, F.R. *Surviving an OSHA Audit: A Manager's Guide.* Lancaster, PA: Technomic Publishing Company, 1998.

USEPA, RCRA Orientation Manual. Washington, DC: United States Environmental Protection Agency, 1990.

USEPA, *Hazardous Waste Management.* Washington, DC. United States Environmental Protection Agency, 2013.

Wentz, C.A. *Hazardous Waste Management.* New York: McGraw–Hill, Inc., 1989.

Chapter 6

Emergency Response and Workplace Security

OSHA AND EMERGENCY RESPONSE

Even though there are no OSHA standards dedicated specifically to the issue of planning for emergencies, all OSHA standards are written for the purpose of promoting a safe, healthy, and accident-free, (hence emergency-free) workplace. Therefore, OSHA standards do play a role in emergency prevention.

Because this is the case, they should be considered when developing emergency plans. A first step when developing emergency response plans is to review these OSHA standards. This can help organizations identify and then correct conditions that might exacerbate emergency situations before they occur.

Emergency Response Plans

Typically, when we think of emergency response plans for the workplace, we often conjure up thoughts about the obvious. For example, the first workplace emergency that might come to mind is *fire*, a major concern because fire in the workplace is something that happens more often than we might want to think and because fire can be particularly devastating—in ways we know too well. Most employees do not need to be informed about the dangers of fire. However, employers have the responsibility to do just this, to inform and train employees on fire, fire prevention, and fire protection. Many local codes go beyond this information requirement, insisting that employers develop and implement a fire emergency response and/or evacuation plan. The primary emphasis has been on the latter: evacuation. However, if the employer equips a workplace with fire extinguishers and other fire fighting equipment and expects employees to respond aggressively to extinguish workplace fires, then not only must the facility have an emergency response plan, but the employer must also ensure that all company personnel called upon to fight the fire are completely trained on how to do so safely [29 CFR 1910.156(c)/.157(g)].

Another commonly considered workplace emergency response plan or scenario is for *medical emergencies.* Many facilities satisfy this requirement simply by directing employees to call 911 or some other emergency number whenever a medical emergency occurs in the workplace. Other facilities, though, may require employees to provide emergency first aid. When the employer chooses the employee-supplied first aid option, certain requirements must be met before any employee can legally administer first aid. First, the first aid responder must be trained and certified to administer first aid. This training aspect must also include training on OSHA's Bloodborne Pathogen Standard. This standard requires that the employee is trained on the dangers inherent with handling and being exposed to human body fluids. The employee must also be trained on how to protect him or herself from contamination. If the first aid responder or anyone else is exposed to and contaminated by body fluids, the employer must make available the hepatitis B vaccine and vaccination series to all employees who have occupational exposure, and conduct post-exposure evaluations and follow-ups to all employees who have had an exposure incident (29 CFR 1910.1030).

The third type of emergency response plan required for implementation in selected (covered) facilities is OSHA's 29 CFR 1910.120 (Hazardous Waste Operations and Emergency Response—HAZWOPER) for releases of *hazardous materials*. Unless the facility operator can demonstrate that the operation does not involve employee exposure or the reasonable possibility for employee exposure to safety or health hazards, the following operations are covered:

1. Clean up operations required by a governmental body that involve hazardous substances and are conducted at uncontrolled hazardous waste sites, state priority site lists, sites recommended by the EPA, NPL, and initial investigations of government-identified sites that are conducted before the presence or absence of hazardous substance has been ascertained.
2. Corrective actions involving clean-up operations at sites covered by the Resource Conservation and Recovery Act of 1976 (RCRA).
3. Voluntary clean up operations at sites recognized by Federal, state, local, or other governmental bodies as uncontrolled hazardous waste sites.
4. Operations involving hazardous waste conducted at treatment, storage, disposal (TSD) facilities regulated by RCRA.
5. Emergency response operations for releases of, or substantial threats of releases of, hazardous substances without regard to the location of the hazard.

The final requirement impacts the largest number of facilities that meet the criteria requiring full compliance with 29 CFR 1910.120 HAZWOPER, because many such facilities do not normally handle, store, treat, or dispose hazardous waste, but do use or produce hazardous materials in their processes.

Because the use of hazardous materials could lead to an emergency from the release or spill of such materials, facilities using these materials must develop and employ an effective site emergency response plan.

Before we discuss the basic goals of an effective emergency response plan, we should define "emergency response." Considering that individual facilities are different, with different dangers and different needs, defining emergency response is not always easy. However, for our purposes, we use the definition provided by CoVan (1995).

"Emergency response is defined as a limited response to abnormal conditions expected to result in unacceptable risk requiring rapid corrective action to prevent harm to personnel, property, or system function (p. 54)."

CoVan makes another important point about emergency response, one critical for the safety and environmental manager. He points out that "although emergency response and engineering tends toward prevention, emergency response is a skill area that safety engineers must be familiar with because of both regulations and good engineering practice" (54).

"Good Engineering Practice": the law by which all competent safety engineers work and live.

Now that we have defined emergency response, let's move on to the basic goals of an effective emergency response plan. Most of the currently available literature on this topic generally lists the goals as twofold:

1. Minimize injury to facility personnel.
2. Minimize damage to facility and then return to normal operation as soon as possible.

Obviously, these goals make a great deal of good sense. However, you may be wondering about key words used: "facility personnel" and "damage to the facility." Remember that we are talking about OSHA requirements here. Under OSHA the primary emphasis is protecting the worker. Protecting the worker's health and safety is OSHA's only focus.

What about people who live off site; the site's neighbors?

What about the environment?

These questions stress the point we emphasize here. Again, OSHA is not normally concerned about the environment, unless contamination of the environment (at the work site) might adversely impact the worker's safety and health. The neighbors? Again, OSHA's focus is the worker. One OSHA compliance office explained to us that if the employer takes every necessary step to protect its employees from harm involving the use or production of hazardous materials, then the surrounding community should have little to fear.

This statement is puzzling. We asked the same OSHA compliance officer about those incidents beyond the control of the employer, accidents that could not only put employees in harm's way, but also endanger the surrounding community. The answer? "Well, that's the EPA's bag. We only worry about the worksite and the worker."

Fortunately, OSHA, in combination with the U.S. Environmental Protection Agency (EPA) have taken steps to overcome this blatant shortcoming (we like to think of it as an oversight). Under OSHA's Process Safety Management (PSM) and EPA's Risk Management Planning (RMP) directive, chemical spills and other chemical accidents that could impact both the environment and the "neighbors" have now been properly addressed. What PSM and RMP really accomplish is changing the typical twofold goal of an effective emergency response plan into a threefold goal.

Let us point out that the accomplishment of these two- or three-fold goals or objectives is essential in emergency response. Accomplishing these goals or objectives requires an extensive planning effort prior to the emergency ("prior" being the keyword, because the attempt to develop an emergency response plan when a disaster is occurring or after one has occurred is both futile and stupid). The safety and environmental manager must never forget that while hazards in any facility can be reduced, risk is an element of everyday existence, and therefore cannot be totally eliminated. The safety and environmental manager's goal must be to keep risk to an absolute minimum. To accomplish this, advance planning is critical and essential. We pointed out earlier that most plans address fire, medical emergencies, and the accidental release or spills of hazardous materials. Note that the development of emergency response plans should also factor in other possible emergencies: natural disasters, floods, explosions, and/or weather-related events that could occur. The site emergency response plans should include:

- Assessment of risk.
- Chain of command for dealing with emergencies
- Assessment of resources
- Training
- Incident command procedures
- Site security
- Emergency teams
- Preplanning
- Evacuation routes
- Shutdown procedures
- Emergency medical treatment
- Decontamination
- Public relations

Typical Contents of an Emergency Response Plan

The Federal Emergency Management Agency (better known as FEMA), the U.S. Army Corps of Engineers, and several other agencies, as well as numerous publications, provide guidance on how to develop a site emergency response plan. Local agencies (such as fire departments, emergency planning commissions/agencies, HazMat teams, and Local Emergency Planning Committees (LEPCs)) also provide information on how to design a site plan. All of these agencies typically recommend that a site's plan contain the elements listed in Table 6.1.

In the site safety and environmental manager's effort to incorporate and manage a facility emergency response plan, and in the response itself, two elements mentioned earlier (security considerations and public relations) must be given special attention. If not handled correctly, the lack of effective security measures and/or improper public relations can turn an already disastrous incident into a mega-disaster.

In planning security considerations, provision should be made to have a well-trained security team limit site access to only those people and that equipment that will assist in coping with and resolving the emergency.

Public relations can be a tricky enterprise. The person identified to interface with the media must have thorough knowledge of the site, process, and personnel involved. The PR person must also have access to the highest levels of site management.

Table 6.1 Site Emergency Response Plan

Emergency Response Notification	List of who to call and information to pass on when an emergency occurs
Record of Changes	Table of changes and dates for them
Table of Content/Introduction	The purpose, objective, scope, applications, policies, and assumptions for the plan
Emergency Response Operations	Details about what actions must take place
Emergency Assistance Telephone Numbers	A current list of people and agencies that may be needed in an emergency
Legal Authority and Responsibility	References the laws and regulations that provide the authority for the plan
Chain of Command	Response organization structure and responsibilities
Disaster Assistance and Coordination	Where additional assistance may be obtained when the regular response organizations are over-burdened
Procedures for Changing or Updating the Plan	Details who makes changes and how they are made and implemented
Plan Distribution	List of organizations and individuals who have been given a copy of the plan
Spill Cleanup Techniques	Detailed information about how response teams should oxganize cleanups
Cleanup/Disposal Resources	List of what is available, where it is obtained and how much is available
Consultant Resources	List of special facilities and personnel who may be valuable in a response
Technical Library/References	List of libraries and other information sources that may be valuable for those preparing, updating, or implementing the plan
Hazard Analysis	Details the kinds of emergencies that may be encountered, where they are likely to occur, what areas of the community may be affected, and the probability of occurrence
Documentation of Spill Events	The various incident and investigative reports on spills that have occurred
Hazardous Materials Information	Listing of hazardous materials, their properties, response data, and related information
Dry Runs	Training exercises for testing the adequacy of the plan, training personnel, and introducing changes

Source: Planning Guide and Checklist for Hazardous Materials Contingency Plans, FEMA-10, Federal Emergency Management Agency, Washington, D.C., July 1981. Adaptation from Brauer, R. L., *Safety and Health for Engineers*. New York: Van Nostrand Reinhold, p. 593, 2005.

Otherwise, he or she will not be able to deal with the public/media in an effective manner.

WORKPLACE SAFETY

The September 11, 2001 terrorist attacks on the World Trade Center and the Pentagon made all of us examine our lives, both at home and at work, in terms of safety. People who have lived out their lives in peace and comfort take a degree of personal safety for granted, and being jarred out of that complacency has advantages and disadvantages.

One of the hardest lessons to learn, for most people, is that none of us can really control how safe we are. We can, however, take some sensible steps in that direction, and at work, some added security is necessary. Remember, the 9/11 attacks were directed at workplaces.

Utilities, public works, and local businesses hold the lives of their communities in their hands. For example, think about how difficult living in a large metropolitan areas becomes when there is flooding that disrupts both the potable water supply and releases wastewater into the floodwaters, contaminating that water. Now, think about the possibilities for disaster with deliberate sabotage of either potable or wastewater systems. Not a pleasant thought, is it? And, of course, water and wastewater facilities also store large quantities of potentially hazardous chemicals, too.

What Not to Do: Judge by Appearances

Stories now circulate about people currently distrustful of those of Middle Eastern appearance, but remember, a person who wishes to cause chaos and destruction can come from any background. Timothy McVey, for example, in appearance would not have attracted unusual attention behind you in the grocery line, or as a coworker on the job. Not only can we not live our lives in constant distrust of everyone around us, the United States was begun on a philosophical idea of individual freedom, and the US legal system founded on the idea of innocent until proven guilty. So what's the answer?

What to Do: Background Checks

The people employed in any workplace are selected under some sort of workplace hiring practices. With turnover in personnel, interviews, background checks, and other hiring processes in many workplaces go on all the time. In the past, with shortages of qualified personnel, finding a job candidate with the right skill sets and limiting or ignoring a thorough background check might have been a possibility. Now, though, it would be sheer negligence. A complete background check must be applied across the board. Examine what your workplace's background check involves. Does someone call and actually check references? Did John Doe actually attend the Dismal Seepage Consolidated High School? Does the background check involve a telephone interview or other direct contact with someone who knows the potential hiree?

Most of us don't have much say in who we are working with. That doesn't mean we are powerless, however. Check out your workplace's security system and see what measures are in place for your protection. If you don't think they're adequate, discuss on-the-job safety with your supervisor, the security officer, or the most appropriate person in management before you go any further. Increases in security are an easy sell these days, and chances are your workplace is already looking into measures to improve workplace security.

What to Do: The Threat Assessment Team

As with every other hazard to safety and environmental health, the first step toward increasing security in the work place is hazard analysis. A Threat Assessment Team can be created to evaluate the facility's vulnerability on several different levels. These include recommending and implementing employee training programs to improve employee response to security threats, implementing facility wide plans for dealing with security threats, and creating communication channels for employees to bring concerns to effective attention. The Threat Assessment Team should employ a three-step process for security analysis: hazard assessment, workplace security analysis, and workplace surveys.

Hazard Assessment

Because threats to a workplace's security can come from many different areas, assessment teams should include representatives from diverse groups within the facility. Management, operations, security, finance, legal, and human resources should all be represented, as should general employees. The hazard assessment should begin with a review of pertinent records.

A records review allows the development of a baseline, which will help the team analyze trends in past security threats. Records that should be examined include OSHA 300 logs, incidence reports, and records or information relating to assault or attempted assault incidents. Insurance, medical, and worker's comp records should also be examined. Police reports and accident investigations may also be useful. Other records that may show trends include grievance records, training records, and other miscellaneous records—minutes of pertinent meetings, for example. Communicating with other similar local businesses or community and civic associations to discuss their own experiences and concerns with security may also prove useful.

Workplace Security Analysis

When the threat assessment team inspects the workplace in a workplace security analysis, they are looking at both facility and work tasks to determine the presence of hazards, conditions, operations, and situations that might place workers at risk from either outside or inside threats. Follow-up inspections should be scheduled with some regularity for continuous improvement.

Workplace Survey

As in any other type of hazard assessment, the employees who face them every day frequently know more about potential problems or threats than anyone else. Areas that make employees feel uncomfortable or insecure are key areas to examine closely and to implement changes. The team may wish to interview people who work in higher risk areas and will often receive more complete responses through questioning.

What to Implement: Hazard Control and Prevention

Control methods can be used to improve, eliminate, or minimize the risks involved with a breach of security. Examine the general workplace design, workstations, and area designs and the existing security measures. Look at the existing security equipment, work practice controls and procedures, and the workplace violence prevention program.

Buildings, workstation, and areas:

- Review all new or renovated facility designs to ensure safe and secure conditions. Design facilities to allow employees to communicate with other staff in emergency situations, via clear partitions, video cameras, speakers or alarms, or other measures appropriate to the workplace.
- Prevent entrapment of the employees and/or minimize potential for assault incidents by the design of work areas and furniture placement.
- Control access to employee work areas by use of keyed entrances, buzzers or key-card access, and security badges.
- Adequate lighting systems increase safety and security, both in indoor areas and the grounds around the facility, especially parking areas. Lighting should meet the requirements of nationally recognized standards (ANSI A-85, ANSI/IES RP-7 1983, ANSI/IES RP-1 1993) and local building codes.

Security Equipment

- Use electronic alarm systems that are activated visually or audibly and that identify the location of the room or employee by an alarm, lighted indicator, or other effective system. Make sure such systems are adequately manned.
- Use closed-circuit televisions to monitor high-risk areas inside and outside the building.
- Use metal detection systems to identify persons with weapons.
- Use cell phones, beepers, CB radios, hand-held alarms, or noise devices for personnel in the field.
- Inspect and repair security equipment regularly to ensure effectiveness.

Work Practice Controls and Procedures

- Provide identification cards for all employees, and require employees to wear them.
- Establish sign-in and sign-out books and an escort policy for non-employees.
- Base staffing considerations on safety and security assessments, for both fixed site and field locations.
- Develop internal communication systems to respond to emergencies.
- Develop a policy on responding to emergency or hostage situations.
- Develop and implement security procedures for employees who work late or off hours; accounting for field staff; when to involve in-house security or local law enforcement in an assault incident; weapons bans in facilities; and employer response to assault incidents.
- Develop written procedures for employees who must enter locations where they feel threatened or unsafe.
- Provide information and give assistance to employees who are victims of domestic violence.
- Develop procedures to ensure confidentiality and safety for affected employees.
- Train employees on awareness, avoidance, and action to take to prevent mugging, robbery, rapes, and other assaults.

Don't Be a Stranger

Here's something to think about: The best on-the-job security for personnel is to know the people you're working with.

Sounds too simple, doesn't it? Yet this is how small towns used to function. Just as your home is more secure when you know your neighbors and they keep an eye on your property and carry in your mail and paper when you go on vacation (and vice versa), exchanging the ordinary information of polite conversation and getting to know something about the people you work with increases everyone's safety. When we pay attention to those around us, when we learn something about them, we have a better idea of their normal state of mind and condition and would be more apt to know if something goes wrong for them. This kind of knowledge, though, must not happen just to be nosey but because we care about those we work with on the most basic human terms.

More subtly, kidnappers and terrorists, as well as others, who open fire on small children in school yards or at crowds in public places, often begin by mentally mapping groups of other people as "the enemy," as "less than human," or as "strangers." Becoming something other than a face in the crowd to your coworkers reinforces your position as a real person to those around you.

Prevent Random Acts of Violence

Companies that handle chemicals have long known that casual trespass is risky—even someone who doesn't mean any harm can cause problems tampering with equipment or materials they don't understand. More and more common is deliberate damage. Many plant security systems limit physical access to the plant and to the chemicals themselves once they come on site. Facility security must be more than warning signage and fencing; facilities must now create serious barriers, whether physical or electronic, securing site, personnel, and community protection.

Protection From Theft

Materials that have intrinsic value must be protected—no need to go any further into why theft is of concern. However, the definition of "theft" is changing, and many people don't think of activities they do as "stealing," although management might. Blatant theft of tangible goods is hard to argue over, but theft of small items and of time are becoming more and more important to those who must keep track of the bottom line. Security systems can help control costly losses.

Expensive equipment (portable-sized power equipment, computer components, and other office machines may have to be kept in place by physically locking down the equipment. Fully document and register equipment information, model numbers and serial numbers, for insurance as well as for tracking of stolen goods.

Tools are another big area of concern. Without some sort of controlled access system, both small hand tools and power tools as well as larger power equipment can be vulnerable to "walking off the site," especially in areas where security cameras are impractical. Guilty parties can include workers, outsider theft, as well as semi-connected folks like sub-contractor crew members or temporary workers. I've seen saw blades mysteriously vanish from the power saws overnight, and a shaper leave with the contractor crew. These job "souvenirs"—often very practical souvenirs—can be kept where they belong with the use of a regulated tag-in/tag-out systems and security cameras.

Equipment that must remain portable presents additional security headaches. For personnel, the tag-in/tag-out system may work very well, but for equipment that would be attractive to outsiders (laptop computers, for example, which are being used more and more in the field), workers may need training on techniques for protecting such equipment from theft. Key to keeping such equipment in the inventory is making sure the employee using it keeps the equipment secured at all times.

Losses in the small office and shop stuff don't seem like much on the surface, but it adds up fast. The half-deliberate theft or absent-minded disappearance of common office supplies or the absolutely deliberate theft of tools and equipment, or coffee service perishables, or cleaning supplies has happened at every facility at one time or another. While management doesn't want to feel like "Big Brother" and workers don't want to feel like "Big Brother Is Watching Them," supplies are expensive, and the time it takes to replace them adds to the cost.

"Theft" can also include time employees spend attending to tasks other than their assigned duties. While employers need to keep a sense of balance about time valued employees spend "on the job" but not on THEIR job, excessive use of personal email or too much time with the solitaire game on the screen costs the company money, just as stolen equipment does. Keeping track of these losses is much more possible now than in the past.

Protection of Equipment and Data

In some respects, more important than the equipment itself, these days, is the data the equipment is used to generate, control, store, format, and distribute. The differences between stand-alone computers and a networked system make some aspects of data protection easier, and others harder. However, networking offers so many benefits that most places are heading in that direction, if they aren't there already. These days, servers and a Management Information System (MIS) staff handle computer operations at many facilities.

Data protection starts with backup. Back up data, and keep duplicate backups off-site as well as on-site. Servers allow automated backup, but automated backup doesn't matter if the place burns down and you don't have a recent backup off site. Stand-alone computers should be outfitted with some means of backing up data, and the backup media should be collected regularly and stored offsite.

Often equipment theft is covered by insurance and can be replaced. Data isn't necessarily so easy to retrieve. If facility data isn't adequately backed up, restoration may be difficult, time consuming, and costly. How much of your computerized data was entered by hand? Then someone may have to go back in and tediously re-enter it, which not only is time consuming but introduces errors as well, making the loss much more expensive. Was it read from an electronic reader of some sort? How much data does that hand held hold, anyway? Loss of equipment or the data on the equipment is tough to handle. The more extensive the system is, the more apt it is to be properly backed up, but the bigger the tangle of data can be to sort out.

Server systems allow MIS staff to keep track of employee Internet and email usage, in part to protect from data theft or loss, and to help diagnose problems. Servers also allow for MIS staff to monitor many other types of employee computer activity. How far management goes in keeping tabs on worker computer use is up to individual facility policy, but the possibility is there. Pulling that information is possible on standalone computers, too, but is much more labor intensive.

MIS personnel should regularly update and share viral protection software, as well as educate employees on the dangers of stranger-generated email. As some folks have to climb a mountain "because it's there," other folks seem to feel obligated to trash computer systems on a widespread basis to prove they can. Some server systems involve solid fire-wall protection, but various Trojan horses, viruses, and bugs are being created and spread all the time. While server systems allow swift distribution of software updates, distribution is more labor intensive on standalone computers. However, if a standalone computer is infected, it is easily quarantined. A really massive infestation of some of the more wicked bugs can bring a business to its knees for days and cost thousands of dollars of wasted work hours, as well as hundreds of hours of MIS staff time to fix.

Protection from Harm

In the case of industrial facilities, security systems also provide protection for the public from hazardous processes and materials. Layers of security that prevent unauthorized people from going where they have no business being can protect outsiders as well as workers from everything from minor annoyances to gunfire. Keycode, password, or ID-card entry systems that limit access to the building, reception areas, and locked access doors to nonpublic areas (crash-barred on the inside, of course) keep unwelcome guests out. Use common sense in allowing former employees access to former work areas; disgruntled former employees have been known to "go postal" in more facilities than USPS offices.

To go along with entry systems, a visible security presence: guards on duty, security cameras, security firm signage, reception, or guard desk entry that requests sign-in and sign-out and proof of identification provides a disincentive for interested outsiders.

Safety for All Concerned

All sorts of materials that plants handle are in and of themselves dangerous; to handle them or work around them, workers must undergo training for their own safety. Waste products in production, treatment, treatment byproducts, and treatment chemicals all pose hazards, as do the physical processes for some treatments. Measures are put into place to ensure that an unprepared worker doesn't walk into a dangerous place. Signs warning that a place might pose an inhalant risk for methane, for example, protects anyone who passes by. Outsiders who might decide to come on facility property might just be curious, or they could be intent on casual or more serious damage, malicious mischief, theft, and physical harm to personnel or equipment. If the idea that some kid might break into facility property on a dare and get killed messing around where he had no business being seems far-fetched, think about how many times we've seen the name of the local high school football team illegally spray-painted onto a water-tower, and think again.

THOUGHT-PROVOKING QUESTIONS

- Define emergency response.
- What are typical causes of industrial emergencies?
- What does a typical Fire Emergency Response plan cover?
- What training should be implemented if workers are expected to fight workplace fires?
- What should an Emergency Medical Response plan cover?
- What training should be implemented if workers are expected to administer first aid?
- What should a HAZWOPER plan cover?
- What operations does HAZWOPER include?
- What are the two initial goals of emergency response?
- What elements should your site Emergency Response Plan include?
- Why is site security important in an emergency?
- Why is PR important before, during, and after an emergency?
- What considerations should the safety and environmental manager take for security?

REFERENCES AND RECOMMENDED READING

Brauer, R.L. *Safety and Health for Engineers,* 2nd ed. New York: Van Nostrand Reinhold, 2005.

Code of Federal Regulations. Hazardous Waste Operations and Emergency Response, title 29, sec 1910.120.

CoVan, J. *Safety Engineering.* New York: John Wiley & Sons, 1995.

Healy, R.J. *Emergency and Disaster Planning.* New York: John Wiley & Sons, 1969.

Henry, K. "New face of security." *Gov. Security* April 2002: 33.

Noll, G.G., et al. *Hazardous Materials: Managing the Incident.* Burlington, MA: Jones & Bartlett Learning, 2012.

Planning Guide and checklist for Hazardous Materials Contingency Plans. FEMA-10. Washington, DC: Federal Emergency Management Agency, 1981.

Safety and Health Requirements Manual, rev. ed., EM 385-1-1. Washington, DC: U.S. Army Corps of Engineers, 1987.

Smith, A.J. *Managing Hazardous Substances Accidents.* New York: McGraw-Hill, 1980.

Spellman, F.R. *A Guide To Compliance for Process Safety Management Planning (PSM/RMP).* Lancaster, PA: Technomic Publishing Company, 1997.

Spellman, F.R. *Occupational Safety and Health Simplified for the Industrial Workplace.* Lanham, MD: Bernan Press, 2015.

U.S. Department of State. "September 11, 2001: Basic Facts." Accessed May 15, 2015. Washington, DC: U.S. Department of State, 2002. http://2001-2009.state.gov/coalition/cr/fs/12701.htm.

Chapter 7

Fire & Hot Work Safety

Although technical knowledge about flame, heat, and smoke continues to grow, and although additional information continues to be acquired concerning the ignition, combustibility, and flame propagation of various solids, liquids, and gases, it still is not possible to predict with any degree of accuracy the probability of fire initiation or consequences of such initiation. Thus, while the study of controlled fires in laboratory situations provides much useful information, most unwanted fires happen and develop under widely varying conditions, making it virtually impossible to compile complete bodies of information from actual unwanted fire situations. This fact is further complicated because the progress of any unwanted fire varies from the time of discovery to the time when control measures are applied.

—Cote & Bugbee, *Principles of Fire Protection*, 1991

The safety and environmental manager (and any other safety professional) must learn to avoid certain so-called "rules of thumb," as well as other gobbledygook commonly accepted as fact. For example, we commonly hear "some people are accident prone." The implication is, of course, that accidents just seem to follow some individuals. No matter what they do, they just seem plagued with "bad luck." In reality, of course, no one is truly accident-prone. Workers have accidents simply because they are careless, indifferent to safety and environmental health regulations and safe work practices, or are required to work under unsafe conditions.

The safety and environmental manager soon finds that safety and environmental health is not something you read about or practice only on occasion. It has to be observed constantly. Most industries place a high premium on safety. They simply can't afford to ignore it.

When on-the-job injuries occur, a causal factor (or factors) is always involved. Typically workers suffer the pain of injury because they have failed to use good judgment. What is the solution to this problem? Experience has shown that well-written safety and health programs can aid in solving this problem. However, experience also indicates that if a well-written safety and environmental program or safe work practice is not followed by workers, then it is of less worth than the paper it is written on. The safety engineer who views his or her job primarily as occupying a desk to write safe work practices and other policies, without ensuring such written procedures and practices are properly disseminated through training, is likely to fail in his and her primary mission: to protect the safety and health of all employees (Spellman, 2015).

Many organizations implement a safe work practice that incorporates the use of a permit procedure for all hot work, except that in normal operations and processes. Hot work is any kind of activity that involves or generates sparks or open flame. It includes heated equipment that might provide an ignition source for a fire. Hot work often involves people from a maintenance department going to other departments to perform activities. The main idea in a hot work procedure is to ensure that supervisors of all departments involved and workers who might be involved in any way in the work participate in the decision to start work and conduct it safely (Brauer, 2005).

INTRODUCTION

Industrial facilities are not immune to fire and its terrible consequences. Each year fire-related losses in the United States are considerable. According to conservative figures reported by the Federal Emergency Management Agency (FEMA) about 1 million fires involving structures and about 3,100 deaths occur each year. The total annual property loss is more than $11.7 billion. Complicating the fire problem is the point that Cote & Bugbee (1991) made earlier: the unpredictability of fire. Fortunately, facility safety and environmental managers are aided in their efforts in fire prevention and control by the authoritative and professional guidance readily available from the National Fire Protection Association (NFPA), the National Safety Council (NSC), Fire Code Agencies, local Fire Authorities, and OSHA regulations. In this chapter, we not only discuss the assistance available from various associations, agencies, and regulatory bodies, we also discuss fire prevention and control, fire protection provided by use of fire extinguisher, hot work permit use, and welding safety.

FIRE SAFETY

Along with providing fire prevention guidance, OSHA regulates several aspects of fire prevention and emergency response in the workplace. Emergency response, evacuation, and fire prevention plans are required under OSHA's 1910.38. The requirement for fire extinguishers and worker training are addressed in 1910.157. Along with state and municipal authorities, OSHA has listed several fire safety requirements for general industry.

All of the advisory and regulatory authorities approach fire safety in much the same manner. For example, they all agree that electrical short circuits or malfunctions usually start fires in the workplace. Other leading causes of workplace fires are friction heat, welding and cutting of metals, improperly stored flammable/combustible materials, open flames, and cigarette smoking.

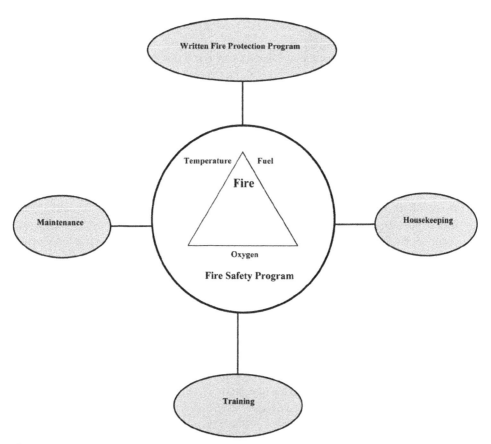

Figure 7.1   Fire Triangle and the Elements Required for a Fire Protection Program

For fire to start, three conditions are necessary: *temperature (heat), fuel,* and *oxygen* (see Figure 7.1). Because oxygen is naturally present in most earth environments, fire hazards usually involve the mishandling of fuel or heat.

The fire triangle helps us understand fire prevention, because the objective of fire prevention and fire fighting is to separate any one of the fire ingredients from the other two. For example, to prevent fires, keep fuel (combustible materials) away from heat (as in airtight containers), thus isolating the fuel from oxygen in the air.

To gain a better perspective of the chemical reaction known as *fire*, remember that the combustion reaction normally occurs in the gas phase; generally, the oxidizer is air. If a flammable gas is mixed with air, there is a minimum gas concentration below which ignition will not occur. That concentration is known as the *lower flammable limit (LFL)*. When trying to visualize LFL and its counterpart, *(upper flammable limit UFL)*, it helps to use an example that most people are familiar with—the combustion process that occurs in the automobile engine. When an automobile engine has a gas/air mixture that is below the LFL, the engine will not start because the mixture is too lean. When the same engine has a gas/air mixture that is above the UFL, it will not start because the mixture is too rich (the engine is flooded). However, when the gas/air mixture is between the LFL and UFL levels, the engine should start (Spellman, 2015).

FIRE PREVENTION AND CONTROL

The best way to try to prevent and control fires in the workplace is to institute a facility fire safety program that includes the elements shown in Figure 7.1. Safety experts agree that the best way to reduce the possibility of fire in the workplace is prevention. For the facility safety engineer this begins with developing a *fire prevention plan*, which must be in writing and must list fire hazards, fire controls, and specify control jobs and personnel responsible and emergency actions to be taken. More specifically, in accordance with OSHA 29 CFR 1910.38, the elements that make up the plan must include:

1. A list of the major workplace fire hazards and their proper handling and storage procedures, potential ignition sources (such as welding, smoking and others), their control procedures, and the type of fire protection equipment or systems that can control a fire involving them.
2. Names or regular job titles of those personnel responsible for maintenance of equipment and systems installed to prevent or control ignitions or fires.
3. Names or regular job titles of those personnel responsible for control of fuel source hazards.
4. Control of accumulation of flammable and combustible waste materials and residues so that they do not contribute to a fire emergency. These housekeeping procedures must be included in the written fire prevention plan.
5. All workplace employees must be apprised of the fire hazards of the materials and processes to which they are exposed.
6. All new employees must be made aware of those parts of the fire prevention plan that the employee must know to protect the employee in the event of an emergency. The written plan must be kept in the workplace and made available for employee review.
7. The employer is required to regularly and properly maintain, according to established procedures, equipment and systems installed on heat-producing equipment to prevent accidental ignition of combustible materials. The maintenance procedure must be included in the written fire prevention plan.

Fire prevention and control measures are those taken *before* fires start and are best accomplished by:

- elimination of heat and ignition sources
- separation of incompatible materials
- adequate means of fire fighting (sprinklers, extinguishers, hoses, etc.)
- proper construction and choices of storage containers
- proper ventilation systems for venting and reducing vapor buildup
- unobstructed means of egress for workers in the event of fire emergency. Adequate aisle and fire-lane clearance for firefighters and equipment must also be maintained.

In the event of a fire emergency, all employees need to know what to do; they need a plan to follow. The fire emergency plan normally is the protocol to follow for fire emergency response and evacuation. Typically, the facility safety engineer is charged with developing fire prevention and emergency response plans that spell out everyone's role. In this effort, the safety engineer's goal should be to make the plan as simple as possible. In addition to a fire emergency response plan, each facility needs to have a well-thought-out fire emergency evacuation plan.

FIRE PROTECTION USING FIRE EXTINGUISHERS

OSHA, under its 29 CFR 1910.157 standard, requires employers to provide portable fire extinguishers that are mounted, located, and identified so they are readily accessible to employees without subjecting the employee to possible injury. OSHA also requires each workplace to institute a portable fire extinguisher maintenance plan. Fire extinguisher maintenance service must take place at least once a year, and a written record must be kept to show the maintenance or recharge date. NOTE: When the facility provides portable fire extinguishers for employee use in the facility, the employee must be provided with training to learn the general principles of fire extinguisher use and the hazards involved in fire fighting.

Employees who are expected to use fire extinguishers in the workplace must be trained on the type(s) of fire extinguishers available to them, the different classes of fire, and where the fire extinguishers are located.

The ABC type fire extinguisher is probably best suited for most industrial applications because it can be used on Class A, B, and C fires. Class A is used for common combustibles (such as paper, wood, and most plastics); Class B is for flammable liquids (such as solvents, gasoline, and oils); and Class C is for fires in or near live electrical circuits. In areas such as electrical substations and switchgear rooms, only Class C (carbon dioxide, CO_2) should be used. Though combination Class A, B, and C extinguishers will extinguish most electrical fires, the chemical residue left behind can damage delicate electrical/electronic components; thus, CO_2-type extinguishers are more suitable for extinguishing electrical fires.

Flammable liquids have a flash point below 100°F. Both flammable and combustible liquids are divided into the three classifications shown below (NFPA, 1981).

Flammable Liquids

Class I-A	Flash point below 73°F, boiling point below 100°F
Class I-B	Flash point below 73°F, boiling point at or above 100°F
Class I-C	Flash point at or above 73°F, but below 100°F

Combustible Liquids

Class II	Flash point at or above 100°F, but below 140°F
Class III-A	Flash point at or above 140°F, but below 200°F
Class III-B	Flash point at or above 200°F

Each employee must know how to use the fire extinguisher. Most importantly, employees mush know when it is *not safe* to use fire extinguishers; that is, when the fire is beyond being extinguishable with a portable fire extinguisher.

Emergency telephone numbers should be strategically placed throughout the workplace. Employees need to know where they are posted. Workers should be trained on the information they need to provide to the 911 operator (or other emergency service number operator) in case of fire.

MISCELLANEOUS FIRE PREVENTION MEASURES

In addition to basic fire prevention, emergency response training and fire extinguisher training, employees must be trained on the hazards involved with flammable and combustible liquids. OSHA Standard 29 CFR 1910.106 addresses this area.

Industrial facilities typically use all types of flammable and combustible liquids. These dangerous materials must be clearly labeled and stored safely when not in use. The safe handling of flammable and combustible liquids is a topic that needs to be fully addressed by the facility safety engineer and workplace supervisor. Worker awareness of the potential hazards that flammable and combustible liquids pose must be stressed. Employees need to know that flammable and combustible liquid fires burn extremely hot, and can produce copious amounts of dense black smoke. Explosion hazards exist under certain conditions in enclosed, poorly ventilated spaces where vapors can accumulate. A flame or spark can cause vapors to ignite creating a flash fire with the terrible force of an explosion.

One of the keys to reducing the potential spread of flammable and combustible fires is to provide adequate containment. All storage tanks should be surrounded by storage dikes or containment systems, for example. Correctly designed and built dikes will contain spilled liquid. Spilled flammable and combustible liquids that are contained are easier to manage than those that have free run of the workplace. Properly installed containment dikes can also prevent environmental contamination of soil and groundwater.

Did You Know?

The plant's workers, supervisors, and safety and environmental manager must be prepared for fire and its consequences. The plant must maintain a fire prevention strategy that will ensure that work areas are clean and clutter free (to ensure fire-lane access). Employees must know how to handle and properly store flammable or combustible chemicals or materials, what they are expected to do in case of a fire emergency, and how and whom to call when fire occurs. If required to use fire extinguishers to fight small workplace fires, employees must know how to properly and safely operate the extinguishers.

HOT WORK PERMIT PROCEDURE

In the performance of hot work in the workplace, various OSHA standards require the following:

The employer shall issue a hot work permit for hot work operations conducted on or near a covered process [including confined spaces].

The permit shall document that the fire prevention and protection requirements in 29 CFR § 1910.252(a) [Fire Prevention and Protection] have been implemented prior to beginning the hot work operations; it shall indicate the date(s) authorized for hot work and identify the object on which hot work is to be performed. The permit shall be kept on file until completion of the hot work operations (29 CFR 1910.119, .134, .252. Code of Federal Regulations, 1995).

Whenever confined space entry is to be made into an entry-by-permit-only confined space, often an important interface between these three standards must exist, especially in the need to ensure safe entry. In this section, we discuss another important procedure, one that also works to ensure confined space operations are conducted safely: hot work permit procedures.

In addition to ensuring that any type of hot work to be performed in confined spaces is accomplished in a safe manner by utilizing hot work permit requirements, other workplace operations might require the use of hot work permit procedures. For example, under OSHA's 29 CFR 1910.119 (Process Safety Management), any time hot work is to be performed on, near, or around covered chemical processes, a hot work permit must be used. Many companies require the use of hot work permits any time hot work is to be performed anywhere within the organization outside normal operations and processes. "Normal operations and processes" might be defined as work normally performed in a welding, brazing, or hot torch cutting shop, or hot work performed as part of an assembly line process, such as that conducted by robots on automobile assembly lines. "Outside normal operations and processes" might be described as performing hot work in work areas where hot work is not typically performed: for example, in office, storage, and/or production areas.

Typically, the organizational safety and environmental manager is responsible for implementing and managing the hot work permitting procedure. As shown in Figure 7.2, the primary elements required to be incorporated into a viable hot work permit system include a standard operating procedure consisting of (1) a written procedure, (2) a permit, (3) worker training, and (4) fire watch provisions.

Exactly what is accomplished by employing the use of a hot work permitting system? A hot work permitting procedure works primarily to ensure that work areas and all adjacent areas to which sparks and heat might be spread (including floors above and below and on opposite sides of walls) are inspected during the work and again 30 minutes after the work is completed, to ensure they are firesafe. For example, during the inspection, work areas and surrounding areas should be inspected to ensure that:

- Sprinklers are in service
- Cutting and welding equipment is in good repair
- Floors are swept clean of combustibles
- Combustible floors are wetted down, covered with damp sand, metal or other shields
- No combustible material or flammable liquids are within 35 feet of the work
- Combustibles and flammable liquids within 35 feet of work are protected with covers, guards, or metal shields
- All wall and floor openings within 35 feet of work are covered

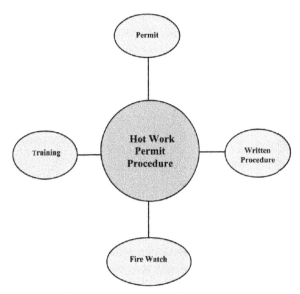

Figure 7.2 Elements Required in a Hot Work Permit Procedure

- Covers are suspended beneath the work to collect sparks
- For work on walls or ceilings, ensure noncombustible construction materials
- Combustibles must be moved away from opposite side of wall
- For work on or in enclosed tanks, containers, ducts, etc., equipment must be cleaned of all combustibles and purged of flammable vapors
- Fire watch is provided during and 30 minutes after operation
- Assigned fire watch is properly trained and equipped

To assist the safety and environmental manager in incorporating a hot work permitting procedure into his or her workplace, in this section, we provide a sample hot work program. We also cover fire watch requirements.

HOT WORK PERMIT PROGRAM (A SAMPLE)[1]

COMPANY'S HOT WORK PERMIT PROGRAM

I. Introduction

OSHA's Process Safety Management Standard (CFR 29 1910.119), Confined Space Entry Program (29 CFR 1910.146), and Respiratory Protection Standard (29 1910.134) requires Company employees, the host facility, and outside contractors to employ safe work practices when performing hot work in or near hazardous materials/ chemicals or in confined spaces. Also required are hot work permits, which describe the proposed work action and the allowable work period.

---

[1] Adapted from F.R. Spellman's *A Guide To Compliance for PSM/RMP/Confined Space Entry*, Technomic Publishing Company, 1997.

II. Hot Work Definition

Hot Work is defined as the use of oxy-acetylene torches, welding equipment, grinders, cutting, and brazing or similar flame-producing or spark-producing operations.

III. Hot Work Permit

1. A hot work permit will be required for contractors and company employees for any hot work performed "in or near" hazardous material/chemical processes, facilities, and confined spaces as follows:

   a. Work on tanks, containers, piping feed systems, or ancillary equipment containing chemicals or fuels, and work in confined spaces.
   b. Work in chemical rooms or on any part of non-diluted chemical systems.

   Note: Special precautions must be used when performing gas welding on this system: *Never use acetylene or propane in the presence of chlorine.*

   c. Work within 25 feet of any flammable/combustible material with NFPA Fire rating of 2 or greater.
   d. Wherever confined space entry testing indicates a hazardous atmosphere.
   e. Wherever a "Hot Work Permit Required" sign is posted.
   f. Work, where in the company's supervisor's judgment, ignition/explosion of chemicals could occur from sparks, hot slag, etc.
   g. (Company Personnel Only) Work anywhere within fence line of company equipment areas.

   Note: Hot Work Permits are not required when the potential for the hazard can be removed throughout the duration of work. This can be accomplished by disconnecting and flushing lines.

   i. Work anywhere within 25 feet of an excavation of a hazardous material piping system/network (no matter the depth, length, width, or other excavation dimension).

2. When contractors perform work on company property and lines, company construction project engineers and/or representatives from the safety and environmental management division will point out to the contractors where hot work permit procedures will be required.
3. Hot Work Permits expire upon completion of each indicated task, and at the end of the workday. A new permit must be completed and issued at the beginning of each workday.

IV. Permit Information/Safe Work Practices

The hot work permit lists required safe work practices that must be documented in the permit and followed during the specified hot work operations. Any of the safe work practice items listed in the permit that are not applicable to a particular work operation must be noted in the appropriate comment area.

***

FIRE WATCH REQUIREMENTS

As stated earlier, a fire watch must be assigned whenever hot work operations are being performed around hazardous materials, in confined spaces, and other times when there is the danger of fire and/or explosion from such work. OSHA has specific requirements regarding fire watch duties.

Firewatchers shall be required whenever welding or cutting is performed in locations where anything other than a minor fire might develop or any of the following conditions exist:

1. Appreciable combustible material, in building construction or contents, are closer than 35 feet (10.7 m) to the point of operation.
2. Appreciable combustibles are more than 35 feet (10.7 m) away but are easily ignited by sparks.
3. Wall or floor openings within a 35-feet (10.7 m) radius expose combustible material in adjacent areas, including concealed spaces in walls or floors.
4. Combustible materials are adjacent to the opposite side of metal partitions, walls, ceilings or roofs, and are likely to be ignited by conduction or radiation.

Firewatchers shall have fire-extinguishing equipment readily available and be trained in its use. They shall be familiar with facilities for sounding an alarm in the event of a fire. They shall watch for fires in all exposed areas, try to extinguish them only when obviously within the capacity of the equipment available, or otherwise sound the alarm. A fire watch shall be maintained for at least a half-hour after completion of welding or cutting operations to detect and extinguish possible smoldering fires.

WELDING SAFETY PROGRAM

Welding is typically thought of as the electric arc and gas (fuel gas/oxygen) welding process. However, welding can involve many types of processes. Some of these other processes include inductive welding, thermite welding, flash welding, percussive welding, plasma welding, and others. McElroy (1980) points out that the most common type of electric arc welding also has many variants including gas-shielded welding, metal arc welding, gas metal arc welding, gas tungsten arc welding, and flux-cored arc welding.

Welding, cutting, and brazing are widely used processes. OSHA's Subpart Q contains the standards relating to these processes in all of their various forms. The primary health and safety concerns are fire protection, employee personal protection, and ventilation. The standards contained in this subpart are as follows:

1910.251   Definitions
1910.252   General Requirements
1910.253   Oxygen-fuel gas welding and cutting
1910.254   Arc welding and cutting

1910.255 Resistance welding
1910.256 Sources of standards
1910.257 Standards organization

In taking a look back on an OSHA study (reported in *Professional Safety*, Feb. 1989) on deaths related to welding/cutting incidents, it is striking to note that of 200 such deaths over an eleven-year period, 80% were caused by failure to practice safe work procedures. Surprisingly, only 11% of deaths involved malfunctioning or failed equipment, and only 4% were related to environmental factors. The implications of this study should be obvious: equipment malfunctions or failures are not the primary causal factor of hazards presented to workers. Instead, the safety and environmental manager's emphasis should be on establishing and ensuring safe work practices for welding tasks. In this section we discuss these safe work practices.

General Welding Safety

Figure 7.3 shows the eight elements required to institute a welding safety program. In the following sections, we discuss each of these elements in detail. NOTE: Much of the information provided is found in OSHA's 29 CFR 1910.252 Welding, Cutting, and Brazing.

Figure 7.3 Elements of a Welding Safety Program

Fire Prevention and Protection

The *Fire Prevention and Protection* element of any welding safety program begins with basic precautions. These basic precautions include the following:

1. **Fire hazards**—if the material or object cannot be readily moved, all movable fire hazards in the area must be moved to a safe location.
2. **Guards**—if the object to be welded or cut can't be moved, and if all the fire hazards can't be removed, then guards are to be used to confine the heat, sparks, and slag, and to protect the immovable fire hazards.
3. **Restrictions**—if the welding or cutting can't be performed without removing or guarding against fire hazards, then the welding and cutting should not be performed.
4. **Combustible material**—wherever floor openings or cracks in the flooring can't be closed, precautions must be taken so that no readily combustible materials on the floor below will be exposed to sparks that might drop through the floor. The same precautions should be taken with cracks or holes in walls, open doorways, and open or broken windows.
5. **Fire extinguishers**—suitable fire extinguishing equipment must be maintained in a state of readiness for instant use. Such equipment may consist of pails of water, buckets of sand, hoses, or portable extinguishers, depending upon the nature and quantity of the combustible material exposed.
6. **Fire watch**—firewatchers are required whenever welding or cutting is performed in locations where other than a minor fire might develop. Firewatchers are required to have fire-extinguishing equipment readily available, and they must be trained in its use. They must be familiar with facilities for sounding an alarm in the event of fire. They must watch for fires in all exposed areas, try to extinguish them only when obviously within the capacity of the equipment available, or otherwise sound the alarm. A fire watch must be maintained for at least a half-hour after completion of welding or cutting operations to detect and extinguish possible smoldering fires.
7. **Authorization**—before cutting or welding is permitted, the individual responsible for authorizing cutting and welding operations must inspect the area. The responsible individual must designate precautions to be followed in granting authorization to proceed, preferably in the form of a written permit (Hot Work Permit).
8. **Floors**—where combustible materials such as paper clippings, wood shavings, or textile fibers are on the floor, the floor must be swept clean for a radius of at least 35 feet (OSHA requirement). Combustible floors must be kept wet, covered with damp sand, or protected by fire-resistant shields. Where floors have been wetted, personnel operating arc welding or cutting equipment must be protected from possible shock.
9. **Prohibited areas**—welding or cutting must not be permitted in areas that are not authorized by management. Such areas include: in sprinklered buildings while such protection is impaired; in the presence of explosive atmospheres, or explosives atmospheres that may develop inside uncleaned or improperly prepared

tanks or equipment that have previously contained such materials, or that may develop in areas with an accumulation of combustible dusts; and in areas near the storage of large quantities of exposed, readily ignitable materials such as bulk sulfur, baled paper, or cotton.

10. **Relocation of combustibles**—where practicable, all combustibles must be relocated at least 35 feet from the work site. Where relocation is impracticable, combustibles must be protected with fireproofed covers, or otherwise shielded with metal of fire-resistant guards or curtains.

11. **Ducts**—ducts and conveyor systems that might carry sparks to distant combustibles must be suitably protected or shut down.

12. **Combustible walls**—where cutting or welding is done near walls, partitions, ceilings or roofs of combustible construction, fire-resistant shields or guards must be provided to prevent ignition.

13. **Noncombustible walls**—if welding is to be done on a metal wall, partition, ceiling or roof, precautions must be taken to prevent ignition of combustibles on the other side from conduction or radiation, preferably by relocating the combustibles. Where combustibles are not relocated, a fire watch on the opposite side from the work must be provided.

14. **Combustible cover**—welding must not be attempted on a metal partition wall, ceilings or roofs that have combustible coverings, nor on any walls or partitions, ceilings or roofs that have combustible coverings, or on walls or partitions of combustible sandwich-type panel construction.

15. **Pipes**—cutting or welding on pipes or other metal in contact with combustible walls, partitions, ceilings, or roofs must not be undertaken if the work is close enough to cause ignition by conduction.

16. **Management**—management must recognize its responsibility for the safe usage of cutting and welding equipment on its property, must establish areas for cutting and welding, and must establish procedures for cutting and welding in other areas. Management must also designate an individual responsible for authorizing cutting and welding operations in areas not specifically designed for such processes. Management must also insist that cutters or welders and their supervisors are suitably trained in the safe operation of their equipment and the safe use of the process. Management has a duty to inform contractors about flammable materials or hazardous conditions of which they may not be aware.

17. **Supervisor**—the supervisor has many responsibilities in welding and cutting operations, including:

- responsibility for the safe handling of the cutting or welding equipment and the safe use of the cutting or welding process.
- determining the combustible materials and hazardous area present or likely to be present in the work location.
- protecting combustibles from ignition by whatever means necessary.
- securing authorization for the cutting or welding operations from the designated management representative.
- ensuring that the welder or cutter secures his or her approval that conditions are safe before going ahead.

- determining that fire protection and extinguishing equipment are properly located at the site.
- where fire watches are required, ensuring that they are available at the site.

18. **Fire prevention precautions**—cutting and welding must be restricted to areas that are or have been made fire safe. When work can't be moved practically, as in most construction work, the area must be made safe by removing combustibles or protecting combustibles from ignition sources.

19. **Welding and cutting used containers**—no welding, cutting, or other hot work is to be performed on used drums, barrels, tanks, or other containers until they have been cleaned so thoroughly as to make absolutely certain that no flammable materials are present and there aren't any substances such as greases, tars, acids, or other materials that when subjected to heat, might produce flammable or toxic vapors. Any pipelines or connections to the drum or vessel must be disconnected or blanked.

20. **Venting and purging**—all hollow spaces, cavities, or containers must be vented to permit the escape of air or gases before preheating, cutting, or welding. Purging with inert gas (e.g., nitrogen) is recommended.

21. **Confined spaces**—to prevent accidental contact in confined space operations involving hot work, when arc welding is to be suspended for any substantial period of time (such as during lunch or overnight), all electrodes are to be removed from the holders and the holders carefully located so that accidental contact can't occur. The machine must be disconnected from the power source. To eliminate the possibility of gas escaping through leaks or improperly closed valves when gas welding or cutting, the torch valves must be closed and the gas supply to the torch positively shut off at some point outside the confined area whenever the torch is not to be used for a substantial period of time (such as during lunch hour or overnight). Where practicable, the torch and hose must also be removed from the confined space.

Note: the safety and environmental manager should use the proceeding information as guidance in preparing the organization's welding safety program.

PPE (Personal Protective Equipment) and Other Protection

Personnel involved in welding or cutting operations must not only learn and abide by safe work practices but also must be aware of possible bodily dangers during such operations. They must learn about the *PPE* (personal protective equipment) and other protective devices/measures designed to protect them.

1. **Railing & welding cable**—a welder or helper working on platforms, scaffolds, or runways must be protected against falling. This may be accomplished by the use of railings, safety harnesses, lifelines, or other equally effective safeguards. Welders must place welding cable and other equipment so that it is clear of passageways, ladders, and stairways.

2. **Eye protection**—helmets or hand shields must be used during all arc welding or arc cutting operations (excluding submerged operations). Helpers or attendants must be provided with the same level of proper eye protection.

Goggles or other suitable eye protection must be used during all gas welding or oxygen cutting operations. Spectacles without side shields with suitable filter lenses are permitted for use during gas welding operations on light work, for torch brazing, or for inspection.

Operators and attendants of resistance welding or resistance brazing equipment must use transparent face shields or goggles (depending on the particular job) to protect their faces or eyes as required.

Helmets and hand shields must meet certain specifications, including being made of a material which is an insulator for heat and electricity. Helmets, shields, and goggles must not be readily flammable and must be capable of sterilization. Helmets and hand shields must be so arranged to protect the face, neck, and ears from direct radiant energy from the arc. Helmets must be provided with filter plates and cover plates designed for easy removal. All parts must be constructed of a material that will not readily corrode or discolor the skin.

Goggles must be ventilated to prevent fogging of the lenses as much as possible. All glass for lenses must be tempered and substantially free from striae, air bubbles, waves and other flaws. Except when a lens is ground to provide proper optical correction for defective vision, the front and rear surfaces of lenses and windows must be smooth and parallel. Lenses must also bear some permanent distinctive marking by which the source and shade may be readily identified. Table 7.1 provides a guide for the section of the proper shade numbers. These recommendations may be varied to suit the individual's needs.

All filter lenses and plates must meet the test for transmission of radiant energy prescribed in ANSI Z87.1 -1968—American National Standard Practice for Occupational and Educational Eye and Face Protection.

Where the work permits, the welder should be enclosed in an individual booth painted with a finish of low reflectivity (such as zinc oxide and lamp black), or

Table 7.1 Recommended Shade Numbers for Welding Operations

Welding Operation	Shade Number
Shielded metal arc welding	10
Gas shielded arc welding (nonferrous)	11
Gas shielded arc welding (ferrous)	12
Shielded metal arc welding:	
3/16-, 7/32-, 1/4-inch electrodes	12
5/16-, 3/8-inch electrodes	14
Atomic hydrogen welding	10–14
Carbon arc welding	14
Soldering	14
Torch brazing	2
Light cutting, up to 1 inch	3 or 4
Medium cutting, 1 inch to 6 inches	4 or 5
Heavy cutting, 6 inches and over	5 or 6
Gas welding (light) up to 1/8 inch	4 or 5
Gas welding (medium) 1/8 inch to 1/2 inch	5 or 6
Gas welding (heavy) 1/2 inch and over	6 or 8

Note: In gas welding or oxygen cutting where the torch produces a high yellow light, use a filter or lens that absorbs the yellow or sodium line in the visible light of the operation.
Source: 29 CFR 1910.252, OSHA.

must be enclosed with noncombustible screens similarly painted. Booths and screens must permit circulation of air at floor level. Workers or other persons adjacent to the welding areas must be protected from the rays by noncombustible or flameproof screens or shields or must be required to wear appropriate eye protection.

3. **Protective clothing**—employees exposed to the hazards created by welding, cutting, or brazing operations must be protected by personal protective equipment, including appropriate protective clothing required for any welding operation.

4. **Confined spaces**—for welding or cutting operations conducted in confined spaces (i.e., in spaces that are relatively small, or restricted spaces such as tanks, boilers, pressure vessels, or small compartments of a ship) personal protective and other safety equipment must be provided.

   Protection of personnel performing hot work in confined spaces includes the following:

   • Proper ventilation
   • Gas cylinders and welding machines must be left on the outside and secured to prevent movement
   • Where a welder must enter a confined space through a manhole or other small opening, means (lifelines) must be provided for quickly removing him or her in case of emergency
   • When arc welding is to be suspended for any substantial period of time, all electrodes must be removed from the holds, the holders carefully located so that accidental contact can't occur, and the machine must be disconnected from the power
   • To eliminate the possibility of gas escaping through leaks of improperly closed valves, when performing gas welding or cutting, the torch valves must be closed and the fuel gas and oxygen supply to the torch positively shut off at some point outside the confined area whenever the torch is not to be used for a substantial period of time.
   • After welding operations are completed, the welder must mark the hot metal or provide some other means of warning others.

Ventilation and Health Protection

All welding should be accomplished in well-ventilated areas. There must be sufficient movement of air to prevent accumulation of toxic fumes or possible oxygen deficiency. Adequate ventilation becomes extremely critical in confined spaces where dangerous fumes, smoke, and dust are likely to collect.

Where considerable hot work is to be performed, an exhaust system is necessary to keep toxic gases below the prescribed health limits. An adequate exhaust system is especially necessary when hot work is performed on zinc, brass, bronze, lead, cadmium, or beryllium-bearing metals. This also includes galvanized steel and metal painted with lead-bearing paint. Fumes from these materials are toxic. They are very hazardous to health.

What does OSHA require for ventilation for hot work operations? The requirements include:

Ventilation must be provided when:

- hot work is performed in a space of less than 10,000 cubic feet per welder
- hot work is performed in a room having a ceiling height of less than 16 feet
- hot work is performed in confined spaces where the hot workspace contains partitions, balconies, or other structural barriers to the extent that they significantly obstruct cross ventilation

The minimum rate of ventilation must be:

- 2,000 cubic feet per minute per welder, except where local exhaust hoods and booths are provided, or where approved airline respirators are provided.

Arc Welding Safety

In 29 CFR 1910.254 (Arc Welding & Cutting), OSHA specifically lists various safety requirements that must be followed when arc welding. For example, in equipment selection, OSHA stipulates the welding equipment must be chosen for safe application to the work to be done. Welding equipment must also be installed safely as per manufacturer's guidelines and recommendations. Finally, OSHA specifies that workpersons designated to operate arc-welding equipment must have been properly trained and qualified to operate such equipment. Training and qualification procedures are important elements that must be included in any welding safety program.

Along with OSHA's requirements above, the safety engineer must ensure the facility's welding safety program includes written safe work practices detailing and explaining safety requirements that must be followed whenever arc welding is performed. In the following section, we summarize OSHA and industry requirements and recommendations for performing arc-welding operations safely.

Note: Arc welding includes shielded metal arc, inert gas shielded arc, and resistance welding. In the following safe work practice, only general safety measures are indicated for these areas, because arc-welding equipment varies considerably in size and type. For example, equipment may range from a small portable shielded metal arc welder to highly mechanized production spot or gas-shielded arc welders. In each instance, specific manufacturer's recommendations should be followed. Along with OSHA requirements, the following work practice includes safety practices which are generally common to all types of arc welding operations.

SAFE WORK PRACTICE: ARC WELDING

1. Ensure all welding equipment is installed according to provisions of the National Electrical Code (NEC) and regulatory bodies.
2. Ensure the welding machine is equipped with a power disconnect switch conveniently located at or near the machine so the power can be shut off quickly.

3. Ensure that the range switch is not operated under load. The range switch, which provides the current setting, should be operated only while the machine is idling and the current is open. Switching the current while the machine is under a load will cause an arc to form between the contact surfaces.

4. Repairs to welding equipment must not be made unless the power to the machine is shut off. The high voltage used for arc welding machines can inflict severe and fatal injuries.

5. Ensure welding machines are properly grounded in accordance with the National Electrical Code or stray current may develop, which can cause severe shock when ungrounded parts are touched. Ensure the ground to your work is securely attached. Grounds are not to be attached to pipelines carrying gases or flammable liquids.

6. Ensure electrode holders do not have loose cable connections. Keep connections tight at all times. Avoid using electrode holders with defective jaws or poor insulation.

7. The polarity switch is not to be changed when the machine is under a load. Ensure you wait until the machine idles and the circuit is open. Otherwise, the contact surface of the switch may be burned and the person throwing the switch may receive a severe burn from the arcing.

8. Ensure welding cables are not overloaded, and do not operate a machine with poor connections.

9. Ensure welding is conducted in dry areas and that hands and clothing are dry.

10. Ensure an arc is not struck whenever someone without proper eye protection is nearby.

11. Ensure pieces of metal that have just been welded or heated are allowed to cool before picking them up.

12. Always wear protective safety glasses.

13. Ensure hollow (cored) castings have been properly vented before welding.

14. Ensure press-type welding machines are effectively guarded.

15. Ensure suitable spark shields are used around equipment in flash welding.

16. When welding is completed, turn off the machine, pull the power disconnect switch, and hang the electrode holder in its designated place.

17. Inspect cables for cuts, nicks, or abrasion.

Safe Work Practice: Gas Welding and Cutting

Specific safety requirements for oxygen-fuel gas welding and cutting are covered under 29 CFR 1910.253 and are listed in the units involving oxyacetylene welding. These safety requirements (precautions) cover proper handling of cylinders, operation of regulators, use of oxygen and acetylene, welding hoses, testing for leaks, and lighting a torch. All of these safety requirements are extremely important and should be followed with the utmost care and regularity.

Along with the normal precautions to be observed in gas welding operations, a very important safety procedure involves the piping of gas. All piping and fittings used to convey gases from a central supply system to work stations must withstand a minimum pressure of 150 psi. Oxygen piping can be of black steel, wrought iron, copper, or brass. Only oil-free compounds should be used on oxygen-threaded connections. Piping for acetylene must be of wrought iron (Note: acetylene gas must never come into contact with unalloyed copper, except in a torch—any contact with it could result in a violent explosion). After assembly, all piping must be blown out with air or nitrogen to remove foreign materials.

According to Giachino & Weeks (1985) five basic rules contribute to the safe handling of oxyacetylene equipment. These are:

1. Keep oxyacetylene equipment clean, free of oil, and in good condition.
2. Avoid oxygen and acetylene leaks.
3. Open cylinder valves slowly.
4. Purge oxygen and acetylene lines before lighting a torch.
5. Keep heat, flame, and sparks away from combustibles.

Cutting Safety

Whenever torch-cutting operations are conducted, the possibility of fire is very real, because proper precautions are often not taken. Torch cutting is particularly dangerous because sparks and slag can travel several feet and can pass through cracks out of sight of the operator. The safety and environmental manager must ensure the persons responsible for supervising or performing cutting of any kind follow accepted safe work practices. Accepted safe work practices for torch-cutting operations typically include:

1. Use of a cutting torch where sparks will be a hazard is prohibited.
2. If cutting is to be over a wooden floor, the floor must be swept clean and wetted before starting the cutting.
3. A fire extinguisher must be kept in reach any time torch-cutting operations are conducted.
4. Cutting operations should be performed in wide-open areas so sparks and slag will not become lodged in crevices or cracks.
5. In areas where flammable materials are stored and cannot be removed, suitable fire-resistant guards, partitions, or screens must be used.
6. Sparks and flame must be kept away from oxygen cylinders and hoses.
7. Never perform cutting near ventilators.
8. Firewatchers with fire extinguishers should be used.
9. Never use oxygen to dust off clothing or work.
10. Never substitute oxygen for compressed air.

THOUGHT-PROVOKING QUESTIONS

- What associations provide professional guidance for fire prevention and control?
- What areas does OSHA regulate that concern fire safety?
- What's the chief cause of workplace fires? Why?
- What are the other causes of workplace fires?

- What are the three conditions necessary for a fire to begin?
- Draw and label the fire triangle.
- At what phase does the chemical reaction usually begin? What's the usual oxidizer?
- What elements should a fire prevention and safety program include?
- What's the first step toward fire prevention?
- What fire prevention and control measures should be taken before any fires occur? Why is each measure important?
- What is a fire emergency plan? What should it include?
- What are the different classes of fire extinguishers, and what is each used for?
- What should employees who might need to use fire extinguishers know?
- Discuss proper chemical storage and how it can affect fire safety.
- Discuss containment and fire safety and prevention.
- How does hotwork fit into confined space entry programs?
- Where and when (other than in confined space entry) could hot work be used?
- What are the safety and environmental manager's responsibilities under the Hot Work Standard?
- What does the viable hot work system entail and include?
- What worker protections does a hot work program provide?
- What do firewatch provisions entail?
- What information should a hot work permit include?
- When does the hot work permit expire?
- What conditions require a hot work permit?
- Who completes which sections of the Hot Work Permit?
- When are firewatchers required?
- What are firewatch duties?
- What are the different types of welding?
- In welding statistics, what percentage of accidents and injuries are caused by failure to follow safe work practices?
- What is the chief concern in welding safety?
- What safeguards for fire prevention and protection should be implemented?
- What are the management and supervisory responsibilities for safe usage?
- What is the proper PPE for safe welding?
- What are the ventilation requirements for safe welding?
- What special requirements should be in place for arc welding?
- What special requirements should be in place for gas welding?
- What are safe-cutting requirements?

REFERENCES AND RECOMMENDED READING

Brauer, R.L. *Safety and Health for Engineers.* New York: Van Nostrand Reinhold, 2005.
Code of Federal Regulations. Welding, Cutting, and Brazing, title 29, sec 1910.251-255.
Code of Federal Regulations. Emergency action plans, title 29, sec 1910.38.
Code of Federal Regulations. Portable fire extinguishers, title 29, sec 1910.157.
Colonna, G., ed. *Fire Protection Guide on Hazardous Materials*, 14th ed. Quincy, MA: National Fire Protection Association, 2010.

Cote, A., and P. Bugbee. *Principles of Fire Protection*. Batterymarch Park, MA: National Fire Protection Association, 1991.

Giachino, J., and W. Weeks. *Welding Skills*. Homewood, IL: American Technical Publications, Inc., 1985.

Goetsch, D.L. *Occupational Safety and Health*, 2nd ed. Englewood Cliffs, NJ: Prentice Hall, 1996.

Hoover, et al. *Health, Safety, and Environmental Control*. New York: Van Nostrand Reinhold, 1989.

Kavianian, H.R., and C.A. Wentz. *Occupational and Environmental Safety Engineering and Management*. New York: Van Nostrand Reinhold, 1990.

McElroy, F.E., ed. *NSC Accident Prevention Manual for Industrial Operations: Engineering and Technology*, 8th ed. Merrifield, VA: International Fire Chiefs Association, 1980.

National Fire Protection Association. *Cutting and Welding Processes—NFPA 51B-1989*. Quincy, MA: NFPA, 1989

National Fire Protection Association. *Fire Protection Handbook*, 16th ed. Quincy, MA: NFPA, 2003.

NFPA 101-1988. *Life Safety Code*. Quincy, MA: National Fire Protection Association, 1988.

Occupational Safety and Health Administration. *Occupational Safety and Health Standards for General Industry, 29 CFR Part 1910, with amendments*. Washington, DC: U.S. Department of Labor, 1995.

"OSHA News: OSHA Studies Workplace Deaths Involving Welding." *Professional Safety* February 8, 1989.

Spellman, F.R. *Confined Space Entry*. Lancaster, PA: Technomic Publishing Company, 1999.

Spellman, F.R. *Occupational Safety and Health Simplified for the Industrial Workplace*. Lanham, MD: Bernan Press, 2015.

Spellman, F.R. *Safe Work Practices for Wastewater Treatment Plants*. Lancaster, PA: Technomic Publishing Company, 1996.

Chapter 8

Lockout/Tagout

When maintenance and servicing are required on equipment and machines, the energy sources must be isolated and lockout/tagout procedures implemented. The terms zero mechanical state or zero energy state have often been used to describe machines with all energy sources neutralized. These terms have been incorporated in many standards. The current term indicating a machine at total rest is energy isolation. Machine energy can be electrical, pneumatic, steam, hydraulic, chemical, thermal, and others. Energy is also the potential energy from suspended parts or springs.

—National Safety Council, 1992

INTRODUCTION

OSHA's 29 CFR 1910.147 states:

Employers are required to develop, document, and utilize an energy control procedures program to control potentially hazardous energy.

The energy control procedures must specifically outline the scope, purpose, authorization, rules, and techniques to be utilized for the control of hazardous energy and the means to enforce compliance including, but not limited to, the following:

- *a specific statement of the intended use of the procedure;*
- *specific procedural steps for shutting down, isolating, blocking, and securing machines and equipment to control hazardous energy;*
- *specific procedural steps for the placement, removal, and transfer of lockout devices or tagout devices and the responsibility for them;*
- *specific requirements for testing a machine or equipment to determine and verify the effectiveness of lockout devices, tagout devices, and other energy control measures.*

—29 CFR 1910.147

Lockout/Tagout Procedures often go hand in hand with workplace maintenance actions that involve machinery repair or installation. In addition, lockout/tagout is also important in Confined Space Entry Procedures to ensure that entry into a confined space can be accomplished safely. One of the major requirements in OSHA's Confined Space Entry Standard is to ensure that the hazards within a permit-required confined space have been removed or isolated. Removing the hazard is always the best way to protect entrants; however, removing all the hazards is, in many cases, impossible. Thus, OSHA requires the control of hazardous energy using isolation, blanking or blinding, disconnection, and/or lockout/tagout procedures. According to Caruey (1991), OSHA estimates that full compliance with the lockout/tagout standard will prevent 120 accidental deaths, 29,000 serious injuries, and 32,000 minor injuries every year.

In our experience, we have found that many workers mistake the results of atmospheric testing that show no hazard exists in a particular confined space as meaning that the space is totally safe for entry. Indeed, this might be the case; however, many other dangers inherent to confined spaces make entry into them hazardous. For example, if the confined space has some type of open liquid stream flowing through it, the chance for engulfment exists. If the space has electrical devices and circuitry inside, an electrocution hazard exists. If hazardous chemicals are stored and taken into the space, the potential for a hazardous atmosphere exists. Many confined spaces contain physical hazards, including piping and other obstructions—for example, rotating machinery is often housed within confined spaces.

To ensure that the confined space is indeed safe, any and all sources of hazardous energy must be isolated before entry is made. The primary method employed to accomplish this is through lockout/tagout procedures.

However, the intent of employing lockout/tagout procedures goes far beyond just providing for safe-confined space entry. The control of hazardous energies by locking or tagging out also applies to most work involved in servicing, adjusting, or maintenance activities involving machines and/or processes that place personnel at elevated risk. In addition to the sources of machine energy mentioned in the opening (electrical, pneumatic, steam, and so forth), of particular concern is inadvertent activation when personnel are in contact with the hazards.

The many safety and environmental managers employed in major industrial groups recognize that the need to incorporate a viable, fully compliant lockout/tagout program (one that includes all elements of 29 CFR 1910.147 (see Figure 8.1)) can't be overstated. Review the historical data. For example, USDL (1988) points out that 7% of all workplace deaths and nearly 10% of serious accidents in many major industrial groups are associated with the failure to properly restrain or de-energize equipment during maintenance. Maintenance workers account for one-third of injuries, even though they are familiar with the machines they are working on. Statistical records show that most injuries involve machines that are still running or that have been accidentally activated. Pasques (et al.) (1989) points out that a sawmill industry study showed start-ups and unwanted movements to be involved about one-third of the time. Surprisingly, no emergency shutoff was available about 50 percent of the time.

In this chapter, we first define the key terms associated with lockout/tagout and then present a sample of a written lockout/tagout program. As with all sample written

Figure 8.1   Requirements of Lockout/Tagout Program

programs presented in this text, this lockout/tagout program has a huge advantage over many other such programs: it has been used in the real world. Both worksite usage and OSHA examination have tested it. Safety and environmental managers can safely employ a written program of this type in their workplace.

LOCKOUT/TAGOUT KEY DEFINITIONS

Affected employee: An employee whose job requires him/her to operate or use a machine or equipment on which servicing or maintenance is being performed under lockout or tagout, or whose job requires him/her to work in an area where such servicing or maintenance is being performed.

Authorized employee: A person who locks out or tags out machines or equipment to perform servicing or maintenance on that machine or equipment. An affected employee becomes an authorized employee when that employee's duties include performing servicing or maintenance covered under the company's lockout/tagout program.

Capable of being locked out: An energy isolating device is capable of being locked out if it has a hasp or other means of attachment to which (or through which) a lock can be affixed, or it has a locking mechanism built into it. Other energy isolating devices are capable of being locked out if lockout can be achieved without the need to dismantle, rebuild, or replace the energy isolating device or permanently alter its energy control capability.

Note: After January 2, 1990, whenever replacement or major repair, renovation, or modification of a machine or equipment is performed, and whenever new machines or equipment are installed, energy isolating devices for such machines or equipment shall be designed to accept a lockout device. See CFR 29 1910.47 for more information.

Energized: Connected to an energy source or containing residual or stored energy.

Energy isolating device: A mechanical device that physically prevents the transmission or release of energy, including (but not limited to) the following: A manually operated electrical circuit breaker, a disconnect switch, a manually operated switch by which the conductors of a circuit can be disconnected from all ungrounded supply conductors, and in addition, in which no pole can be operated independently; a line valve; a block; and any similar device used to block or isolate energy. Push buttons, selector switches, and other control circuit type devices are not energy isolating devices.

Energy source: Any source of electrical, mechanical, hydraulic, pneumatic, chemical, thermal, or other energy.

Hot tap: A procedure used in repair, maintenance, and service activities, which involves welding on a piece of equipment (pipelines, vessels, or tanks) under pressure, to install connections or appurtenances. Commonly used to replace or add sections of pipeline without the interruption of service for air, gas, water, steam, and petrochemical distribution systems.

Lockout: The placement of a lockout device on an energy isolating device, in accordance with an established procedure, ensuring that the energy isolating device and the equipment being controlled cannot be operated until the lockout device is removed.

Lockout device: A device that utilizes a positive means (such as a lock, either key or combination type) to hold an energy isolating device in the safe position and prevent the energizing of a machine or equipment. Included are blank flanges and bolted slip blinds.

Normal production operation: The utilization of a machine or equipment to perform its intended production function.

Selecting and/or maintenance: Workplace activities such as constructing, installing, setting up, adjusting, inspecting, modifying, and maintaining and/or servicing machines or equipment. These activities include lubrication, cleaning, or unjamming of machines or equipment, and making adjustments or tool changes, where the employee may be exposed to the unexpected energization or start-up of the equipment or release of hazardous energy.

Setting up: Any work performed to prepare a machine or equipment to perform its normal production operation.

Tagout: The placement of a tagout device on an energy isolating device, in accordance with an established procedure, to indicate that the energy isolating device and the equipment being controlled may not be operated until the tagout device is removed.

Tagout device: A prominent warning device, such as a tag and a means of attachment, which can be securely fastened to an energy isolating device in accordance with an established procedure, to indicate that the energy isolating device and the equipment being controlled may not be operated until the tagout device is removed.

LOCKOUT/TAGOUT PROGRAM (A SAMPLE)

I. Purpose

This program has been developed to ensure protection of employees and to maintain compliance with OSHA Standard 1910.147, Control of Hazardous Energy.

These instructions establish the minimum requirements for the lockout or tagout of energy isolating devices whenever maintenance or servicing is done on machines or equipment. It shall be used to ensure that the machine or equipment is stopped, isolated from all potentially hazardous energy sources and locked out before employees perform any servicing or maintenance. These procedures apply:

1. Whenever an employee has to remove or bypass a guard or other safety device;
2. Where such servicing results in the employee placing all or part of their body in a danger zone;
3. Where the unexpected energizing or start-up of the machine or equipment or release of stored energy could cause injury or death.

II. Authorized Employees

Authorized company employees shall be trained in lockout/ tagout procedure. Retraining will be done whenever an authorized employee's job responsibility changes or when the periodic inspection identifies procedural deficiencies. All affected (and other) employees whose work operations are or may be in the area shall be instructed in the purpose and use of lockout/tagout procedure.

III. Application and Compliance

This procedure applies to all company employees and shall be enforced. All employees must signify that they have read and understand each part of this procedure by signing the training record (see Section IX of this procedure).

All employees are required to comply with the restrictions and limitations imposed upon them during the use of lockout or tagout. The employee, upon observing a machine that is locked or tagged out for servicing **shall not** attempt to start or use that machine.

IV. Tags

Tags are used to prevent the unexpected energization of equipment and generally are restricted to those controls that cannot be locked or safeguarded by any other method.

1. When the energy isolating device(s) cannot be locked out, use a tag to prevent inadvertent actuation.
2. Tags will be attached to equipment or machinery at the control panel and at other points of operation.

3. Tags must bear the words "DANGER" and "DO NOT OPERATE" or "DO NOT USE" printed on both sides of the tag.
4. Tags shall bear the name, department, date, and telephone number of the employee (or department) performing the work.

V. Locks

Locks are employee-identifiable, key-operated padlocks. Lockout devices are those openings in equipment control handles or switches that can accept a padlock. Multiple lockout devices are those devices which have spaces for the application of more than one lock. This type of device is locked into the lockout opening, and subsequent locks prevent removal of the device.

Authorized employees will be assigned their own individual lock with one unique key. A master key will generally not be available except under very unusual circumstances. If a master key is necessary, it will be under the strict control of the work center supervisor. Locks will not be removed by anyone other than the authorized employee who placed it on the control, unless the procedure described in the "Special Condition" section of this program is followed. Employees working on the same machine will each use their own lock. When work continues from one shift to the next, employees leaving must remove their locks and employees beginning work must apply their own locks.

VI. Energy Analysis

Note: Example Statement
You should include a similar statement in your own lockout/tagout program.

In 20__, Company's Safety Division surveyed each company facility and identified all machines or equipment covered by this program. All energy sources were identified, along with their methods of control. Each time a new machine is installed, the Company Safety Division is responsible for performing an energy analysis. This energy analysis procedure is ongoing and must be updated as required.

A. Equipment Energy Analysis Survey

Starting in 20__, and anytime a new machine or new configuration to an existing machine has been installed since then, the company's safety director conducted a survey of all the company's properties and made the following findings:

List all equipment that must be locked/tagged out here
Equipment _____
Equipment _____
Equipment _____
Etc. _____

Various equipment/machinery identified in the company's maintenance management system (preventive maintenance program), have equipment-specific lockout/

tagout procedures; Company is not required to document the equipment-specific procedure for each particular machine or equipment, because:

1. the machine and/or equipment were found to have no potential for stored or residual energy or re-accumulation of stored energy after shut down that could endanger employees;
2. the machine or equipment has a single energy source, which can be readily identified and isolated;
3. the isolation and locking out of that energy source will completely de-energize and deactivate the machine or equipment;
4. the machine or equipment is isolated from the energy source, and locked out during servicing or maintenance;
5. a single lockout device will achieve a locked-out condition;
6. the lockout device is under the exclusive control of the authorized employee performing the servicing or maintenance;
7. the service or maintenance does not create hazards for other employees; and
8. the company, in utilizing this exception, has had no accidents involving the unexpected activation or re-energization of the machine or equipment during servicing or maintenance.

For those machines and/or equipment that do not require equipment-specific lockout/tagout procedures, the lockout/tagout procedure in Section VII is to be used at the company.

VII. Lockout/Tagout Procedure

Lockout/tagout procedures for company equipment other than equipment-specific lockout procedures for the incinerators and other covered equipment are:

1. Notify appropriate operations and maintenance supervisors of lockout/tagout.
2. Place the main switch, valve, control, or operating lever in the "off," "closed," or "safe" position.
3. **CHECK** and **TEST** to **MAKE CERTAIN** that the proper controls have been identified and deactivated.
4. Place a lock to secure the disconnection whenever possible. If a lock cannot be used on electrical equipment, an electrician shall remove fuses or disconnect the circuit.
5. If a system cannot be locked out with a lock, attach a HOLD-OFF, DO NOT ENERGIZE or other such tag to the switch, valve, or lever. If the company work center does not use employee-identifiable locks, a lock and tag must be used together.
6. When auxiliary equipment or machine controls are powered by separate supply sources, such equipment or controls shall also be locked or tagged to prevent any hazard that may be caused by operating the equipment or exposure to live circuits.

7. When equipment uses pneumatic or hydraulic power, pressure in lines or accumulators shall be checked. Using whatever safe means possible, this pressure shall be disconnected or pressure lines disconnected.

8. When stored energy is a factor as a result of position, spring tension, or counterweighting, the equipment shall be placed in the bottom or closed position, or it shall be blocked to prevent movement.

9. When the work involves more than one person, additional employees shall attach their locks and tags as they report.

10. When outside contractors are involved, the equipment shall be locked out and tagged in accordance with this procedure by the project manager supervising the work. Only in emergency cases is equipment to be shut down by other than a company representative.

11. When the servicing or maintenance is completed and the machine or equipment is ready to return to normal operating condition, the following steps will be used:

    a. Check the machine or equipment and the immediate area around the machine to ensure that tools, materials, and other nonessential items have been removed and that the machine or equipment components are operationally intact. Ensure that all guards have been replaced.

    b. Check the work area to ensure that all employees have been safely positioned or removed from the area.

    c. Verify that the controls are in neutral.

    d. Remove the lockout devices and re-energize the machine or equipment. The employee who applied the device will remove each lockout/tagout device from each energy isolating device.

    e. Notify affected employees that the servicing or maintenance is completed and that the machine or equipment is again ready for use.

VIII. Periodic Inspections

A periodic inspection of the energy control procedure(s) will be conducted at least quarterly to ensure that the requirements of the program and the standard are being followed to ensure full employee protection.

The work center supervisor and/or safety division personnel will perform the periodic inspection.

The inspection will be conducted to identify any program inadequacies that need correcting. The inspection will review the employee's responsibilities under the procedure being inspected.

The company will provide certification that the inspection of a lockout/tagout has been performed. The certification will identify the machine on which the lockout/tagout is being utilized, the date of the inspection, names of employees involved, and the name of the individual performing the inspection.

IX. Training

Training will be provided by the company's safety division to ensure that all employees understand the purpose and function of the lockout/tagout program and that employees acquire the knowledge and skills required for safe application, usage, and removal of energy controls. The training shall include:

1. Each (person who actually locks/tags out) employee shall receive training in the recognition of hazardous energy sources, the type and magnitude of the energy available in the workplace, and the methods and means necessary for energy isolation and control.
2. Each (person affected by the lockout/tagout) employee shall be instructed in the purpose and use of the energy control procedure.
3. All employees whose work operations are or may be in areas where energy control procedures may be used, shall be instructed about the procedure, and about the prohibitions relating to attempts to restart or re-energize machines or equipment that are locked out or tagged out.

When tagout is used, employees shall receive additional training on the limitations of tags. A training record will be used to record each employee's training.

X. Special Conditions

LOCKOUT/TAGOUT REMOVAL WHEN AUTHORIZED EMPLOYEE IS ABSENT

When the authorized employee who applied the lockout or tagout device is not available to remove it, that device may be removed under the direction of the supervisor, provided that specific procedures and training for such removal have been developed, documented, and incorporated into the lockout/tagout program. Specific procedures include the following elements:

1. Verifying that the authorized employee who applied the device is not at the facility;
2. Making all reasonable efforts to contact the authorized employee to inform him/her that his or her lockout or tagout device has been removed;
3. Ensuring that the authorized employee has this knowledge before he/she resumes work at that facility; and
4. Completing the "Lockout/Tagout Removal When Authorized Employee is Absent" form—the form shall be kept by the work center supervisor.

XI. Methods of Informing Outside Contractors of Procedures

Whenever outside servicing personnel are to be engaged in lockout/tagout activities covered by this program, the company and the outside employer will inform each other of their respective lockout or tagout procedures.

Company employees are to be trained to understand and comply with the restrictions and prohibitions of the outside contractor's lockout/tagout energy control program.

The Safety Division will provide to the company work center supervisor a contractor briefing form that will inform the contractor on precautionary measures involved in performing lockout/tagout on the site.

THOUGHT-PROVOKING QUESTIONS

- What should effective lockout/tagout accomplish?
- How does lockout/tagout relate to confined space entry?
- How and why is lockout/tagout used, other than for safe-confined space entry?
- What is the importance of the individually designated lock and keys? When are they used and why?
- What is the difference between lockout and tagout?
- What is an energy analysis, and what should it accomplish?
- What is the proper procedure for energy control?
- What purpose are periodic inspections?
- What kinds of paperwork should be kept for lockout/tagout?
- What steps must be followed if an employee fails to remove his lock or tag?
- How does lockout/tagout affect outside contractors?

REFERENCES AND RECOMMENDED READING

Caruey, A. "Lock Out the Chance for Injury." *Safety & Health* 143, no. 5 (1991): 46.

Code of Federal Regulations. Occupational Safety and Health Administration, Department of Labor, title 29, sec. 1900-1910.147.

National Safety Council. *Accident Prevention Manual for Business & Industry: Engineering & Technology*, 10th ed. Chicago: National Safety Council, 1992.

Paques, J.J., S. Masse, and R. Belanger. "Accidents Related to Lockout in Quebec Sawmills." *Professional Safety,* September 1989: 17–20.

Spellman, F.R. *Confined Space Entry.* Lancaster, PA: Technomic Publishing Company, 1999.

Spellman, F.R. *Occupational Safety and Health Simplified for the Industrial Workplace.* Lanham, MD: Bernan Press, 2015.

Chapter 9

Confined Space Safety

You may have seen headlines in a newspaper or on the news related to rescue attempts that result in fatalities. The headline leads the reader into a story that is tragic but all too familiar to those who work in the safety and health profession. In 1998, four men were assigned to work inside an empty sewage tank. One man entered the tanks with a safety harness but no safety line. When he uncoupled a hose that was used to drain the tank's contents, residual sewage and gases in the line flowed back into the tank.

The mixture of methane and hydrogen sulfide overcame the man inside the tank, and he lost consciousness. One by one, his coworkers (who were observing the operation from outside the tank) entered the tank to attempt a rescue. All were overcome within a minute. They drowned in sewage.

***

The Occupational Safety and Health Administration fined the company $125,000 and ordered it to conduct hands-on training about working in confined spaces.

1993 CONFINED SPACE TRAGEDY

Date of Incident: October 13, 1993
Time of Incident: 9:30 a.m.
Location of Incident: Seventh Chemical Plant, #3 Toxaphene Stripper Tank
Losses Incurred: Bodily injury, two fatalities

Summary Description of Event

Two men were injured and two men were killed in a confined space incident triggered by the improper use of an airline respirator. On October 13, 1993, maintenance worker John Joe arrived at work at 7:30 a.m. and was discussing his daily tasks with his supervisor. His task for this day was to remove builtup residue from the #3 Toxaphene Stripper Tank, which was 13 feet in diameter and 12 feet deep.

At 8:40 a.m. John donned a full-faced airline respirator and entered the tank through the 18-inch manhole. Once he was in the tank, his assistant removed the ladder so the sludge bucket could pass through the manhole. After emptying several buckets, John used his assistant to replace the ladder so he could climb out with the last bucket. As he was climbing out, the airline got caught and separated a quick-connect fitting. While John was trying to reconnect it, the ladder slipped, and he attempted to hold on but fell to the bottom of the tank.

His assistant placed another ladder in the space and attempted to enter but was unable to do so because of the overpowering odor. A nearby maintenance mechanic entered to attempt the rescue but suffocated. The supervisor also entered the tank but escaped before becoming overcome. Rescue squads were called in to assist. They ventilated the space with a blower but were able to remove the two men. Both men were pronounced dead on arrival at the hospital. The assistant and supervisor were treated and released.

Post-response Assessment

This incident suggests there was a serious deficiency with the employer's confined-space entry program. OSHA's permit-required confined space entry standard (29 CFR 1910.146 (d) (4) (iv)) requires entrants to be provided with proper personal protective equipment, presuming that the equipment is the correct type and is properly maintained and working.

In this case, the airline should have had a double-action connector preventing it from being accidentally disconnected. The entrant should have also been equipped with an emergency cylinder to facilitate self-rescue in the event of a problem. Analysis also suggests other OSHA violations, such as failure to conduct a thorough assessment, inadequate ventilation, inadequate rescue plans, the absence of a retrieval harness and mechanical lifting devices, and the absence of an attendant and entry supervisor.

***

INTRODUCTION[1]

The tragic events that occurred inside that empty sewage tank and the Toxaphene Stripper Tank (confined spaces), provide us with several important considerations related to the danger involved with confined space entry. First, confined spaces can be very unforgiving. Secondly, entering confined spaces without the proper equipment and training amplifies and acerbates the inherent danger. The first two considerations are rather obvious, the third (why four men died, not one) is less obvious. Specifically, the fact that the rescuers risked and lost their lives while attempting to save their fellow worker is actually a common occurrence—one so common that experienced safety and health professionals and those who have actually experienced such tragic

---

[1] Based on information in F.R. Spellman (2015). *Occupational Safety and Health Simplified for the Industrial Workplace*. Lanham, MD: Bernan Press.

occurrences think and worry about it any time a confined space entry is made. But it is a concern not shared by many workers, especially untrained workers.

A common occurrence? Not normally a concern of many workers? Yes. Absolutely. When confined space fatalities occur, multiple fatalities are the common result, the norm. Why? Because rescuers leap right in—and become victims themselves.

Just how often does this occur? Statistics on fatalities in the workplace provided by the Centers for Disease Control report that 92 fatalities occurred from confined space entries in 2014. Out of that number, over 60 percent of the victims were rescuers— *over 60 percent!*

This tendency to leap into a confined space with total disregard for one's own safety is what the author calls the "John Wayne" syndrome. John Wayne, who frequently played larger-than-life heroes, rushed into dangerous situations to rescue victims in movie after movie, with no regard for his own safety or well-being. Workers often disregard their own safety to attempt to rescue fellow workers, which too often ends in tragedy.

Should workers who disregard their own safety by attempting to rescue workers in jeopardy be considered heroes? Arguing against the rescuers' hero status is difficult. The fact is some workers do risk and give their lives in the valiant attempt to rescue fellow workers. Thus, they are heroes—albeit dead ones.

The point we make here (and throughout this text) is that serious problems exist with posthumous heroism and that workers should never be placed in the position where such life-threatening decisions are made.

Let's get back to our last question and answer it. Yes, the rescuers were obviously heroes. It is not our intention to state otherwise. To the first question: Did they have to die? Absolutely not; no one ever "has" to die in such a manner. Were they misled? Again, we feel they absolutely were.

What does all this have to do with fatalities that are the result of confined space entries? Actually, everything. There is absolutely no excuse for such fatalities to occur in the first place. One thing is unequivocally certain: When fatalities occur as a result of confined space entry, someone is responsible; someone did not do his or her job; someone was negligent. In confined space, negligence kills.

OSHA took punitive action against the responsible person in charge (the company) for the tragic occurrence and consequences. Did the $125,000 fine compensate for the death of four workers? Was the additional requirement of conducting hands-on training about working in confined spaces a stiff enough penalty for the responsible party? We leave the answers to these two questions to you.

Without question, when an organization's safety engineer's duties include compliance with OSHA's 29 CFR 1910.146 Confined Space Entry Standard, his or her hands are full, on a full-time, continuous basis. There is no easy way out when it comes to ensuring full compliance with this vital requirement. Full compliance is completely possible, but it requires exceptional attention to regulatory compliance, to detail, and to the ongoing management of the program, as it should be managed, to lead, not to mislead, workers.

In this chapter, we describe and explain the elements needed (see Figure 9.1) to ensure full compliance with OSHA's Confined Space Entry requirements. From Figure 9.1, you see that the main program elements are attached to the semicircle by solid lines, while ancillary or interfacing OSHA Standards (Lockout/Tagout and

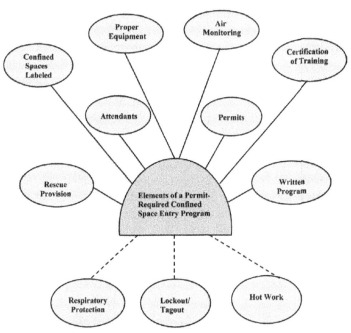

Figure 9.1   Confined Space Entry Elements Attached to the Semicircle by Solid Lines Indicate Major OSHA Requirements that must be Part of an Organization's Confined Space Entry Program. Elements Attached by Dashed Lines Indicate Standards or Requirements that may have a Direct Interface with Confined Space Entry

Respiratory Protection) and the Hot Work Permitting requirement are attached below by dashed lines. Lockout/Tagout, Respiratory Protection, and Hot Work Permitting are essential in protecting workers from hazards that are sometimes present in confined spaces. However, not all confined spaces present such risks.

OSHA'S CONFINED SPACE ENTRY PROGRAM[2]

OSHA has a specific standard that mandates compliance with its requirements for making confined space entries. However, in light of those four dead workers, let us point out that no matter how many standards and regulations OSHA and other regulators write, promulgate, and attempt to enforce, if employers and employees do not abide by their responsibilities under the act, the requirements are not worth the paper they are printed on.

OSHA's Confined Space Entry Program (CSEP) is a good one—a vital guideline to protect workers and others. CSEP was issued to protect workers who must enter confined spaces. It is designed and intended to protect workers from toxic, explosive, or asphyxiating atmospheres and from possible engulfment from small

---

[2] This section is adapted from F.R. Spellman, *Confined Space Entry*. Lancaster, PA: Technomic Publishing Company, 1999.

particles such as sawdust and grain (e.g., wheat, corn, and soybeans normally contained in silos). It focuses on areas with hazards that could potentially cause death or injury, areas or spaces classified as *permit-required* confined spaces. Under the standard, employers are required to identify all permit-required spaces in their workplaces, prevent unauthorized entry into them, and protect authorized workers from hazards through an entry-by-permit only program.

CSEP covers all of general industry (**NOTE**: this rule does not apply to agriculture 29 CFR 1928, construction 29 CFR 1926, or shipyard employment 29 CFR 1915), including agricultural services (the keyword here is "services" and not agriculture), manufacturing, chemical plants, refineries, transportation, utilities, wholesale and retail trade, and miscellaneous services. It applies to manholes, vaults, digesters, contact tanks, basins, clarifiers, boilers, storage vessels, furnaces, railroad tank cars, cooking and processing vessels, tanks, pipelines, and silos, among other spaces.

Confined Space Entry: Definitions

Most rules, regulations, and standards have their own set of terms essential for communication between managers and the workers required to comply with the guidelines. Therefore, key terms that specifically pertain to OSHA's Confined Space Entry Program are defined and presented here in alphabetical order. The definitions are from OSHA's *Occupational Safety and Health Standards for General Industry* (29 CFR 1910 Subpart J—General Environment 29 CFR 1910.146 Confined Space Entry, 1995). The bottom line: Understanding any rule or regulation is difficult unless you have a clear understanding of the terms used.

Acceptable entry conditions: the conditions that must exist in a permit space to allow entry and to ensure that employees involved with a permit-required confined space entry can safely enter into and work within the space.

Attendant: an individual stationed outside one or more permit spaces who monitors the authorized entrants and who performs all attendant's duties assigned by the employer's permit space program.

Authorized entrant: an employee who is authorized by the employer to enter a permit space.

Blanking and blinding: the absolute closure of a pipe, line, or duct by the fastening of a solid plate (such as a spectacle blind or a skillet blind) that completely covers the bore and that is capable of withstanding the maximum pressure of the pipe, line, or duct with no leakage beyond the plate.

Confined space: A space that:

1. is large enough and so configured that an employee can bodily enter and perform assigned work; and
2. has limited or restricted means for entry or exit (e.g., tanks, vessels, silos, storage bins, hoppers, vaults, and pits are spaces that may have limited means of entry); and
3. is not designed for continuous employee occupancy.

Double block and bleed: the closure of a line or pipe by closing and locking or tagging.

Emergency: any occurrence or event (including any failure of hazard control or monitoring equipment) internal or external to the permit space that could endanger entrants.

Engulfment: the surrounding and effective capture of a person by a liquid or finely divided (flammable) solid substance that can be aspirated to cause death by filling or plugging the respiratory system or that can exert enough force on the body to cause death by strangulation, constriction, or crushing.

Entry: the action by which a person passes through an opening into a permit-required confined space. Entry includes ensuing work activities in that space and is considered to have occurred as soon as any part of the entrant's body breaks the plane of an opening into the space.

Entry permit (permit): the written or printed document provided by the employer to allow and control entry into a permit space that contains the information shown in an approved entry permit.

Entry supervisor: the person (such as the employer, foreperson, or crew chief) responsible for determining whether acceptable entry conditions are present at a permit space where entry is planned, for authorizing entry and overseeing entry operations, and for terminating entry as required by the Confined Space Entry Standard.

Note: In practice (in the real world of performing confined space entry operations), common routine often designates the entry supervisor as the "competent" or "qualified" person. So designated in writing, the competent or qualified person is that entry supervisor who has had the appropriate training and experience and possesses the knowledge required to supervise and bring about safe, correct confined space entries. An entry supervisor may also serve as an attendant or as an authorized entrant, as long as that person is trained and equipped as required by the Confined Space Entry Standard for each role he or she plays. Also, the duties of entry supervisor may be passed from one qualified individual to another qualified individual during the course of an entry operation.

Hazardous atmosphere: an atmosphere that may expose employees to the risk of death, incapacitation, impairment of ability to self-rescue (i.e., to escape unaided from a permit space), injury, or acute illness from one or more of the following causes:

1. Flammable gas, vapors, or mist in excess of 10% of its lower explosive or lower flammable limit (LEL or LFL, which basically mean the same thing);
2. Airborne combustible dust at a concentration that meets or exceeds its LFL/LEL; (**NOTE:** This concentration may be approximated as a condition in which the dust obscures vision at a distance of 5 feet (1.52 m) or less);
3. Atmospheric oxygen concentration below 19.5 percent or above 23.5 percent;
4. Atmospheric concentration of any substance for which a dose or a permissible exposure limit (PEL) is published in Subpart G (of the 1910 General Industry Standard), Occupational Health and Environmental Control, or in Subpart Z, Toxic and

Hazardous Substances, which could result in employee exposure in excess of its dose or permissible exposure limit (PEL). (**NOTE:** An atmospheric concentration of any substance that is not capable of causing death, incapacitation, impairment of ability to self-rescue, injury, or acute illness due to its health effects is not covered by this provision);

5. Any other atmospheric condition that is immediately dangerous to life and health (IDLH).

Note: For air contaminants for which OSHA has not determined a dose or permissible exposure limit, other sources of information [Safety Data Sheets (SDS) that comply with the Hazard Communication Standard (commonly known as HazCom), §1910.1200 of the General Industry Standard, published information, and internal documents] can provide guidance in establishing acceptable atmospheric conditions.

Hot work permit: the employer's written authorization to perform operations (e.g., riveting, welding, cutting, brazing, burning, and heating) capable of providing a source of ignition.

Immediately dangerous to life or health (IDLH): any condition that poses an immediate or delayed threat to life, that would cause irreversible adverse health effects, or that would interfere with an individual's ability to escape unaided from a permit space.

Note: Some materials—hydrogen fluoride gas and cadmium vapor, for example—may produce immediate transient effects that, even if severe, may pass without medical attention, but are followed by sudden, possibly fatal collapse 12–72 hours after exposure. The victim "feels normal" after recovery from these transient effects until collapse. Such materials in hazardous quantities are considered to be "immediately" dangerous to life or health.

Inerting: the displacement of the atmosphere in a permit space by a noncombustible gas (such as nitrogen) to such an extent that the resulting atmosphere is noncombustible. (**NOTE:** This procedure produces an IDLH oxygen-deficient atmosphere.)

Isolation: the process by which a permit space is removed from service and completely protected against the release of energy and material into the space by such means as blanking or blinding; realigning or removing sections of lines, pipes, or ducts; a double block and bleed system; lockout or tagout of all sources of energy; or blocking or disconnecting all mechanical linkages.

Line breaking: the intentional opening of a pipe, line, or duct that is or has been carrying flammable, corrosive, or toxic material, an inert gas, or any fluid at a volume, pressure, or temperature capable of causing injury.

Non-permit confined space: a confined space that does not contain (or with respect to atmospheric hazards) have the potential to contain any hazard capable of causing death or serious physical harm.

Oxygen-deficient atmosphere: an atmosphere containing less than 19.5 percent oxygen by volume.

Oxygen-enriched atmosphere: an atmosphere containing more than 23.5 percent oxygen by volume.

Permit-required confine space (permit space): a confined space that has one or more of the following characteristics:

1. Contains or has a potential to contain a hazardous atmosphere;
2. Contains a material that has the potential for engulfing an entrant;
3. Has a configuration such that an entrant could be trapped or asphyxiated by inwardly converging walls or by a floor that slopes downward and tapers to a smaller cross-section; or
4. Contains any other recognized serious safety or health hazard.

Permit-required confined space program (permit space program): the employer's overall program for controlling (and where appropriate, for protecting employees from) permit space hazards and for regulating employee entry into permit spaces.

Permit system: the employer's written procedure for preparing and issuing permits for entry and for returning the permit space to service following termination of entry.

Prohibited condition: any condition in a permit space that is not allowed by the permit during the period when entry is authorized.

Rescue service: the personnel designated to rescue employees from permit spaces.

Retrieval system: the equipment (including a retrieval line, chest or full-body harness, wristlets (if appropriate) and a lifting device or anchor—usually a tripod and winch assembly) used for non-entry rescue of persons from permit spaces.

Testing: the process by which the hazards that may confront entrants in a permit space are identified and evaluated. Testing includes specifying the tests that are to be performed in the permit space. (**NOTE:** Testing enables employers both to devise and implement adequate control measures for the protection of authorized entrants and to determine if acceptable entry conditions are present immediately prior to and during entry.)

Evaluating the Workplace

The employer shall evaluate the workplace to determine if any spaces are permit-required confined spaces (The Office of the Federal Register, 1995). The organization's safety and environmental manager needs to ask, "Does my organization need to comply with OSHA's Confined Space Entry Standard?" It depends, and OSHA wants all safety and environmental managers to make that determination by evaluating their workplaces.

So, how do we go about evaluating our workplaces to determine if we must comply? Before we answer this question, a note of caution. In the evaluation procedure that you must follow to evaluate your workplace, you must take every care and caution that you do not walk into, climb into, or crawl into any space unless you are absolutely certain that it is safe to do so. In short, for safety, you must assume any unfamiliar confined space presents hazards until you have determined by examination and testing that it does not.

How do you go about evaluating a worksite for compliance? To determine if a particular work site must comply with OSHA's Confined Space Standard we must take certain steps. First, we must be familiar with what a confined space is. A confined space:

1. is large enough and so configured that an employee can bodily enter and perform assigned work; and
2. has limited or restricted means for entry or exit (e.g., tanks, vessels, silos, storage bins, hoppers, vaults, and pits are spaces that may have limited means of entry); and
3. is not designed for continuous employee occupancy.

The next step is to survey the plant site, the facility, the factory, or other type of work site to determine if any spaces or structures fall under OSHA's definition of a confined space. While performing such a survey, you must record on paper the name and location of each space or structure identified for evaluation later. You should also have a list of all worksite confined spaces. This list should be distributed to all employees, placed in plain view on employee bulletin boards, and inserted into your site's written confined space program. One thing is certain—when OSHA audits your facility, they will want to see your list of confined spaces.

During the evaluation survey process, if confined spaces are identified, then the determination must be made whether or not they are "permit-required" or "non-permit" confined spaces. To do this, you must be familiar with OSHA's definitions for both. Recall from the definitions list that a:

A. **Non-permit confined space** is a confined space that does not contain or (with the respect to atmospheric hazards) have the potential to contain any hazard capable of causing death or serious physical harm.
B. **Permit-required confined space (permit space)** is a confined space that has one or more of the following characteristics:

1. Contains or has a potential to contain a hazardous atmosphere;
2. Contains a material that has the potential for engulfing an entrant;
3. Has an internal configuration such that an entrant could be trapped or asphyxiated by inwardly converging walls or by a floor that slopes downward and tapers to a smaller cross-section; or
4. Contains any other recognized safety or health hazard.

For a space that is obviously a permit-required confined space (for any of the reasons stated above), you are required to label such a space. If you prefer, the label can also be stenciled on the entrance to, or near the entrance to a confined space. The point is the permit-required confined space must be clearly labeled to inform employees of the location and the danger posed by the permit space.

After identifying and labeling all site permit-required confined spaces, the employer has two choices: (1) to designate such spaces as "off limits" to entry by any employee (unauthorized entry must be prevented) or (2) the employer must develop a written confined space program.

Permit-Required Confined Space Written Program

In this section, we assume that the safety and environmental manager has identified worksite permit-required confined spaces. He or she can either prohibit the entry of any organizational personnel from entering such spaces or develop a written permit-required confined space program.

The first step the employer must take in implementing a permit-required confined space program is to take the measures necessary to prevent unauthorized entry. Typically this is accomplished by (first) labeling all confined spaces. The next step is to list all confined spaces and clearly communicate to employees that the listed spaces are not to be entered by organizational personnel under any circumstances.

Remember that the *employer* is responsible for identifying, labeling, and listing all site permit-required confined spaces, and must also identify and evaluate the *hazards* of each confined space.

Once the hazards have been identified and evaluated, the identity and hazard(s) of each site's confined space must be listed in the organization's written confined space entry program (obviously, it is important that employees are made well aware of all the hazards). The next step is to develop written procedures and practices for those personnel who are required to enter, for any reason, permit-required confined spaces.

The procedures and practices used for permit-required confined space entry must be *in writing* and at the very least must include directions for:

- Specifying acceptable entry conditions
- Isolating the permit space
- Purging, inerting, flushing, or ventilating the permit space as necessary to protect entrants from external hazards
- Providing pedestrian, vehicle, or other barriers as necessary to protect entrants from external hazards
- Verifying that conditions in the permit space are acceptable for entry throughout the duration of an authorized entry.

Under OSHA's program, the employer must also provide specified equipment to employees involved in confined space entry. The requirements under this specification and the required equipment are covered in the following section.

Permit-Required Confined Space Entry: Equipment

OSHA, in its Confined Space Entry Standard (1910.146), specifies the equipment required to make a "safe" and "legal" confined space entry into permit-required confined spaces. Note that the employer must provide this equipment to the employee at no cost. The employer is also required not only to procure this equipment at no cost to the employee, but also to maintain the equipment properly. Most importantly, the employer is required to ensure that employees use the equipment properly. We will come back to this important point later. For now, let's take a look at the type of equipment required for making a safe and legal permit-required confined space entry.

Note: "Equipment" means approved, listed, labeled, or certified as conforming to applicable government or nationally recognized standards, or to applicable scientific principles. It does not mean jerry-rigged or "pulled off the wall" devices that might (or might not) be suitable for use by employees—it means that only safe and approved equipment in good condition is to be used—period.

Testing and Monitoring Equipment

Numerous makes and models of confined space air monitors (gas detectors or sniffers) are available on the market, and selection should be based on your facility's specific needs. For example, if the permit-required confined space to be entered is a sewer system, then the specific need is a multiple-gas monitor. This type of instrument is best suited for sewer systems, where toxic and combustible gases and oxygen-deficient atmospheres are prevalent.

No matter what type of air monitor is selected for a specific use in a particular confined space, any user must be thoroughly trained on how to effectively use the device. Users must also know the monitor's limitations and how to calibrate the device according to manufacturer's requirements. Having an approved air monitor is useless if workers are not trained in its operation or proper calibration.

When choosing an air monitor for use in confined space entry, you must ensure that the monitor selected is not only suitable for the type of atmosphere to be entered, but also that it is equipped with audible and visual alarms that can be set, for example, at 19.5% or lower for oxygen, and preset for levels of the combustible or toxic gases it is used to detect.

Ventilating Equipment

In many cases, you can eliminate, reduce, or modify atmospheric hazards in confined spaces by ventilating—using a special fan or blower to replace the bad air inside a confined space with good air from outside the enclosure. Whatever blower or ventilator type you chose to use, a certain amount of common sense and a consideration of the depth of the manhole, size of the enclosure, and number of openings available is required. Keep in mind that the blower must be equipped with a vapor-proof, totally enclosed electrical motor, or a non-sparking gas engine. Obviously, the size and configuration of the confined space dictates the size and capacity of the blower to be used. Typically, a blower with a large-diameter flexible hose (elephant trunk) is most effective.

Personal Protective Equipment

OSHA requires PPE for confined space entries. The entrant must be equipped with the standard personal protective equipment (PPE) required to make a vertical entry into a permit-required confined space (a full-body harness combined with a lanyard or lifeline), and also the PPE required to protect him or her from specific hazards.

For example, an employee who is to enter a manhole is typically equipped with (1) an approved hard hat to protect the head; (2) approved gloves to protect the hands; (3) approved footwear (safety shoes) to protect the feet; (4) approved safety eye wear or face protection to protect the eyes and face; (5) full-body clothing (long sleeve shirt and trousers) to protect the trunk and extremities; and (6) a tight-fitting NIOSH (National Institute for Occupational Safety and Health) approved self-contained breathing apparatus (SCBA), or supplied-air hose mask with emergency escape bottle, for IDLH atmospheres.

Lighting

Many confined spaces could be described as nothing more than dark (and sometimes foreboding) holes in the ground—often a fitting description. As you might guess, typically many confined spaces are not equipped with installed lighting. To ensure safe entry into such a space, the entrant must be equipped with intrinsically safe lighting.

Intrinsically safe? Absolutely.

Think about it. The last thing you want to do is to send anyone into a dark space (filled with methane) with a torch in his or her hand, and a light source that emits sparks might as well be a torch. Confined spaces present enough dangers on their own without adding to the hazards. However, even after the space has been properly ventilated (with copious and continuous amounts of outside fresh air), and with the source of methane shut off (blinded or blanked, etc.), we still, obviously, have a space that has the potential for an extremely explosive atmosphere. Do not underestimate the hazards such a confined space presents!

So, what do we do? Good question. If lighting is required in a confined space, we need to ensure that it is provided to the entrant—for his or her safety, as well as to enable work to be done. For confined space entries, explosion-proof lanterns or flashlights (intrinsically safe devices) are recommended. These devices (if NIOSH and OSHA-approved) are equipped with spring-loaded bulbs that, upon breaking, eject themselves from the electrical circuit, preventing ignition of hazardous atmospheres.

Another safe, low-cost, instant light source now readily available for confined space entry are lightsticks. They can be used safely near explosive materials because they contain no source of ignition. Lightsticks are available with illumination times from 0.5 to 12 hours. Lightsticks are activated by simply tossing the lightstick on the ground or against a wall, which breaks the inner glass capsule—illumination is immediate.

Another common work light used for confined space entry is the droplight. UL-approved droplights that are vapor-proof, explosion-proof, and equipped with ground fault circuit interrupter (GFCIs) are the recommended type for confined space entry.

Note: If you have a confined space that has the potential for an explosive atmosphere with permanently installed light fixtures in place, you must remember that these lights must be certified for use in hazardous locations and maintained in excellent condition.

Barriers and Shields

As safety and health professionals, we are concerned with not only the safety of the confined space entrant, but also the safety of those outside the confined space. For example, an open manhole presents a pedestrian and traffic hazard. To prevent accidents in areas where manhole work is in progress, we can use several safety devices—manhole guard rail assemblies, guard rail tents, barrier tape, fences, and manhole shields, for example. Remember that we not only want to prevent someone from falling into a manhole (or other type of confined space opening) but also we want to prevent unauthorized entry. Occasionally, manholes or ordinarily inaccessible areas, when open for work crews, present an attractive nuisance. Even ordinary curiosity may lead people (especially children) to put themselves at risk by attempting to enter a confined space.

Along with protecting the confined space opening from someone falling into it or entering it illegally, we must also control traffic around or near the opening. To do this we may need to employ the use of cones, signs, or stationed guard personnel.

Don't forget the nighttime hours. After dark, it is difficult to see an open confined space opening or guard device; these devices should be lighted with vehicle strobes or beacon lights.

Ingress and Egress Equipment: Ladders

Have you ever peered inside a 40-foot deep, 24-inch diameter vertical manhole? Not a pleasant sight? Maybe—maybe not. It depends on your point of view. If the manhole has no lighting (as most do not) then you are peering into what appears to be a bottomless pit (and maybe it is). Have you been there? If so, no further explanation is needed. You know that at best, entering any manhole such as the one just described can be a perilous undertaking.

If you have never faced entering a manhole such as the one just described, let's consider an important point. If you are tasked to enter such a confined space, you will be interested in entering it (ingressing) safely (taking all required precautions) and returning (egressing) safely.

Experience with assessing safety considerations in confined space areas has shown that many of the installed ladders (in place to allow entry and exit inside confined spaces) are not always in the best material condition. Why? Consider the environment they are constantly exposed to year after year.

Confined spaces may be shrouded in moist, chemical-laden atmospheres, conditions that are excellent for corroding most metals. Most ladders installed in confined spaces are made of metal. Not only do we require our workers to enter dangerous permit-required confined spaces, but without properly evaluating all of the confined space's conditions, we may also be asking them to enter them in a totally unsafe manner on equipment that may fail.

Installed ladders within confined spaces must be inspected on a periodic basic to ensure their integrity. Don't forget about the devices used to hold the ladders in place. Most securing or attachment bolts or screws are also made of metal as well, metal that will corrode and weaken with time. We have found ladders that were attached to the wall by rust and rust alone, simply waiting for a victim. Adding weight would send the ladder and passenger on a less than thrilling ride that would almost certainly result in death. *Don't let this happen to you!*

How about those spaces that do not have installed ladders? For confined spaces not equipped with ladders, stairways, or some other installed means of ingress and egress, we often employ the use of portable ladders. One way or another, we are required to provide a safe way in and out of a confined space, and ladders often fit this need.

Upon occasions (more frequent than we would like), however, ladders or stairways for safe entry or exit are not available, practical, or practicable. When such a situation arises, winches and hoisting devices are commonly used to raise and lower entrants. Remember that any lowering and lifting devices must be OSHA-approved as safe to use. Using a rope attached to the bumper of a vehicle to lower or raise an entrant, for example, is strictly prohibited. Only hand-operated lifting/hoisting devices should be employed. Motorized devices are unforgiving, especially whenever the entrant gets caught up in an obstruction (machinery, pipe, angle iron, etc.) that prevents his or her body from moving. The motorized device doesn't care, it just continues to pull the entrant out. On a motorized device, a person stuck in a confined space could literally be pulled apart. OSHA regulations were created to prevent just such gruesome incidents from occurring. But gruesome and fatal events (sometimes involving multiple fatalities) do occur.

Rescue Equipment

When confined space rescue is to be effected by any agency other than the facility itself (emergency rescue service, fire department, etc.), the facility is not required to provide the rescue equipment. However, when confined space rescue is to be performed by facility personnel, proper rescue equipment is required.

What is proper rescue equipment?

Proper rescue equipment basically consists of the equipment needed to remove personnel from confined spaces in a safe manner. "In a safe manner" means "to prevent further injury to the entrants and any injury to the rescuers."

Confined space rescue equipment (commonly called retrieval equipment) typically consists of three components: a safety harness, rescue and retrieval line, and means of retrieval.

Let's take a closer look at each of these components.

A *full-body harness* combined with a lanyard or lifeline evenly distributes the fall-arresting forces among the worker's shoulders, legs, and buttocks, reducing the chance of further internal injuries. A harness also keeps the worker upright and more comfortable while awaiting rescue.

The full-body harness used for confined space rescue should consist of flexible straps that continually flex and give with movement, conforming to the wearer's body and eliminating the need to frequently stop and adjust the harness. Usually constructed of a combination of nylon, polyester, and specially formulated elastomer, the proper harness resists the effects of sun, heat, and moisture to maintain its performance on the job. The full-body harness should include a sliding back D-ring (to attach the retrieval line hook), and a non-slip adjustable chest strap.

The heavy-duty *rescue and retrieval line* is usually a component of a winch system. Both ends of the retrieval lines should be equipped with approved locking mechanisms of at least the same strength as the lines for attaching to the entrant's harness and anchor point.

The winch systems used today are either an approved *two-way system or a three-way system*. The two-way system is used for raising and lowering rescue operations whenever a retractable lifeline is not needed. Typical systems feature three independent braking systems, a tough, two-speed gear drive and approximately 60 feet of steel cable. Three-way systems offer additional protection when a self-retracting lifeline is used. The winch is usually a heavy-duty model (usually rated at 500 lb. or 225 kg) with disc brake to stop falls within inches and is equipped with a shock-absorption feature to minimize injuries. The proper winch should allow the user to raise and lower loads at an average speed of 10–32 feet per minute in an emergency.

The *means of retrieval* usually includes the proper winch with built-in fall protection attached to a 7- or 9-foot tripod. The tripod should be of sufficient height to allow the victim to be brought above the rim of the manhole or other opening and placed on the ground.

Other Equipment

If tools are to be used during a confined space entry or rescue, it may be necessary to use non-sparking tools if flammable vapors or combustible residues are present. These non-sparking, nonmagnetic, and corrosion-resistant tools are usually fashioned from copper or aluminum.

A fire extinguisher, additional radios for communication, spare oxygen bottles (both for SCBAs and cascade systems as needed), a first aid kit, or other equipment necessary for safe entry into and rescue from permit spaces may also be necessary.

Pre-Entry Requirements

Before anyone is allowed to enter a permit-required confined space, certain space conditions must first be evaluated. The first step taken should be to determine whether workers must enter the permit-required space to complete the task at hand. You should ask yourself, "Do we really need to enter the permit-required confined space?" If the answer is yes, then before initiating a confined space entry, the space should be tested with a calibrated air monitor to determine if acceptable entry conditions exist before entry is authorized.

If air monitoring indicates that entry can be made safely without respiratory protection, or if appropriate respiratory protection must be worn, then the supervisor (qualified or competent person) must decide how to bring about the entry in the safest manner possible.

Whether the atmosphere is safe or unsafe (without proper respiratory protection), you must ensure that monitoring is continuous. Taking only one reading and basing your decisions on that reading is not wise. In fact, it's unsafe. Conditions can change within a confined space at any time. It is critical to the well-being of the entrant to know when these changes take place, and what the changes are.

When conducting the air test for atmospheric hazards, a standard testing protocol should be followed:

First—test for oxygen
Then—test for combustible gases and vapors
Then—test for toxic gases and vapors.

You should also test the atmosphere within a confined space at different levels. For example, if you are about to authorize the entry of workers into a manhole that is 30 feet in depth, you should test, top to bottom, for a stratified atmosphere. Remember, some toxic gases (methane, for example) are lighter than air. They tend to accumulate at the higher levels within the manhole. If the manhole may contain carbon monoxide (which has a vapor density similar to air) you should test at the middle level. Hydrogen sulfide (a deadly killer) is heavier than air; therefore, you should test close to the bottom of the manhole. Along with testing at different levels for stratification of toxic gases, you should also check in all directions, to the point possible.

The key point to remember is that atmospheric testing should be continuous, especially when entrants are inside the confined space.

To ensure that continuous atmospheric testing is conducted while an entrant is inside the confined space, an attendant (at least one) must be stationed outside the space to conduct the testing.

In addition to continuously monitoring the atmosphere of the permit-required confined space, the attendant or some other designated person must be familiar with the procedure for summoning rescue and emergency services.

Note: For those facilities having fully trained and equipped on-site rescue teams, it is common practice (and prudent practice) to have the rescue team standing outside the confined space to be on immediate call if required.

Another important function of the attendant or other designated person involved in permit-required confined space entry is to ensure that unauthorized entry into the confined space is prevented.

Before any permit-required confined space entry can be brought about, a proper confined space entry permit must be used.

When employees from more than one work center (e.g., electricians, machinists, painters, and others from different work centers), or more than one employer are involved in confined space entry, an entry procedure to ensure the safety of all entrants must be developed and implemented.

After the confined space entry is completed, procedures must be in place and used to ensure that the space has been closed off and the permit canceled.

The final step that should be taken after any confined space entry has been affected and is completed is to critique the procedure. Questions should be asked and answers given. For example: Did anything go wrong during the entry procedure? Did an unauthorized person make an entry into the space? Did any of the equipment used fail? Was anyone injured? Were there any employee complaints about the procedure? Other questions might arise. If questions do come up, steps must be taken to make sure they are answered or that corrections are made to ensure the next entry into a permit-required confined space is a safer one.

At least once each year, the permits accumulated during the year (confined space permits must be retained by the employer for one year) should be reviewed. If it is apparent from the review that the procedure should be changed, then change it as needed.

Permit System

A permit system for permit-required confined space entry is required by the Confined Space Standard. An entry supervisor (qualified or competent person) must authorize entry, prepare and sign written permits, order corrective measures if necessary, and cancel permits when work is completed. Permits must be available to all permit space entrants at the time of entry and should extend only for the duration of the task. They must be retained for a year to facilitate review of the confined space program (29 CFR 1910.146 (e)(f), 1995).

The information above sums up OSHA's requirements under its Confined Space Entry Standard (29 CFR 1910.146) and in particular for sections (e) Permits System and (f) Entry Permit.

The gist of OSHA's requirements under these sections includes:

1. Ensuring that a permit is actually used for entry into permit-required confined spaces
2. Ensuring that an entry supervisor (the qualified or competent person) authorizes the entry
3. Ensuring that the entry permit is signed
4. Ensuring that any corrective measures are taken if found necessary
5. Ensuring the permit is canceled when work is completed

Confined space entry permits must be available to all permit space entrants at the time of entry and should extend only for the duration of the task. As we stated previously, the permits must be retained for a year to facilitate review of the confined space program.

Permit Requirements

What does a confined space permit require, and what does it look like? These are standard questions that arise any time confined space training is being conducted and at those times when a facility is developing a permit-required confined space program for use and for compliance with OSHA. OSHA, in its 1910 Standard, has published sample permits (listed in Appendix D to § 1910.146). These samples can also aid you in fashioning your own permit.

According to OSHA, an entry permit must include the following:

1. Identification of the permit space to be entered;
2. The purpose of the entry;
3. The date and authorized duration of the entry permit;
4. The authorized entrants within the permit space by name, or by such other means as will enable the attendant to determine quickly and accurately, for the duration of the permit, which authorized entrants are inside the permit space;

5. The personnel, by name, currently serving as attendants;
6. The individual, by name, currently serving as the entry supervisor (qualified or competent person), with a space for the signature or initials of the entry supervisor who originally authorized entry;
7. The hazards of the permit space to be entered;
8. The measures used to isolate the permit space and to eliminate or control permit space hazards before entry (what this really means is that lockout/tagout must be completed);
9. The acceptable entry conditions;
10. The results of initial and periodic tests performed, accompanied by the names or initials of the testers and by an indication of when the tests were performed;
11. The rescue and emergency services that can be summoned and the means (such as the equipment to use and the numbers to call) for summoning those services;
12. The communication procedures used by authorized entrants and attendants to maintain contact during the entry;
13. Equipment, such as personal protective equipment, testing equipment, communications equipment, alarm systems, and rescue equipment:
14. Any other information that needs to be included, given the circumstances of the particular confined space, to ensure employee safety;
15. Any additional permits, such as for hot work, that have been issued to authorize work in the permit space.

Confined Space Training

"The employer shall provide training so that all employees whose work is regulated by this [standard] acquire the understanding, knowledge, and skills necessary for the safe performance of the duties assigned. ..." (CFR 1910.146 (g), 1995). Any work requirement is easier to perform if the person doing the task is fully trained on the proper way to accomplish it. Training offers another advantage as well: increased safety. In accomplishing any work task safely, proper training is critical. Confined space entry operations are extremely dangerous undertakings. We stated earlier that confined spaces are very unforgiving. This is the case even for those workers who have been well trained. However, training helps to reduce the severity of any incident. When something goes wrong (as is often the case) it is better to have fully trained personnel standing by than to have people standing by who are not trained and do not know how to properly rescue an entrant, let alone how to rescue themselves. When you get right down to it, having fully trained workers for any job just makes good common sense.

Training Requirements for Confined Space Entry

OSHA is very clear on its requirement to train confined space entry personnel. Both initial and refresher training must be provided. This training must provide employees with the necessary understanding, skills, and knowledge to perform confined space entry safely. Refresher training must be provided and conducted whenever an employee's duties change, when hazards in the confined space change, or whenever an evaluation of the confined space entry program identifies inadequacies in the employee's knowledge. The training must establish employee proficiency in the duties required and shall introduce new or revised procedures as necessary for compliance with the standard.

OSHA also requires the employer to certify *in writing* that the employee has been trained. This certification must include the employee's name, the signature of the trainer, and the dates of training. Typically, employers certify this training by conducting written and practical examinations (including training dry runs or drills). When an employee meets the certification requirements, the employee is normally awarded a certificate stating that he or she has been trained and certified (by whatever means). These written certifications should be filed in the employee's personnel record and training records.

Any time you conduct safety training, you must keep accurate records of the training. OSHA will want to see these records when they audit your facility (for whatever reason). Any supervisor or training official that provides critically important (possibly life-saving) training would be foolish not to keep and maintain accurate training records. They may be needed in a legal action.

To facilitate the recordkeeping process, a form or roster with a statement like the one shown in Figure 9.2 is highly recommended.

Remember, not only does OSHA require training on its Confined Space Standard and other associated standards (i.e., Lockout/Tagout, Respiratory Protection, and Hot Work Permits), this training is critically important to the well-being of workers. By

---

ATTENDANCE ROSTER

TRAINER: _____ DATE: _____

CONFINED SPACE TRAINING

In accordance with the recordkeeping and training requirements of the Confined Space Entry Standard, I have received training on Confined Space Entry Procedures. I have agreed to verify my understanding and training on 29 CFR 1910.146 OSHA's Confined Space Entry Standard by signing this roster. This training meets the requirements as specified by 29 CFR 1910.146.

Name: Work Center:

_____   _____
_____   _____
_____   _____
_____   _____
_____   _____

Figure 9.2 Typical Training Roster Form

making sure they know that their work organization is taking all possible steps to ensure their safety, they should buy in to the required safe work practices themselves.

Workplace Confined Space Training Program

Are you at a loss as to what the actual training program should entail for the worker? Exactly what should you include in your workplace confined space training program? It depends. Any workplace training program on just about any OSHA-requirement is somewhat site-specific. For example, confined space training for wastewater workers might be different from the training given to telephone repair persons who have to enter underground vaults, because the hazards might not be the same.

As a rule of thumb, it is hard to go wrong on any OSHA-required training if the requirements spelled out in the applicable standard are explained to all workers involved. In addition, for confined space entry training it is important, at a minimum, to cover the following:

1. Explain and point out the requirements of 29 CFR 1910.146 (OSHA's Confined Space Standard).
2. Clearly explain who is responsible for what under the program.
3. Explain key definitions.
4. Inform each trainee of the exact location of the worksite's permit-required confined spaces.
5. Explain how to use the worksite's confined space permit.
6. Explain the potential for engulfment.
7. Explain and demonstrate how to use air-monitoring equipment.
8. Explain and demonstrate how to use required confined space entry equipment.
9. Explain the potential for hazardous atmospheres.
10. Explain the worksite's procedures for confined space rescue.
11. Explain the interface between confined space entry and lockout/tagout, respiratory protection, and hot work permits.
12. Explain how to properly use the worksite's pre-entry checklist.

Confined Space Certification Exam (A Sample)

We stated earlier that measuring the employee's level of knowledge of Confined Space Entry Procedures is important. One way to accomplish this is to administer a written proficiency examination such as the sample exam that follows.

Note from Author: You may want to look at and analyze the questions (especially the types of questions) asked in this examination. A prudent course of action on your part would be to ensure that your confined space training program provides the knowledge necessary to enable the workers to correctly answer all these questions. After administering an exam on safety and health topics, make sure the trainees do not walk away from any safety exam with any wrong answers in their memories. If they answer any question incorrectly, let them know the correct answer before they leave the exam area. Also, it is interesting to note that this particular test was used in the past by OSHA instructors to test OSHA Compliance Officers' Confined Space Entry knowledge.

PERMIT-REQUIRED CONFINED SPACE TEST[3]

1. What is one of the first questions that should be answered before planning entry into a permit-required confined space?

Answer: Can this job/task be accomplished without entering the permit space?

2. OSHA addresses confined space hazards in two specific comprehensive standards. One of the standards covers General Industry and the other covers:
   A. Agriculture
   B. Longshoring
   C. Construction
   D. Shipyards

Answer: D. Shipyards

3. OSHA's definition of confined spaces in general industry includes:
   A. The space being more than 4 feet deep
   B. Limited or restricted means for entry and exit
   C. The space being designed for short-term occupancy
   D. Having only natural ventilation

Answer: B. Limited or restricted means for entry and exit

4. Which of the following would *not* constitute a hazardous atmosphere under the permit-required confined space standard?
   A. Less than 19.5% oxygen
   B. More than the IDLH of hydrogen sulfide
   C. Enough combustible dust that obscures vision at a distance of 5 feet
   D. 5% of LEL

Answer: D. 5% of LEL

5. OSHA's review of accident data indicates that most confined space deaths and injuries are caused by the following three hazards:
   A. Electrical, Falls, Toxics
   B. Asphyxiants, Flammables, Toxics
   C. Drowning, Flammables, Entrapment
   D. Asphyxiants, Explosions, Engulfment

Answer: B. Asphyxiants, Flammables, Toxics (Federal Register, 1-14-93, pg. 4465, Third column under "1. *Atmospheric Hazards*")

6. Toxic gases in confined spaces can result from:
   A. Products stored in the space and the manufacturing processes
   B. Work being performed inside the space or in adjacent areas
   C. Desorption from porous walls and decomposing organic matter
   D. All of the above

Answer: D. All of the above

---

[3] The Confined Space Examination presented here is an adaptation from the examination used by the U.S. Department of Labor—Occupational Health Administration Office of Training and Education—OSHA Training Institute, Des Plaines, IL.

7. Oxygen deficiency in confined spaces does **not** occur by:
   A. Consumption by chemical reactions and combustion
   B. Absorption by porous surfaces such as activated charcoal
   C. Leakage around valves, fittings, couplings, and hoses of oxygen-fuel gas welding equipment
   D. Displacement by other gases

Answer: C. Leakage around valves, fittings, couplings, and hoses of oxygen-fuel gas welding equipment

8. What reading (in %O2) would you expect to see on an oxygen meter after an influx of 10% nitrogen into a permit space?
   A. 5.0%
   B. 11.1%
   C. 18.9%
   D. 90.0%

Answer: C. 18.9% [100% air – 10% nitrogen = 90% air; 90% air × 0.21% O_2 = 18.9% O_2]

9. An attendant is which of the following?
   A. A person who makes a food run to the local 7–11 store for refreshments for the crew inside the confined space.
   B. A person who often enters a confined space while other personnel are within the same space.
   C. A person who watches over a confined space while other employees are in it and only leaves if he or she must use the restroom.
   D. A person with no other duties assigned other than to remain immediately outside the entrance to the confined space and who may render assistance as needed to personnel inside the space. The attendant never enters the confined space and never leaves the space unattended while personnel are within the space.

*Answer: D. A person with no other duties assigned other than to remain immediately outside the entrance to the confined space and who may render assistance as needed to personnel inside the space. The attendant **never** enters the confined space and never leaves the space unattended while personnel are within the space.*

10. Per 1910.146, an atmosphere that contains a substance at a concentration exceeding a permissible exposure limit intended solely to prevent long-term (chronic) adverse health effects is **not** considered to be a hazardous atmosphere on that basis alone.
    A. True
    B. False

Answer: A. True [FR, 1-14-93, pg. 4474, top of 3rd column]

11. Of the following chemical substances, which one is a simple asphyxiant **and** is flammable:
    A. Carbon monoxide (CO)
    B. Methane (CH4)

C. Hydrogen Sulfide (H2S)
D. Carbon dioxide (CO2)

Answer: B. Methane (CH4)

12. Entry into a permit-required confined space is considered to have occurred:
    A. When an entrant reaches into a space too small to enter
    B. As soon as any part of the body breaks the plane of an opening into the space
    C. Only when there is clear intent to fully enter the space (therefore, reaching into a permit space would not be considered entry)
    D. When the entrant says, "I'm going in now."

Answer: B. As soon as any part of the body breaks the plane of an opening into the space

13. If the LEL of a flammable vapor is 1% by volume, how many parts per million is 10% of the LEL?
    A. 10 ppm
    B. 100 ppm
    C. 1,000 ppm
    D. 10,000 ppm

Answer: C. 1,000 ppm [1% = 10,000 ppm x10% = 1,000 ppm]

14. The principal of operation of most combustible gas meters used for permit entry testing is:
    A. Electric arc
    B. Double displacement
    C. Electrochemical
    D. Catalytic combustion

Answer: D. Catalytic combustion

15. The LEL for methane is 5% by volume and the UEL is 15% by volume. What reading should you get on a combustible gas meter when you calibrate with a mixture of 2% by volume methane with a balance of nitrogen?
    A. 10,000 ppm (1% LEL)
    B. 40% LEL
    C. Zero
    D. 80% of the flash point

Answer: C. Zero [**NOTE:** If balance had been air: %Volume divided by %LEL −2/5 = 40% LEL]

16. The proper testing sequence for confined spaces is the following:
    A. Toxics, Flammables, Oxygen
    B. Oxygen, Flammables, Toxics
    C. Oxygen, Toxics, Flammables
    D. Flammables, Toxics, Oxygen

Answer: B. Oxygen, Flammables, Toxics

17. Circle the following true statement(s).
    A. Employers must document that they have evaluated their workplace to determine if any spaces are permit-required confined spaces.
    B. If employers decide that their employees will enter permit spaces, they shall develop and implement a written permit space program.
    C. Employers do not have to comply with any of 1910.146 if they have identified the permit spaces and have told their employees not to enter those spaces.
    D. The employer must identify permit-confined spaces by posting danger signs.

Answer: B. If employers decide that their employees will enter permit spaces, they shall develop and implement a written permit space program. (**NOTE:** *see CPL 2.100, Section (c) Question #2 to rule out "A"*)

18. Circle the following true statement(s).
    A. Under paragraph (c)(5), (i.e., alternate procedures), continuous monitoring can be used in lieu of continuous forced air ventilation if no hazardous atmosphere is detected.
    B. Continuous forced air ventilation eliminates atmospheric hazards.
    C. Continuous atmospheric monitoring is required if employees are entering permit spaces using alternate procedures under paragraph (c)(5).
    D. Periodic atmospheric monitoring is required when making entries using alternate procedures under paragraph (c)(5).

Answer: D. Periodic atmospheric monitoring is required when making entries using alternate procedures under paragraph (c)(5).

19. OSHA's position allows employers the option of making a space eligible for the application of alternate procedure for entering permit spaces, paragraph (c)(5), by first temporarily "eliminating" all non-atmospheric hazards, then controlling atmospheric hazards by continuous forced air ventilation.
    A. True
    B. False

Answer: B. False

20. Respirators allowed for entry into and escape from immediately dangerous to life or health (IDLH) atmospheres are _____.
    A. Airline
    B. Self-contained breathing apparatus (SCBA)
    C. Gas mask
    D. Air-purifying
    E. A and B

Answer: B. Self-contained breathing apparatus (SCBA) (**NOTE:** *a "combination airline with auxiliary SCBA" would be approved, but not an airline*).

21. Circle the following *false* statement(s):
    A. If all hazards within a permit space are eliminated without entry into the space, the permit space may be reclassified as a non-permit confined space under paragraph (c)(7).

B. Minimizing the amounts of regulation that apply to spaces where hazards have been eliminated encourages employers to actually remove all hazards from permit spaces.

C. A certification containing only the date, location of the space, and the signature of the person making the determination that all hazards have been eliminated shall be made available to each employee entering a space that has been reclassified under paragraph (c)(7).

D. An example of eliminating an engulfment hazard is requiring an entrant to wear a full-body harness attached directly to a retrieval system.

Answer: D. An example of eliminating an engulfment hazard is requiring an entrant to wear a full-body harness attached directly to a retrieval system.

22. Circle the following *false* statement(s):
    A. Compliance with OSHA's Lockout/Tagout Standard is considered to eliminate electromechanical hazards.
    B. Compliance with the requirements of the Lockout/Tagout Standard is not considered to eliminate hazards created by flowable materials such as steam, natural gas and other substances that can cause hazardous atmospheres or engulfment hazards in a confined space.
    C. Techniques used in isolation are blanking, blinding, misaligning or removing sections of line soil pipes, and a double block and bleed system.
    D. Water is considered to be an atmospheric hazard.

Answer: D. Water is considered to be an atmospheric hazard. [**NOTE:** see CPL 2.100, pg. 18, Question #11 & Question #12 for additional discussion about water in a confined space.]

23. Circle the following true statement(s):
    A. "Alarm only" devices, which do not provide numerical readings, are considered acceptable direct-reading instruments for initial (pre-entry) or periodic (assurance) testing.
    B. Continuous atmospheric testing must be conducted during permit space entry.
    C. Under alternate procedures, OSHA will accept a minimal "safe for entry" level as 50 percent of the level of flammable or toxic substances that would constitute a hazardous atmosphere.
    D. The results of air sampling required by 1910.146, which show the composition of an atmosphere to which an employee is exposed are **not** exposure records under 1910.1020.

Answer: C. Under alternate procedures, OSHA will accept a minimal "safe for entry" level as 50 percent of the level of flammable or toxic substances that would constitute a hazardous atmosphere. [**NOTE:** see CPL 2.100, pp. 19-20, Question #6. To rule out "A," see CPL 2.100, pg. 22, Question #16 and to rule out "D," see CPL 2.100, pg. 24, Question #3]

24. Example(s) of simple asphyxiants are:
    A. Nitrogen (N2)
    B. Carbon monoxide (CO)

    C. Carbon dioxide (CO2)

    D. A and C

Answer: D. A and C [**NOTE:** if some workers choose only "A" because CO2 has a PEL, give them credit]

25. Which statement(s) is/are true about combustible gas meters (CGMs)?
    A. CGMs can measure all types of gases.
    B. The percent of oxygen will affect the operation of CGMs.
    C. Most CGMs can measure only pure gases.
    D. CGMs will indicate the lower explosive limit for explosive dusts.

Answer: B. The percent of oxygen will affect the operation of CGMs.

26. Circle the following true statement(s):
    A. An off-site rescue service has to have a permit space program before performing confined space rescues.
    B. The only respirator that a rescuer can wear into an IDLH atmosphere is a self-contained breathing apparatus.
    C. Only members of in-house rescue teams shall practice making permit space rescues at least once every 12 months.
    D. Each member of the rescue team shall be trained in basic first aid and CPR.
    E. To facilitate non-entry rescue, with no exceptions, retrieval systems shall be used whenever an authorized entrant enters a permit space.

Answer: D. Each member of the rescue team shall be trained in basic first aid and CPR. (See .146(k)(1)(iv) for correct answer. See CPL 2.100, pg. 24, Question #1 under section (k) to rule out "A"; pg. 25, Question #3 to rule out "B"; FR 1-14-93, pg. 4527, Third column, second paragraph to rule out "C"; and see 1146(k)(3) to rule out "E.")

27. The Permit-Required Confined Space standard requires the employer to initially:
    A. Train employees to recognize confined spaces
    B. Measure the levels of air contaminants in all confined spaces
    C. Evaluate the workplace to determine if there are any confined spaces
    D. Develop an effective confined space program

Answer: C. Evaluate the workplace to determine if there are any confined spaces.

28. If an employer decides that he/she will contract out all confined space work, then the employer:
    A. Has no further requirement under the standard
    B. Must label all spaces with a keep out sign
    C. Must train workers on how to rescue people from confined spaces
    D. Must effectively prevent all employees from entering confined spaces

Answer: D. Must effectively prevent all employees from entering confined spaces.

29. Not required on a permit for confined space entry is:
    A. Names of all entrants
    B. Name(s) of entry supervisor(s)

C. The date of entry
D. The ventilation requirements of the space

Answer: D. The ventilation requirements of the space

30. Circle the following training requirement that is identical for entrant, attendant, and entry supervisor.
    A. Know the hazards that may be faced during entry
    B. The means of summoning rescue personnel
    C. The schematic of the space to ensure all can get around in the space
    D. The proper procedure for putting on and using a self-contained breathing apparatus.

Answer: A. Know the hazards that may be faced during entry

31. Attendants can:
    A. Perform other activities when the entrant is on break inside the confined space.
    B. Summon rescue services as long as he/she does not exceed a 200-feet radius around the confined space.
    C. Enter the space to rescue a worker but only when wearing an SCBA and connected to a lifeline.
    D. Order evacuation if a prohibited condition occurs.

Answer: D. Order evacuation if a prohibited condition occurs.

32. An oxygen-enriched atmosphere is considered by 1910.146 to be:
    A. Greater than 22% oxygen
    B. Greater than 23.5% oxygen
    C. Greater than 20.9% oxygen
    D. Greater than 25% oxygen when the nitrogen concentration is greater than 75%.

Answer: B. Greater than 23.5% oxygen

33. The following confined space, which would be permit-required is:
    A. A grain silo with inward sloping walls
    B. A ten-gallon methylene chlorine reactor vessel
    C. An overhead crane cab that moves over a steel blast furnace
    D. All of the above

Answer: A. A grain silo with inwardly sloping walls

34. A written permit space program requires:
    A. That the employer purchase SCBAs and lifelines, but the employees purchase safety shoes and corrective lens safety glasses.
    B. That the employer test all permit-required confined spaces at least once per year or before entry, whichever is most stringent.
    C. That the employer provide one attendant for each entrant up to five and one for each two entrants when there are more than five.
    D. That the employer develops a system to prepare, issue, and cancel entry permits.

Answer: D. That the employer develop a system to prepare, issue, and cancel entry permits

35. Of the following, which is *not* a duty of the entrant?
    A. Properly use all assigned equipment
    B. Communicate with the attendant
    C. Exit when told to
    D. Continually test the level of toxic chemicals in the space

Answer: D. Continually test the level of toxic chemicals in the space

36. Of the following, which is *not* a duty of the entry supervisor?
    A. Summon rescue services
    B. Terminate entry
    C. Remove unauthorized persons
    D. Endorse the entry permit

Answer: A. Summon rescue services

37. When designing ventilation systems for permit space entry:
    A. The air should be blowing into the space
    B. The air should always be exhausting out of the space
    C. The configuration, contents, and tasks determine the type of ventilation methods used
    D. Larger ducts and bigger blowers are better.

Answer: C. The configuration, contents and tasks determine the type of ventilation methods used

38. Of the following, which is *not* a duty of the attendant?
    A. Know accurately how many entrants are in the space
    B. Communicate with entrants
    C. Continually test the level of toxic chemicals in the space
    D. Summon rescue services when necessary.

Answer: C. Continually test the level of toxic chemicals in the space

39. Circle the following true statement(s):
    A. Carbon monoxide gas should be ventilated from the bottom
    B. The mass of air going into a space equals the amount leaving
    C. Methane gas should be ventilated from the bottom
    D. Gases flow by the inverse law of proportion

Answer: B. The mass of air going into a space equals the amount leaving

40. Hot work is going to be performed in a solvent reactor vessel that is 10 feet high and 6 feet in diameter. Which of the following is the *preferred* way to do this?
    A. Use submerged arc-welding equipment
    B. Inert the vessel with nitrogen and provide a combination airline with auxiliary SCBA respirator for the welder
    C. Fill the tank with water and use underwater welding procedures

D. Pump all the solvent out, ventilate for 24 hours, and use non-sparking welding sticks

E. Clean the reactor vessel, then weld per 1910.252

Answer: E. Clean the reactor vessel, then weld per 1910.252

41. The certification of training required for attendants, entrants, and entry supervisors must contain (circle all that apply):
   A. The title of each person trained
   B. The signature or initials of each person trained
   C. The signature or initials of the trainer
   D. The topics covered by the training

Answer: C. The signature or initials of the trainer

42. Paragraph 1910.146 (g) requires that training of all employees whose work is regulated by the permit-required confined space standard shall be provided:
   A. On an annual basis
   B. When the employer believes that there are inadequacies in the employee's knowledge of the company's confined space procedures
   C. When the union demands it
   D. All of the above

Answer: B. When the employer believes that there are inadequacies in the employee's knowledge of the company's confined space procedures

Training is mandatory when an employee is first assigned confined space entry duties, when those duties change, whenever a change in permit-required confined space entry operations presents a new hazard, or whenever an employer believes an employee needs additional procedural assistance. The preceding point cannot be overstated. Here is another crucial point: The primary message sent by the employer in training his or her workers for confined space entry should be: Look before you leap.

On-site personnel including entrants, attendants, and entry supervisors assigned to affect permit-required confined entries must be fully aware of their duties under the OSHA Standard. OSHA, under part (h) entrants, (i) attendants, and (j) entry supervisors of 1910.146 clearly defines these duties. Again, training is the key ingredient to effecting safe permit-required confined space entry. Obviously, assigning anyone specific duties is easy, but ensuring that these duties are performed in the correct manner—especially when training has not been conducted—is much more difficult. Supervisors and workers must know their duties and must know how to complete their duties in a safe and correct manner.

Duties of Authorized Entrants

The key responsibility of any permit-required confined space entrant is to gain knowledge of the hazards that may be faced during entry. The entrant must also be knowledgeable enough to understand the mode, signs or symptoms, and consequences of exposure to immediate or potential hazards. This knowledge requirement is central to the critical importance that training plays in compliance with this program or any safety program.

Knowledge of the hazards and/or potential hazards is just part of the requirements involved in being a "qualified" entrant. The entrant must also know his or her equipment: how to use it, what it is to be used for, and its limitations.

The entrant must know how to communicate with the attendant. Communication can be via radio/walkie-talkie (which must also be intrinsically safe, capable of producing no sparks), by hand signals (visual contact must be maintained), and/or by voice, whistle, or some other prearranged and practiced sound-making device.

The entrant must alert the attendant whenever he or she recognizes any warning sign or symptom of exposure to a dangerous situation. He or she must also communicate to the attendant any changing condition that could make the entry more hazardous than it already is.

The entrant must know when to exit the confined space and do so without hesitation, without prompting, without delay. He or she must maintain a position within the space whereby he or she can exit quickly if necessary. When ordered to exit by the attendant, the entrant must not delay, think about it, or pause for any reason. When ordered to exit, the entrant must exit immediately.

Duties of Attendants

The employer has the responsibility of ensuring that the permit-required confined space attendant is fully trained and knowledgeable. The attendant must know the hazards that may be faced during entry, including information on the mode, signs or symptoms, and consequences of the exposure. The attendant must be aware of the behavioral effects of hazard exposure to which the entrants may be subjected.

That the attendant plays a critical role in confined space entry should be apparent. This critical role cannot be filled by just anyone. The attendant must be fully trained and qualified to perform his or her assigned responsibilities.

The attendant is responsible for maintaining an accurate count of authorized entrants in the permit space and must ensure that the means used to identify authorized entrants accurately identifies who is in the permit space.

The attendant remains outside the permit space until properly relieved by another qualified attendant. When the employer's permit entry program allows attendant entry for rescue, attendants may enter a permit space to attempt a rescue **IF** they have been trained and equipped for rescue operations.

The attendant maintains communication between him/herself and the entrant(s) at all times. The attendant also monitors conditions within and outside the space that might endanger the entrants and orders the entrants to exit if necessary.

If the attendant detects a hazardous situation, the behavioral effects of hazard exposure in an authorized entrant, and/or determines that he or she cannot (for whatever reason) perform their attendant duties, the attendant must order the immediate evacuation of the permit space.

The attendant is also responsible for summoning rescue and other emergency services as soon as he or she determines that authorized entrants may need assistance to escape from permit space hazards.

The attendant prohibits unauthorized persons from entering a permit space and/or from interfering with an entry in progress.

The attendant has one responsibility and one responsibility only: to perform the duties of a permit space attendant without allowing any distraction.

Duties of Entry Supervisors

As with any other work activity, supervisors play a key role in permit-required confined space entry. In permit space entry, the supervisor is responsible for issuing confined space permits. To do this according to the standard, the entry supervisor must know the hazards of the confined spaces, verify that all tests have been conducted and all procedures and equipment are in place before endorsing a permit, terminate entry if necessary, cancel permits, and verify that rescue services are available and the means for summoning them are operable. In addition, entry supervisors are to remove unauthorized individuals who attempt to enter the confined space. They also must determine, at least when shifts and entry supervisors change, that acceptable conditions, as specified by the permit, continue.

Remember, the entry supervisor signs "the bottom line" on the permit. Before signing that bottom line on any safety document, supervisors should use good judgment, along with care and caution. If and when anything goes wrong in a confined space entry, the first item that the OSHA investigator will want to see is the permit. When lawyers are involved (as is often the case when workers are killed or badly injured on the job) the permit becomes an important document that will end up in a court of law, along with the supervisor in charge of the confined space entry operation.

Confined Space Rescue

Of the more than 1.6 million workers who enter confined spaces each year, approximately 63 die from asphyxiation, burns, electrocution, drowning and other tragedies related to confined space entry operations. But more alarming is the fact that 60 percent of those who die in confined spaces are untrained rescuers who not only fail to save a coworker, but also are killed during the rescue attempt (The John Wayne Syndrome). OSHA requires that a trained, equipped rescue team be available whenever employees work in confined spaces (Coastal Video, 1993).

Confined space accident accounts always seem to read the same way: there is a victim in a confined space, and a topside hero (or heroes) abandons common sense and concern for their own safety and enters a dangerous environment to save their coworker, their friend, and like the victim whose peril sparked their action, they also die.

Those heroes chose to risk (and some of them lose) their lives when they entered a confined space that contained an immediate and apparent risk. The victims' deaths affect their families, their friends, and their coworkers.

Rescue Services

The employer who engages in permit-required confined space entry has the option of whether to use an off-site or in-plant rescue service. If the decision is made to use an off-site rescue service, a number of factors must be considered.

The first factor to consider is: Is such a rescue service readily available? This is a logical, straightforward question. However, when you seek to answer it you may find that the question is easier to ask than it is to obtain an answer that ends your search. Why? Let's take a look at what typically occurs when this option is chosen.

The natural inclination is to list **dial 911** or another local emergency number on your confined space permit as your rescue service. However, is such rescue service really available to you from the local fire department or some other emergency service? You need to find out. In our experience, when we call local fire departments and explain to them that we are about to make a confined space entry and are giving them a heads up to be aware of the operation, they are usually puzzled. "We fight fires and make some rescues. But, confined space rescue? Sorry, we are not trained for that." If you try to locate off-site rescue service, you probably will hear that response, because it is typical.

We simply cannot list 911 as the standby emergency rescue service (and hope that whoever responds will be able to effect rescue) unless we are absolutely certain they will respond in less than four minutes (remember, a victim in a confined space cannot live without air for more than four minutes) and are fully trained to effect the rescue.

The second factor that you must take into consideration (once you have identified a rescue service that can respond in four minutes or less) is: Is this service familiar with your facility? Have you invited the members of the service into your facility for familiarization with your facility?

Another factor to consider is on-site training. Has the rescue service actually practiced making confined space rescues in your confined spaces? Are they willing to spend the time to acquire the information they need to handle a crisis situation at your facility? This is an important point that an OSHA auditor will be certain to verify if and when your facility is audited.

On-site rescue teams have considerations as well. If you decide to employ the services of an on-site rescue team, OSHA requires that:

1. The employer shall ensure that each member of the rescue team is provided with, and is trained to use properly, the personal protective equipment and rescue equipment necessary for making rescues from permit spaces.
2. Each member of the rescue team shall be trained to perform the assigned rescue duties. Each member of the rescue team shall also receive the training required of authorized entrants stated in the standard.
3. Each member of the rescue team shall practice making permit space rescues at least once every 12 months, by means of simulated rescue operations in which they remove dummies, mannequins, or actual persons from the actual permit spaces or from representative permit spaces. Representative permit spaces shall, with respect to opening size, configuration, and accessibility, simulate the types of permit spaces from which rescues are to be performed.
4. Each member of the rescue team shall be trained in basic first aid and in cardiopulmonary resuscitation (CPR). At least one member of the rescue team holding current certification in first aid and in CPR shall be available.

In the OSHA standard, the above requirements describe the rescue team as a "rescue service." From experience, calling this rescue service a rescue team is better (and more appropriate) because a team is what it is. To properly effect confined space rescue, the rescue service must be a "team," individuals who work together seamlessly. Each member must have good endurance, enthusiasm, and willingness to learn and possess a team-oriented attitude.

Rescue Service Provided by Outside Contractors

When an employer arranges to have persons other than the host employer's employees perform permit space rescue, the host employer shall:

1. Inform the rescue service of the hazards they may confront when called on to perform rescue at the host employer's facility.
2. Provide the rescue service with access to all permit spaces from which rescue may be necessary so that the rescue service can develop appropriate rescue plans and practice rescue operations.

Non-Entry Rescue

The rescue services we've discussed to this point all involved making external (non-entry rescue) confined space rescues—the preferred method of rescue recommended by this text, even though it may not be feasible on all occasions. The rule of thumb that we use is that if external rescue via a tripod, winch, retrieval line, and body harness cannot be made, then the confined space entry should not be made in the first place. When such retrieval systems are used, they shall meet the following requirements:

1. Each authorized entrant shall use a chest or full-body harness, with a retrieval line attached at the center of the entrant's back near shoulder level, or above the entrant's head. Wristlets may be used in lieu of the chest or full-body harness if the employer can demonstrate that the use of a chest or full-body harness is unfeasible or creates a greater hazard, and that the use of wristlets is the safest and most effective alternative.
2. The other end of the retrieval line shall be attached to a mechanical device or fixed point outside the permit space in such a manner that rescue can begin as soon as the rescuer becomes aware that rescue is necessary. A mechanical device (such as a tripod and winch assembly) is used to retrieve personnel from vertical type permit spaces more than five feet (1.52m) deep.

A final word on permit-required confined space rescue: In the event of a rescue where the entrant is exposed to a hazardous material for which a Safety Data Sheet (SDS) or other similar written information is required to be kept at the worksite, that SDS or written information must be made available to the medical facility treating the exposed entrant.

Alternative Protection Methods

Minimizing the amount of regulation that applies to spaces where hazards have been eliminated encourages employers to actually removal all hazards. OSHA has specified alternative protection procedures that may be used for permit spaces where the only hazard is atmospheric and ventilation alone can control the hazard. Let's take a brief look at these alternative protection procedures.

"Hierarchy" of Permit-Required Confined Space Entry

The following hierarchy of permit-required confined space entry is useful to anyone involved in designing a worksite confined space entry program.

1. 1910.146(c)(7) Reclassification - Hazards Eliminated
   * Requires certification, (c)(7)(iii)

   The employer may reclassify a permit-required confined space as a non-permit confined space under the following procedures:
   - If the permit space poses no actual or potential atmospheric hazard and if all hazards within the space are eliminated without entry into the space, the permit space may be reclassified as a non-permit confined space for as long as the non-atmospheric hazards remain eliminated.
   - If it is necessary to enter the permit space to eliminate hazards, such entry shall be performed under the guidelines presented in the standard. If testing and inspection during that entry demonstrate that the hazards within the permit space have been eliminated, the permit space may be reclassified as a non-permit confined space for as long as the hazards remain eliminated.

Note: OSHA points out that control of atmospheric hazards through forced air ventilation does not constitute elimination of the hazards.

   - The employer shall document the basis for determining that all hazards in a permit space have been eliminated, through a certification that contains the date, the location of the space, and the signature of the person making the determination. The certification must be made available to each employee entering the space.

Note: Great care and caution should be exercised before anyone signs certification stating that a particular confined space is not hazardous (for any reason). Remember that the person who signs such a document is responsible, and therefore liable for his or her decision.

   - If hazards arise within a permit space that has been declassified to a non-permit space, each employee in the space must exit the space. The employer must then reevaluate the space and determine whether it must be reclassified as a permit space.

2. 1910.146(c)(5)(i)(E) Alternate Entry - Hazards Controlled (by continuous forced air ventilation)
   - Requires documentation and supporting data, (c)(5)(i)(E)
   - Requires training, (g)
   - Requires a "min-program," (c)(5)(ii)
   - Requires certificate, (c)(5)(ii)(H)

An employer may use the alternate procedures specified in the Standard (c)(5)(ii) for entering a permit space under the conditions set forth in the following:

- The employer can demonstrate that the only hazard posed by the permit space is an actual or potential hazardous atmosphere.
- The employer can demonstrate that continuous forced air ventilation alone is sufficient to maintain that permit space safe for entry.
- The employer develops monitoring and inspection data that supports his or her reclassification decision.
- If an initial entry of the permit space is necessary to obtain the data required, the entry must be made by the requirements set forth for entry into a permit-required confined space.
- The determinations and supporting data are documented by the employer and are made available to each employee who enters the permit space.

Let's summarize these requirements. To qualify for alternative procedures employers must satisfy all the following conditions:

- Ensure that it is safe to remove the entrance cover (e.g., a manhole filled with methane might explode if, when removing the metal manhole cover, the cover and/or tools used causes a spark)
- Determine that ventilation alone is sufficient to maintain the permit space safe for entry and that work to be performed within the permit-required space will introduce no additional hazards.
- Gather monitoring and inspection data to support the above requirements.
- If entry is necessary to conduct initial data gathering, perform such entry under the full permit program.
- Document the determination and supporting data and make them available to employees.

3. Permit Space Entry—Hazards can neither be Eliminated nor Controlled
   If this is the case, then the following is required:

   - written program, (d) as required by (c)(4)
   - permits, (e) & (f)
   - training, (g)
   - attendant, (d)(6)
   - testing, (d)(5)
   - rescue, (k)

Maintaining your Confined Space Entry program (once it is established for your facility) will require regular attention on your part in evaluation and analysis of your facilities and their spaces. As your facility changes, grows, and ages, the confined spaces on your site may change, too, demanding reassessment. Meeting the requirements of OSHA's standards for your facility is an ongoing process, not a once-and-done event.

Procedures for Atmospheric Testing

You can never trust your senses to determine if the air in a confined space is safe! You cannot see or smell many toxic gases and vapors, nor can you determine the level of oxygen present.

Personnel involved in permit-required confined space entry must understand that some vapors or gases are heavier than air and will settle to the bottom of a confined space. Other gases are lighter than air and will be found around the top of the confined space. Because of the behaviors of various toxic gases, you must test all areas (top, middle, bottom) of a confined space with properly calibrated testing instruments to determine what gases are present.

Testing Procedures

Atmospheric testing is required for two distinct purposes: evaluation of the hazards of the permit space, and verification that acceptable entry conditions for entry into that space exist.

1. **Evaluation testing**. The atmosphere of a confined space should be analyzed using equipment of sufficient sensitivity and specificity to identify and evaluate any hazardous atmospheres that may exist or arise, so that appropriate permit entry procedures can be developed and acceptable entry conditions stipulated for that space. Evaluation and interpretation of these data and development of the entry procedure should be done by, or reviewed by, a technically qualified professional (e.g., OSHA consultation service, Certified Safety Professional (CSP), Certified Industrial Hygienist (CIH), registered safety Professional Engineer, etc.) based on evaluation of all serious hazards.
2. **Verification testing**. The atmosphere of a permit space that may contain a hazardous atmosphere should be tested for residues of all contaminants identified by evaluation testing using permit specified equipment to determine that residual concentrations at the time of testing and entry are within the range of acceptable entry conditions. Results of testing (i.e., actual concentration, etc.) should be recorded on the permit in the space provided adjacent to the stipulated acceptable entry condition.
3. **Duration of testing**. Measurement of values for each atmospheric parameter should be made for at least the minimum response time of the test instrument specified by the manufacturer.
4. **Testing stratified atmospheres**. When monitoring for entries involving a descent into atmosphere that may be stratified, the atmospheric envelope should be tested at a distance of approximately four feet (1.22m) in the direction of travel and to each side. If a sampling probe is used, the entrant's rate of progress should be slowed to accommodate the sampling speed and detector response.

Air Monitoring and OSHA

When an OSHA compliance officer audits your facility, if you have permit-required confined spaces that are entered by your employees, the auditor will pay particular attention to your air-monitoring procedures.

Typically, the OSHA auditor will want to see copies of your confined space permits for the past year. From these permits, the auditor will chose one and set it aside. Later, the auditor will ask to interview those involved in making that confined space entry. The auditor may ask the confined space personnel several different questions related to their knowledge of confined space entry. The auditor may desire to see these personnel perform the entry again (if possible).

During the OSHA auditor's interview process, air monitoring will be discussed. The auditor will want to see the instrument used during the confined space entry. The auditor will note the condition of the instrument, looking specifically for any damage, dirt, or dead batteries (Are they using the right batteries or have they "jerry-rigged" a battery pack?), and will test to determine if any sensors are malfunctioning, etc.

The OSHA auditor almost always asks one of the confined space entry personnel to demonstrate both how to calibrate and how to use the instrument.

In addition, the OSHA auditor typically asks several questions related to air monitoring to determine the knowledge level of the workers. These may include:

1. Has the operator been trained?
2. Who gave them the training? What was covered? How long did the training last? Any hands-on or on-the-job training?
3. What types of instruments are used?
4. Where is the manufacturer's instruction manual? Have they read the manual?
5. How often do they use the instrument?
6. Do they have calibration data, logbook, etc.
7. What calibration gas do they use? Why did they choose this gas (a question mainly for % LEL—are they using propane, methane, pentane. . .)?
8. Do they zero the instrument as part of the calibration?
9. Who calibrates the equipment? How often? How is it done?
10. Do they have a calibration curve or correction factor chart?
11. What are the interferences for the toxic sensors?
12. Is the meter intrinsically safe for the environment they are monitoring?
13. Are they waiting long enough for the sensors to respond (and for remote sampling some manufacturers suggest 1 second per foot of tubing)?
14. Are they testing all levels and areas where entrants will be working?
15. If using several individual instruments, are they testing in the right sequence (oxygen, flammables, toxics)?
16. What do the numbers on the instrument mean? Are they exact?
17. What are you comparing the numbers to? What is considered safe for entry?
18. Have they replaced any sensors? Any batteries? Any other parts? Do they have maintenance logs?
19. Do they send the instrument back to manufacturer on a regular basis for complete calibration and maintenance?
20. Do they field check?

Note: If you use your portable gas detector for sewer entry, the OSHA auditor will check your detector to see if it complies with OSHA's May 19, 1994 technical amendment to the confined space rule CFR 1910.146, where Federal OSHA modified the Appendix E language to read as follows:

> The oxygen sensor/broad range sensor is best suited for initial use in situations where the actual or potential contaminants have not been identified, because broad range sensors, unlike substance-specific sensors, enable employers to obtain an overall reading of the hydrocarbons (flammables) present in the space.

Other OSHA Permit-Required Confined Space Audit Items

Earlier, we discussed the types of queries that an OSHA auditor would make concerning a typical worksite's air-monitoring practices used in performing permit-required confined space entry. In this section, we discuss the "other" OSHA audit items—ones dealing specifically with permit-required confined space entry procedures.

When an OSHA auditor audits your confined space entry program, you can be assured that they will look at most (if not all) of the items listed below.

1. Are aisles in the vicinity of the confined space marked?
2. Are aisles and passageways properly illuminated?
3. Are aisles kept clean and free of obstructions?
4. Are fire aisles, access to stairways, and fire equipment kept clear?
5. Is there safe clearance for equipment through aisles and doorways?
6. Have all confined spaces and permit-required confined spaces been identified?
7. Are danger signs posted (or other equally effective means of communication) to inform employees about the existence, location, and dangers of permit-required confined spaces?
8. Is the written permit-required confined space entry program available to employees?
9. Is the permit-required confined space sufficiently isolated? Have pedestrian, vehicle, or other necessary barriers been provided to protect entrants from external hazards?
10. When working in permit-required confined spaces, are environmental monitoring tests taken and means provided for quick removal of workers in case of an emergency?
11. Are confined spaces thoroughly emptied of any corrosive or hazardous substances (such as acids or caustics) before entry?
12. Are all lines to a confined space containing inert, toxic, flammable, or corrosive materials valved off and blanked or disconnected and separated before entry?
13. Is each confined space checked for decaying vegetation or animal matter, which may produce methane?
14. Is the confined space checked for possible industrial waste, which could contain toxic properties?
15. Before permit space entry operations begin, has the entry supervisor identified on the permit signed the entry permit to authorize entry?
16. Has the permit been made available at the time of entry to all authorized entrants (by being posted at the entry portal or by other equally effective means) so that entrants can confirm that pre-entry preparations have been completed?

17. Is necessary personal protective equipment (PPE) available?
18. Has necessary lighting equipment been provided?
19. Has equipment (such as ladders) needed for safe ingress and egress by authorized entrants been provided?
20. Is rescue and emergency services equipment available?
21. Is it required that all agitators, impellers, or other rotating equipment inside confined spaces be locked out if they present a hazard?
22. Is all portable electrical equipment used inside confined spaces either grounded and insulated or equipped with ground fault protection?
23. Is at least one attendant stationed outside the confined space for the duration of the entry operation?
24. Is there at least one attendant whose sole responsibility is to watch the work in progress, sound an alarm if necessary, and render assistance?
25. Is the attendant trained and equipped to handle an emergency?
26. Is the attendant and/or are other workers prohibited from entering the confined space without lifelines and respiratory equipment if there is any question as to the cause of an emergency?
27. Is communications equipment provided to allow the attendant to communicate with authorized entrants as necessary to monitor entrant status and to alert entrants of the need to evacuate the permit space?

If your worker training brings your workers to the level that they can provide reasonable answers for the sample OSHA questions, and if your facility is compliant with the list of OSHA auditor questions, you are well on your way to a successful Confined Space Entry Program.

THOUGHT-PROVOKING QUESTIONS

- Describe and define "John Wayne Syndrome."
- What is the ultimate cause of all confined space fatalities?
- What other safety programs work hand-in-hand with Confined Space Entry?
- What are the differences between permit-required and non-permit-required confined spaces?
- Define confined space.
- Define entry.
- What are the steps in workplace evaluation?
- What do OSHA's equipment specifications cover?
- What is the difference between approved and unapproved equipment?
- Who is responsible for equipment maintenance?
- What training should cover testing and monitoring equipment?
- How should you select PPE, and what is required?
- What considerations should you be concerned about in lighting for confined spaces?
- What special considerations should be taken with open manholes?
- Why should you examine built-in ingress and egress equipment?

- What is the OSHA standard concerning mechanized equipment for worker entry and egress?
- Describe the three components of proper rescue equipment.
- What should you look for in tools for confined space rescue?
- In pre-entry testing, what order should you test in? For what should you test? What are the proper testing techniques?
- What is the ideal position for the rescue team during a confined space entry?
- What are the requirements if people from several work centers are working together on a confined space entry?
- How long should permits be kept on file? How often should they be reviewed?
- Define acceptable entry conditions.
- Discuss OSHA's training requirements.
- Why is record-keeping such a critical practice?
- What can a trainer do to ensure and record how effectively training sessions worked?
- How often should training be planned and presented?
- What are the problems inherent in finding offsite rescue services?
- Describe OSHA's requirements for an on-site rescue team.
- What does OSHA require of employees who use outside contractors for rescue services?
- What is the bottom line on confined space entry and rescue?
- When should you use SDSs in confined space entry?
- What is the procedure for reclassifying a permit-required space into a non-permit-required space?
- How can hazards be controlled or eliminated?
- How can a facility qualify for alternative confined space entry requirements?
- What are the two purposes of atmospheric testing?
- Why test? How long and how often should you test?
- What is OSHA going to want to know and see concerning your air monitoring?
- How important is proper monitoring equipment calibration? Why?
- What role does company management play in Confined Space Entry?
- What use are checklists? Why are they effective?

REFERENCES AND RECOMMENDED READING

Coastal Video. *Confined Space Rescue Booklet*. Virginia Beach, VA: Coastal Video Communication Corporation, 1993.

Code of Federal Regulations. Occupational Safety and Health Administration, Department of Labor, title 29, sec. 1900-1910.

Sayers, D.L., and J.L. Walsh. *Thrones, Dominations*. New York: St. Martin's Press, 1998: 263–264.

Spellman, F.R. *Safe Work Practices for Wastewater Operators*. Lancaster, PA: Technomic Publishing Company, 1996.

Spellman, F.R. *Confined Space Entry: A Guide To Compliance*. Lancaster, PA: Technomic Publishing Company, 1999.

The Bureau of National Affairs. *The Job Safety and Health Act of 1970*. Washington, DC: Bureau of National Affairs, 1971.

Water Environment Federation. *Confined Space Entry.* Alexandria, VA: Water Environment Federation, 1994.

SUGGESTED READINGS

The following publications are available from American Industrial Hygiene Association Publications, 3141 Fairview Park Dr., Suite 777, Falls Church, VA 22042, (703) 849-8888, www. aiha.org

- Chelton, C.F. *Manual of Recommended Practice for Combustible Gas Indicators and Portable Direct-Reading Hydrocarbon Detectors,* 2nd ed, 1993. ISBN 0-932627-48-X; 55 pages.
- Perper, J., and B. Dawson. *Direct-Reading Colorimetric Indicator Tubes Manual,* 2nd ed., 1993. 60 pages.
- Vernon, R., and T. King. *Confined Space Entry: An AIHA Protocol Guide,* 1995. ISBN 0-932627-67-6; 53 pages.

The following book can be purchased from the American Society of Safety Engineers, 520 N. Northwest Hwy, Park Ridge, IL 60068, (847) 699-2929, www.asse.org

- Rekus, J.F. *Complete Confined Spaces Handbook,* 1994. ISBN: 1-87371-487-3; 400 pages.

The following publications are available from American Conference of Governmental Industrial Hygienists Publications, 1330 Kemper Meadow Dr., Cincinnati, OH 45240, (513) 742-2020, www.acgih.org

- *Industrial Ventilation: A Manual of Recommended Practice,* 22nd ed., 1995. ISBN: 1-882417-09-7; 470 pages.
- Maslansky, C.J., and S.P. Maslansky. *Air Monitoring Instrumentation,* 1993. ISBN 0-442-00973-9; 304 pages.
- Ness, S.A. *Air Monitoring for Toxic Exposures,* 1991. ISBN 0-442-20639-9; 534 pages.
- *1997 TLVs and BEIs,* 1997. ISBN: 1-882417-19-4; 156 pages.

The following are available from the American National Standards Institute (ANSI), 25 West 43rd St., Fourth Floor, New York, NY 10036, (212) 642-4900, www.ansi.org

- *American National Standard for Respiratory Protection* ANSI Z88.2-1992.
- *Requirements for Safety Belts, Harnesses, Lanyards and Lifelines for Construction and Demolition Use* ANSI A10.14-1991.
- *Safety Requirements for Confined Spaces,* ANSI Z117.1-1995.
- *Safety Requirements for Personal Fall Arrest Systems, Subsystems and Components* ANSI Z359.1-1992.

Chapter 10

Personal Protective Equipment

The primary objective of any health and safety program is worker protection. It is the responsibility of management to carry out this objective. Part of this responsibility includes protecting workers from exposure to hazardous materials and hazardous situations that arise in the workplace. It is best for management to try to eliminate these hazardous exposures through changes in workplace design or engineering controls. When hazardous workplace exposures cannot be controlled by these measures, personal protective equipment (PPE) becomes necessary. When looking at hazardous workplace exposures, keep in mind that government regulations consider PPE the last alternative in worker protection because it does not eliminate the hazards. PPE only provides a barrier between the worker and the hazard. If PPE must be used as a control alternative, a positive attitude and strong commitment by management is required.

—S.Z. Mansdorf, 1993

INTRODUCTION

Above, Mansdorf makes a number of important statements concerning personal protective equipment (PPE), ones worth taking some time to look at carefully.

1. It is best for management to try to eliminate these hazardous exposures through changes in workplace design or engineering controls.

   Sound familiar? We consistently make this same point throughout this text. A hazard, any hazard, if possible, should be "engineered out" of the system or process. Determining when and how to engineer out a hazard is one of the safety engineer's primary functions. However, the safety engineer can much more effectively accomplish this if he or she is included in the earliest stages of design. Remember, it does little good (and is often very expensive) to attempt to engineer out any hazard once the hazard is in place.

2. When hazardous workplace exposures cannot be controlled by these measures, personal protective equipment (PPE) becomes necessary.

While the goal of the safety and environmental manager is certainly to engineer out all workplace hazards, we realize that this goal is virtually impossible to achieve. Even in this day of robotics, computers, and other automated equipment and processes, the man-machine-process interface still exists. When people are included in the work equation, the opportunity for their exposure to hazards is very real. As injury statistics make clear, it happens.

3. . . . consider PPE the last alternative in worker protection because it does not eliminate the hazards.

This is extremely important for two reasons: First, the safety and environmental manager's primary goal is (as we have said before) to engineer out the problem. If this is not possible, the second alternative is to implement administrative controls. When neither is possible, PPE becomes the final choice. The key words here are "the final choice." Secondly, PPE is sometimes incorrectly perceived—by both the supervisor and/or the worker—as their first line of defense against all hazards. This, of course, is incorrect and dangerous. The worker must be made to understand (by means of enforced company rules, policies, and training) that PPE affords only minimal protection against most hazards. IT DOES NOT ELIMINATE THE HAZARD.

4. PPE only provides a barrier between the worker and the hazard.

Our experience shows us that when some workers put on their PPE, they also don a "Superperson" mentality. What does this mean? Often, when workers use eye, hand, foot, head, hearing, or respiratory protection, they also adopt an "I can't be touched" attitude. They feel safe, as if the PPE somehow magically protects them from the hazard, so they act as if they are protected, are invincible, are beyond injury. . . . They feel, however illogically, that they are well out of harm's way. Nothing could be further from the truth. Let's look at an example.

***

CASE STUDY 10.1

A work crew was assigned to clear trees, shrubs, and undergrowth from a densely wooded area to provide clear access for valve checkers, who routinely (on a semiannual basis) inspect the operation of mechanically operated values on an underground wastewater interceptor line in the area. Because the pipeline transited a rural forested area, this clearing assignment was both routine and necessary. Many of the workers used chain saws in this clearing operation. All of them had been trained on the proper operation and safety considerations involved in using chain saws, and each worker had been issued the appropriate PPE to use for this assignment: gloves, safety shoes, safety glasses, and hard hats with wire mesh face shields and ear muffs attached.

During the clearing operation one of the workers inadvertently cut his left leg quite severely on the inner calf with the chain saw he was using. The victim was transported to the nearest medical facility and received extensive treatment for the deep and ragged wound.

During the accident follow-up investigation phase, we asked the victim to explain what happened, how he got injured. The answer he gave us did not surprise us, but his honesty did. He stated that he had all his PPE on. "Sure, it was uncomfortable to wear," but he had worn it anyway. And while he was cutting away, he really didn't consider the hazards involved with operating the 20" chain saw. "I knew I was well-protected with my PPE and all, so I just let the ol' saw rip away." And, of course, that is just what happened, the saw ripped away, right into his leg. "Just felt like I was fully protected," he had said, shaking his head in disbelief at his own negligence.

Sound ridiculous? Such incidents happen many times every day. Workers tend to forget that PPE is only a barrier between themselves and the hazard, one that works to dissipate force and keep hazardous materials from contacting vulnerable parts of the body. The hazard is still there, behind the barrier that PPE provides. Workers forget how easily most barriers can be circumvented or torn away. Unless the hazard is engineered out, it is always there. All the PPE in the world cannot fully protect a worker who is not also aware and vigilant.

***

OSHA'S PPE STANDARD

In the past, many OSHA standards have included PPE requirements, ranging from very general to very specific. It may surprise you to know, however, that not until the 1990's (1993–1994) did OSHA incorporate a stand-alone PPE Standard into its 29 CFR 1910/1926 Guidelines. This *Personal Protective Equipment* standard is covered (General Industry) under 1910.132-.138, but you can find PPE requirements elsewhere in the General Industry Standards. For example, 29 CFR 1910.156, OSHA's Fire Brigades Standard has requirements for firefighting gear. In addition, 29 CFR 1926.95-106 covers the construction industry. As shown in Figure 10.1, the PPE standard focuses on head, feet, eye, hand, respiratory, and hearing protection.

Common PPE classifications and examples include:

1. Head protection (hard hats, welding helmets)
2. Eye protection (safety glasses, goggles)
3. Face protection (face shields)
4. Respiratory protection (respirators)
5. Arm protection (protective sleeves)
6. Hearing protection (ear plugs, muffs)
7. Hand protection (gloves)
8. Finger protection (cots)
9. Torso protection (aprons)
10. Leg protection (chaps)
11. Knee protection (kneeling pads)
12. Ankle protection (boots)

Figure 10.1   Elements Required in a Personal Protective Equipment (PPE) Program

13. Foot protection (boots, metatarsal shields)
14. Toe protection (safety shoes)
15. Body protection (coveralls, chemical suits)

Both respiratory and hearing protection have had their own standards for quite some time and are discussed later in this text. Respiratory protection is covered under 1910.134 and hearing protection under 1910.95.

Using PPE is often essential, but it is generally the last line of defense after engineering controls, work practices, and administrative controls. Engineering controls involve physically changing a machine or work environment. Administrative controls involve changing how or when employees do their jobs, such as scheduling work and rotating employees to reduce exposures. Work practices involve training workers how to perform tasks in ways that reduce their exposure to workplace hazards.

OSHA's PPE Requirements

OSHA mandates several requirements for both the employer and the employee under its PPE Standard. OSHA's requirements include:

1. Employers are required to provide employees with personal protective equipment that is sanitary and in good working condition.
2. The employer is responsible for examining all PPE used on the job to ensure that it is of a safe (and approved) design and in proper condition.

3. The employer must ensure that employees use PPE.
4. The employer must provide a means for obtaining additional and replacement equipment; defective and damaged PPE is not to be used.
5. The employer must ensure that PPE is inspected on a regular basis.
6. The employee must ensure that he or she dons PPE when required.
7. Where employees provide their own PPE, the employer must ensure that it is adequate, including properly maintained and sanitized.

Note: While the employer must ensure the employee wears PPE when required, both the employer and employee should factor in three things: (1) The PPE used must not degrade performance unduly; (2) it must be reliable; and (3) it must be suitable for the hazard involved.

Hazard Assessment

How does a safety and environmental manager determine when and where an employer should provide PPE and when the employee should use it? This can be determined in three ways: (1) From the manufacturer's guidance (when it comes to equipment and processes produced by a manufacturer, the manufacturer is considered the "expert" on the equipment or process and is normally best suited to determine the hazards associated with the equipment and/or processes they manufacture); (2) if the process or equipment the employee is working on/with involves chemicals, the Material Safety Data Sheets (MSDS) for the chemicals involved list the required PPE to be used; and (3) OSHA mandates that the employer perform a hazard assessment of the workplace.

The purpose of the *hazard assessment* is to determine if hazards are present or likely to be present that necessitate the use of PPE. If a facility presents such hazards, the employer is required to (1) select and have each affected employee use the types of PPE that will protect the affected employee from the hazards identified in the hazard assessment; (2) communicate selection decisions to each affected employee; and (3) select PPE that properly fits each affected employee.

The employer is required to verify that the workplace hazard assessment has been conducted through a written certification that identifies the workplace evaluated, the person certifying that the evaluation has been performed and the date of the hazard assessment and that also identifies the document as a certification of hazard assessment.

Note: The safety and environmental manager must maintain up-to-date copies of the PPE Hazard Assessment forms. During a recent OSHA audit, the auditor wanted to see copies of the assessments conducted at our workcenters.

PPE Training Requirements (TOC)

OSHA requires the employer to provide training to each employee required to use PPE. This training must inform the employee on when the PPE is necessary, what PPE is necessary, how to properly don, doff, adjust, and wear PPE, the limitations of the PPE, and the proper care, maintenance, useful life, and disposal of the PPE.

Note: During an OSHA audit of your facility, the auditor may want to look at a copy of your facility's PPE training program. Almost certainly, the auditor will want to review your training records for PPE training. Remember this: You can conduct all the training in the world and have it performed by well-known experts in the field, but if you did not document the training, in OSHA's eyes, it never occurred. You **must** have proof of training conducted.

After workers complete PPE training, OSHA requires each employee to demonstrate his or her understanding of the training. This is usually best accomplished by conducting a written examination (make sure you keep records of this, too).

If the employer has reason to believe that any affected employee who has already been trained does not have the understanding and skill required, the employer must retrain each such employee. In this retraining requirement, remember that everything in life is dynamic (constantly changing), including the workplace and work assignments. OSHA understands this dynamic trend, and thus requires the employer to retrain employees who install new processes, equipment or requirements—any new element in a job task that might render previous training obsolete. Changes also occur in PPE itself. Maybe a new type or model of PPE is introduced and used in the workplace. If this is the case, the employer must ensure that employees using such PPE are fully trained on the PPE.

***

SAMPLE PPE TRAINING GUIDE

This sample PPE Training Guide has been successfully used for more than twelve years. While other training guides on the subject may be more inclusive, this guide helps the safety engineer formulate his or her facility's PPE training requirement.

PERSONAL PROTECTIVE EQUIPMENT (PPE) TRAINING GUIDE

Introduction

This training guide is designed to be used by company work center supervisors to provide required OSHA/VOSH PPE training on 29 CFR 1910.132-138. OSHA's PPE Standard is a "Performance Standard," meaning that employers/employees must meet the minimum requirements in the standard and/or other requirements, as determined by performance and experience and specified by organizational safety officials.

Presentation

1. REQUIREMENTS of PPE STANDARD:
   The OSHA PPE Standard mandates several requirements:

   a. Employer designated safety officials must conduct both a work center and employee job classification hazard assessment to determine who must use PPE and where it must be used.

b. Employers must provide approved PPE to employees who are required to use it in the normal performance of their duties.
c. Employers must train employees on where, how, and when to use PPE. Employers must also train employees on the limitations of PPE.
d. Employees are required to make continued assessments of actual PPE usage to ensure that PPE actually works as designed. Most of the above requirements are to be performed by the company safety and environmental manager.

The safety and environmental manager will train work center supervisors.
Training of work center employees will be provided by work center supervisors.

Employee Information

PPE is designed to protect you from health and safety hazards that cannot be removed from your work environment. PPE is specifically designed to protect many parts of your body, including your eyes, face, head, hands, feet, and hearing.

PPE such as respirators are designed to protect your pulmonary function. Respirators are covered under the company's Respiratory Protection Program and are not required to be presented in this training session under the PPE Standard.

Eye and Face Protection

1. Eye and/or face protection must be worn any time that it is required by SDS.
2. Eye and/or face protection must be worn any time that company Safe Work Practices require it.
3. Both eye and face protection are now required any time you work with:
   a. Chemicals
   b. Hazardous gases
   c. Flying particles
   d. Molten metals
   e. Whenever deemed necessary and appropriate by the supervisor.
4. Welder's eye and face protection must be worn when welding. Ensure proper UV-rated protective glass is used for welding as follows:
   • For welding operations using less than 60 amps, shade 7 is required.
   • For welding operations using 60-160 amps, shade 8 is required.
   • For welding operations using 160-250 amps, shade 10 is required.
   • For welding operations using more than 250 amps, shade 11 is required.
   • For torch soldering use shade 2.
   • For torch brazing use shade 3.
5. When wearing safety glasses, coverage from front and sides is required.
   If employee prescription glasses meet ANSI standards for safety glasses, the employer must provide employee with protective side shields, and the employee must use them.

   If the employee's prescription glasses do not meet ANSI Safety Glass requirements, the employer is not required to provide prescription glasses but must provide oversize safety goggles. The oversize safety goggles must fit and seal properly over prescription glasses.

6. If the employee wears contact lenses, he/she may face additional hazards from chemicals or dust. Dust caught under the lens can cause painful abrasions. Chemicals can react with contacts to cause permanent injury. Under no circumstances are contact lenses to be considered protective devices. Eye protection must be worn in addition to or instead of contact lenses.

7. Face and eye protection devices must be distinctly marked to facilitate identification of manufacturer.

Head Protection

1. Employees are required to wear protective helmets (hard hats) when working in areas where there is a potential for injury to the head from falling objects.

2. Employees are required to wear protective helmets (hard hats) when working in or around construction projects.

3. Employees are required to wear protective helmets (hard hats) when working in or around areas where flying debris could cause potential head injuries.

4. Employees are required to wear protective helmets (hard hats) when working near exposed electrical conductors, which could contact the head.

5. Employees are required to wear protective helmets (hard hats) when organizational safety officials and/or supervisors determine the need.

6. The hard hat suspension must be designed to absorb some impact. It must be adjusted to fit the wearer and to keep the shell a minimum distance of one-and-one-fourth inches above the wearer's head.

7. Company employees are required to wear Class B hard hats. Class B hard hats are made from insulating material designed to protect you from impact and from electric shock by voltages of up to 20,000 volts.

8. Employees are to periodically check their hard hat suspension. Look for loose or torn cradle straps, loose rivets, broken sewing lines or other defects.

9. Hard hats are to be dated when issued and must be replaced every two to five years or after major impact.

Hand Protection

1. Employees must be issued hand protection when they are exposed to hazards such as those from skin absorption of harmful substances, severe cuts or lacerations, severe abrasions, punctures, chemical burns, thermal burns, and harmful temperature extremes.

2. Extreme caution is to be exercised whenever employees are wearing gloves while working on moving machinery.

3. The supervisor is to provide the employee with the correct type of hand protection required for the job. Check SDS if unsure.

4. Whatever gloves are selected and provided by the supervisor, make sure they fit. The supervisor is responsible for ensuring that gloves selected are the most appropriate gloves for a particular application, for determining how long they can be worn, and whether they can be reused.

5. Instruct employees on how to inspect gloves.

Foot Protection

1. Employees must wear approved safety shoes when the possibility of foot injury could occur from heavy or sharp objects that fall on the feet.
2. Employees must wear approved safety shoes when something could roll over their feet.
3. Employees must wear approved safety shoes when something could pierce the sole of the shoe.
4. Employees must wear approved safety shoes whenever directed by supervisor or designated safety officials.
5. Employees who work around exposed electrical wires or connections need to wear metal-free nonconductive shoes or boots.
6. Employees who are required to work in a static-free environment, for example, when working with computers or other electronic equipment, should wear a conductive shoe designed to drain static charges into a mat or the floor.
7. Employees who might have to work in or around chemical spills are required to be supplied with rubber or synthetic footwear (Type based on SDS).

Cleaning and Maintenance

All PPE must be properly cleaned and maintained. Cleaning is particularly important for eye and face protection, where dirty or fogged lenses could impair vision. PPE is to be inspected, cleaned, and maintained at regular intervals so that the PPE provides the requisite protection.

Summary

It is the employer's responsibility to teach employees about the PPE they require. However, it is the employee's responsibility to wear it. No one can use it for the employee except the employee.

When properly and regularly used, PPE effectively provides the needed barrier for workers against the hazard. But PPE is useless unless the worker wears it and uses it properly.

THOUGHT-PROVOKING QUESTIONS

- What should a safety engineer do if a task's risks can't be "engineered out" or administratively controlled?
- What's the chief problem with PPE?
- How does PPE function?
- What are employer responsibilities for PPE?
- What are employee responsibilities for PPE?
- What three factors must apply for required PPE?
- How does a safety engineer decide when and what PPE is needed?
- What are OSHA's documentation requirements concerning hazard assessment?
- How should workers learn about their required PPE?

- When should eye protection be used?
- When should face protection be used?
- When should head protection be used?
- When should hand protection be used?
- When should foot protection be used?
- How should PPE be maintained?

REFERENCES AND RECOMMENDED READING

Mansdorf, S.Z. *Complete Manual of Industrial Safety.* Englewood Cliffs, NJ: Prentice-Hall, Inc., 1993.

Spellman, F.R., *Surviving an OSHA Audit: A Management Guide.* Lancaster, PA: Technomic Publishing Company, 1998.

Chapter 11

Respiratory Protection

The basic purpose of any respirator is, simply, to protect the respiratory system from inhalation of hazardous atmospheres. Respirators provide protection either by removing contaminants from the air before it is inhaled or by supplying an independent source of respirable air. The principal classifications of respirator types are based on these categories.

—NIOSH Guide to Industrial Respiratory Protection, 1987

INTRODUCTION

Written procedures shall be prepared covering safe use of respirators in dangerous atmospheres that might be encountered in normal operations or in emergencies. Personnel shall be familiar with these procedures and the available respirators [OSHA 29 CFR 1910.134 (c)].

Respirators Defined: Respirators are devices that can allow workers to safely breathe without inhaling particles or toxic gases. Two basic types are (1) *air-purifying*, which filter dangerous substances from the air; and (2) *air-supplying*, which deliver a supply of safe breathing air from a tank (SCBA), or group of tanks (cascade system), or an uncontaminated area nearby via a hose or airline to your mask.

You should recall that respiratory protection might be a requirement in effecting safe-confined space entry. We also stated that often the organization's safety engineer holds the responsibility for making this determination. If indeed the safety engineer determines that respiratory protection is required, then it is incumbent upon him or her to implement a written respiratory protection program that is in compliance with OSHA's Respiratory Protection Standard (29 CFR 1910.134).

Remember, though, that respiratory protection is often necessary to protect workers who may not ever be called upon to enter a confined space with an atmosphere containing airborne contaminants. Workers may need protection from airborne contaminants in any workplace or worksite situation where airborne contaminants are health hazards.

We have continuously stressed the vital need to attempt first to engineerout such hazards (any hazard). However, when engineering and other methods of control cannot eliminate airborne hazards, proper selection and use of respiratory protection are part of the safety and environmental manager's responsibility.

Unlike past practices, where respiratory protection entailed nothing more than providing respirators to workers who could be exposed to airborne hazards and expecting workers to use the respirator to protect themselves, today, supplying respirators without the proper training, paperwork, and testing is illegal. Employers are sometimes unaware that by supplying respirators to their employees without having a comprehensive respiratory protection program, they are making a serious mistake, because by issuing respirators, they have implied that a hazard actually exists. In a lawsuit, they then become fodder for the lawyers.

OSHA mandates that an effective program must be put in place. This respiratory protection program must not only follow OSHA's guidelines, but must also be well planned and properly managed. A well-planned, well-written respiratory protection program must include the eleven elements shown in Figure 11.1. In this chapter, we discuss these eleven elements and explain what they require. This information will enable the safety engineer to implement a respiratory protection program that complies with OSHA requirements.

Figure 11.1   Elements Required for Compliance with OSHA's Respiratory Protection Standard (29 CFR 1910.134)

Note: For permit-required confined space entry operations, respiratory protection is a key piece of safety equipment, one always required for entry into an Immediately Dangerous to Life or Health (IDLH) space and one that must be readily available for emergency use and rescue if conditions change in a non-IDLH space. Remember, however, that *only air-supplying respirators should be used in confined spaces where there is not enough oxygen.*

Selecting the proper respirator for the job, the hazard, and the worker is very important, as is thorough training in the use and limitations of respirators. Compliance with OSHA's Respiratory Standard begins with developing written procedures covering all applicable aspects of respiratory protection. Because this requirement is important, in this chapter we present a written Respiratory Protection Program that includes ten of OSHA's eleven required elements. Again, while this sample program is designed for a fictitious organization named the "Company," in reality it has been successfully used for several years (along with the Respirator Program Evaluation Checklist) and has proven its worth and effectiveness through worksite testing and OSHA evaluation.

WRITTEN RESPIRATORY PROTECTION PROGRAM (A SAMPLE)

Company's Respiratory Protection Program

I. INTRODUCTION

The Occupational Safety and Health Act (OSH Act) requires that every employer provide a safe and healthful work environment. This includes ensuring workers are protected from unacceptable levels of airborne hazards. While most air is safe to breathe, certain work operations and locations have characteristic problems of air contamination. Control measures are required to reduce airborne hazard concentrations to safe levels. When controls are not feasible, or while they are being implemented, workers must wear approved respiratory protection.

The company has adopted this "Respiratory Protection Program" to comply with OSHA regulations (as set forth in 29 CFR 1910.134) and to do all that is possible to protect those employees who are filling a job classification that requires respirator use in the performance of their duties. All departments and work centers are included and must adhere to the requirements set forth in this program. Company's "Respiratory Protection Program" is an organized approach for assuring employees a safe work place by providing specific requirements in these areas:

1. Designation of individual departmental responsibilities.
2. Definition of various terms used in the "Respiratory Protection Program."
3. Designation of types of respirators and their applications.
4. Designation of procedures for respirator selection and distribution.
5. Designation of procedures to be used for inspection and maintenance of respirators.
6. Designation of procedures for employee respirator fit testing.
7. Designation of a procedure for medical surveillance.

8. Designation of a training program for personnel participating in Company's "Respiratory Protection Program."
9. Documentation procedure for personnel participating in Company's "Respiratory Protection Program."

II. RESPONSIBILITIES

A. Department directors will be responsible for the following:

1. Implement and ensure compliance of departmental personnel with Company's "Respiratory Protection Program."
2. Specify the job classifications that use respirators, and ensure this job requirement is included in job descriptions for these classifications.

B. Company's safety division has the following responsibilities under Company's "Respiratory Protection Program."

1. Develop and modify as necessary Company's written "Respiratory Protection Program."
2. Check and review quarterly all work center programs, including the work center respirator inspection record.
3. Compile and maintain a master respirator inventory list for Company.
4. Implement an ongoing respirator training program.
5. Conduct initial and annual employee fit testing.
6. Provide initial and annual spirometric evaluation to ensure that employees are capable of wearing a respirator under their given work conditions.
7. Provide technical assistance in determining the need for respirators and in the selection of appropriate types of respirators.
8. Forward training, fit test, initial/annual spirometric evaluation, and medical doctor's evaluation for suitability to wear a respirator to human resources manager for inclusion into employee's personnel record.
9. Inspect quarterly the accuracy and proper maintenance of records specified in this program.
10. Conduct air quality tests annually on internal combustion engine-driven airline respirator compressors to ensure proper air quality.

C. Company supervisory personnel are responsible for the following:

1. Ensure that respirators are available to employees as needed.
2. Ensure that employees wear appropriate respirators as required.
3. Ensure inspection of cartridge-type respirators on a monthly basis, and self-contained breathing apparatus (SCBA's) and airline hose mask systems on a weekly and monthly basis. Ensure records of respirator inspections are maintained.
4. Ensure employees are fit tested and receive initial/annual spirometric evaluation prior to using a respirator.

D. The employee is responsible for the following:

1. Use supplied respirators in accordance with instructions and training.
2. Clean, disinfect, inspect, and store assigned respirator(s) properly.
3. Perform self-fit test prior to each use, and ensure that manageable physical obstructions such as facial hair (mustaches only) do not interfere with respirator fit.
4. Report respirator malfunctions to supervision and conduct "After Use Inspection" of SCBA-type respirator.
5. Report any poor health conditions that may preclude safe respirator usage.

E. Company human resources manager is responsible for the following:

1. Schedule required initial medical examination and spirometric, evaluation for all new employees who fill job classifications requiring the use of respirators.
2. Maintain records of employee medical, spirometric and fit test results.

III. DEFINITION OF TERMS

Company's "Respiratory Protection Program" defines various terms as follows:

Aerosol: A suspension of solid particles or liquid droplets in a gaseous medium.

Asbestos: A broad mineralogical term applied to numerous fibrous silicates composed of silicon, oxygen, hydrogen, and metallic ions like sodium, magnesium, calcium, and iron. At least six forms of asbestos occur naturally. Types of asbestos that are currently regulated—actinolite, amosite, anthophylite, chrysotile, crocidolite, and tremolite.

Banana oil: A liquid that has a strong smell of bananas, used to check for general sealing of a respirator during fit testing.

Blasting abrasive: A chemical contaminant composed of silica, silicates, carbonates, lead, cadmium, or zinc and classified as a dust.

Breathing resistance: The resistance that can build up in a chemical respirator cartridge that has become clogged by particulates.

Chemical hazard: Any chemical that has the capacity to produce injury or illness when taken into the body.

Cleaning respirators: Cleaning respirators involves washing with mild detergent and rinsing with potable water.

Dust: A dispersion of tiny solid airborne particles produced by grinding or crushing operations.

Forced Vital Capacity (FVC): The maximal volume of air which can be exhaled forcefully after a maximal inhalation.

Fit testing: An evaluation of the ability of a respiratory device to interface with the wearer in such a manner as to prevent the work place atmosphere from entering the worker's respiratory system.

Forced Expiratory Volume (FEV1): That volume of air which can be forcibly expelled during the first second of expiration.

Fume: Solid particles generated by condensation from the gaseous state.

Gas: A substance that is in the gaseous state at ordinary temperature and pressure.

IDLH (Immediately Dangerous to Life and Health): Any condition that poses an immediate threat to life, or that is likely to result in acute or immediately severe health effects.

Irritant smoke (stannic oxychloride): A chemical used to check for general sealing of a respirator during a fit test.

Mist: A dispersion of liquid particulates.

Oxygen deficiency: Any level below the PEL of 19.5%

Particulates: Dusts, mists, and fumes.

Permissible Exposure Limit (PEL): The maximum time-weighted average concentration of a substance in air that a person can be exposed to during an 8-hour shift.

PEL/IDLH CHART

Chemical Name	PEL (8 hr. average)	IDLH
1. Ammonia	50 ppm	300 ppm
2. Carbon Dioxide	5,000 ppm	50,000 ppm
3. Carbon Monoxide	50 ppm	1,200 ppm
4. Sodium Hydroxide	Must use SCBA to enter	
5. Sulfur Dioxide	5 ppm	100 ppm
6. Chlorine	1 ppm	10 ppm
7. Hydrogen Chloride	5 ppm	100 ppm
8. Hydrogen sulfide	10 ppm	100 ppm
9. Propane	1,000 ppm	2,100 ppm
10. Oxygen	19.5% (Min)	——
11. Flammable	10% LEL	

PEL—Permissible Exposure Limit.
IDLH—Immediately Dangerous to Life and Health Limit.

Respirator: A face mask that filters out harmful gases and particles from air enabling a person to breathe and work safely.

Respiratory hazard: Any hazard that enters the human body by inhalation.

Saccharin: A chemical sometimes used to check for general sealing of a respirator during fit testing.

Smoke: Particles that result from incomplete combustion.

Spirometric evaluation: A test used to measure pulmonary function. A measurement of FVC and FEV1 of 70% or greater is satisfactory. A measurement of less than 70% may require further pulmonary function evaluation by a medical doctor.

Vapor: The gaseous state of a substance that is liquid or solid at ordinary temperature and pressure.

IV. TYPES OF RESPIRATORS

A. Chemical Cartridge Respirators

1. Description: Chemical cartridge respirators may be considered low-capacity gas masks. They consist of a facepiece, which fits over the nose and mouth of the wearer. Attached directly to the facepiece is a small replaceable filter-chemical cartridge.
2. Application: Usually this type of respiratory protection equipment is used where there is exposure to solvent vapors or dust and particulate matter, as with sandblasting, spray coating, or degreasing. They may not be worn in IDLH atmospheres.

B. Airline respirators (helmet, hoods, and masks) Cascade-fed or Compressor-fed

1. Description: These devices provide air to the wearer through a small-diameter, high-pressure hose line from a source of uncontaminated air. The source is usually derived from a compressed airline with a valve in the hose to reduce the pressure. A filter must be included in the hose line (between the compressed airline and the respirator) to remove oil and water mists, oil vapors, and any particulate matter that may be present in the compressed air. Internally lubricated compressors require that precautions be taken against overheating since the heated oil will break down and form carbon monoxide. Where the air supply for airline respirators is taken from the compressed airline, a carbon monoxide alarm must be installed in the air supply system. Completion of prior-to-operation preventive maintenance check on the carbon monoxide alarm system is critical.
2. Application: Airline respirators used in industrial application for confined space entry (IDLH atmosphere) must be equipped with an emergency escape bottle.

C. Self-Contained Breathing Apparatus (SCBA)

1. Description: This type of respirator provides Grade D breathing air (not pure oxygen), either from compressed air or breathing air cylinders, or by chemical action in the canister attached to the apparatus. It enables the wearer to be independent of any outside source of air. This equipment may be operable for periods between one-half to two hours. The operation of the self-contained breathing apparatus is fairly complex, and it is therefore necessary that the wearer have special training before being permitted to use it in an emergency situation.
2. Application: Because the oxygen-producing mechanism is self-contained in the apparatus, it is the only type of equipment that provides complete protection and at the same time permits the wearer to travel for considerable distances from a source of respirable air. SCBAs (with the exception of hot work activities) can be used in many industrial applications.

V. RESPIRATOR SELECTION & DISTRIBUTION PROCEDURES

Work center supervisors select respirators. Selection is based on matching the proper color-coded cartridge with the type of protection desired. Selection is also dependent upon the quality of fit and the nature of the work being done. Cartridge-type respirators are issued to the individuals who are required to use them. Each individually assigned respirator is identified in a way that does not interfere with its performance. Questions about the selection process are to be referred to the safety division.

VI. RESPIRATOR INSPECTION, MAINTENANCE, CLEANING, AND STORAGE

To retain their original effectiveness, respirators should be periodically inspected, maintained, cleaned, and properly stored.

Note: In the following sections, several references are made to various inspection records. You should design site-specific standard record forms and inspection records for use with your respiratory protection program.

A. Inspection

1. Respirators should be inspected before and after each use, after cleaning, and whenever cartridges or cylinders are changed. Appropriate entries should be made in a respirator "Inspection After Each Use" record.
2. If a 1/2-face air-purifying respirator is taken out of use, indicate it on the inspection records. The respirator must be inspected thoroughly before it is put back in use.
3. Work center supervisors shall ensure all cartridge-type respirators are inspected once per month and make appropriate entries in a "Supervisor's Monthly Respirator Inspection Checklist." The work center supervisor or designated person shall inspect all SCBAs and airline respirators weekly and monthly, and make appropriate entries in a "SCBA/Air Line Respirator Weekly and Monthly Inspection and Maintenance Checklist" record. These records are to be kept by each work center for a period of three years.
4. Safety Division personnel will inspect these records quarterly.

B. Maintenance

1. Respirators that do not pass inspection must be replaced or repaired prior to use. Respirator repairs are limited to the changing of canisters, cartridges, cylinders, filters, head straps, and those items as recommended by the manufacturer. No attempt should be made to replace components, or make adjustments, modifications, or repairs beyond the manufacturer's recommendations.

C. Cleaning

Individually assigned cartridge respirators are cleaned as frequently as necessary by the assignee to ensure proper protection is provided. SCBA respirators are cleaned after each use.

The following procedure is used for cleaning respirators:

1. Filters, cartridges, or canisters are removed before washing the respirator, and discarded and replaced as necessary.
2. Cartridge-type and SCBA respirator facepieces are washed in a detergent solution, rinsed in clean potable water, and allowed to dry in a clean area. A clean brush is used to scrub the respirator to remove adhering dirt.

D. Storage

After inspection, cleaning, and necessary repairs, respirators are stored to protect against dust, sunlight, heat, extreme heat, extreme cold, excessive moisture, or damaging chemicals. Respirators are to be stored in plastic bags or the original case. Individuals assigned respirators are to store their respirator in an assigned personal locker. General-use SCBAs are to be stored in designated cabinets, racks, or lockers with other protective equipment. Respirators are not to be stored in toolboxes or left in the open. Individual cartridges or masks with cartridges are to be sealed in plastic bags to preserve their effectiveness.

VII. RESPIRATOR FIT TESTING

The "Respiratory Protection Program" provides standards for respirator fit-testing. The goal of respirator fit testing is to (1) provide the employee with a face seal on a respirator that exhibits the most protective and comfortable fit and (2) to instruct the employee on the proper use of respirators and their limitations.

There are three levels of fit testing: **Initial, Annual,** and **Pre-Use Self-Testing**.

A. The Initial and Annual fit tests are rigorous procedures used to determine whether the employee can safely wear a respirator

The Initial and Annual tests are conducted by the safety division. Both tests utilize the Cartridge and SCBA-type respirator to check each employee's suitability for wearing either type. Fit testing requires special equipment and test chemicals such as banana oil, irritant smoke, or saccharin. In general, any change to the face or mouth may alter respirator fit and may require the use of a specially fitted respirator; Company's safety division will make this determination. Upon completion of initial fit testing, the safety division forward the original of the employee's Fit Test Record to the human resources manager for inclusion in the employee's file. A copy will be forwarded to the affected work center supervisor.

B. Pre-Use Self-Testing is a routine requirement for all employees who wear respirators

Each time the respirator is used, it must be checked for positive and negative seal. The Safety Division will train supervisors on this procedure. Supervisors are responsible for training employees in their individual work centers.

1. Positive Pressure Check Procedure (cartridge style respirator): After the respirator has been put in place and straps adjusted for firm but comfortable tension, the exhalation valve is blocked by the wearer's palm. He or she takes a deep breath and gently exhales a *little* air. Hold the breath for 10 seconds. If the mask fits properly, it will feel as if it wants to pop away from the face, but no leakage will occur.
2. Negative Pressure Check Procedure (cartridge style respirator): While still wearing the respirator, cover both filter cartridges with the palms, and inhale slightly to partially collapse the mask. Hold this negative pressure for 10 seconds. If no air leaks into the mask, it can be assumed the mask is fitting properly.

Note: Self-test fit testing can be conducted, for both positive and negative pressure checks, on the SCBA-type respirator by crimping the hoses with fingers, and vice blocking airways with palm of hands.

If either test shows leakage, the following procedure should be followed:

1. Ensure mask is clean. A dirty or deteriorated mask will not seal properly, nor will one that has been stored in a distorted position. Proper cleaning and storage procedures must be used.
2. Adjust the head straps to have snug, uniform tension on the mask. If only extreme tension on the straps will seal the respirator, report this to the supervisor. Note that a mask with uncomfortably tight straps rapidly becomes obnoxious to the wearer.

1910.134 (g)(1)(A) states: Personnel with facial hair that comes between the sealing surface of the facepiece and the face, or that interferes with valve function shall not be permitted to wear tight-fitting respirators. Thus, respirator wearers with beards or side burns that interfere with the face seal are prohibited from wearing tight-fitting respirators on the job.

Dental changes—loss of teeth, new dentures, braces, and so forth—may affect respirator fit and may require a new fitting with a different type mask.

Note: Any change to the face or mouth that may alter respirator fit must be brought to the immediate attention of the work center supervisor.

VIII. MEDICAL SURVEILLANCE

OSHA states that no one should be assigned a task requiring use of respirators unless they are found medically fit to wear a respirator by competent medical authorities. Company's "Respiratory Protection Program" will include a medical surveillance procedure that includes:

A. Pre-employment Physical/Spirometric Evaluation/Five-Year Follow-Up Physical Exam

All new and regular employees who fill job classifications that require respirator use in the performance of their duties are required to pass an initial medical examination to determine fitness to wear respiratory protection on the job. Annual spirometric evaluations will be conducted to ensure that employees covered under this program meet the OSHA requirements for fitness to wear respirators. On a continuous five-year basis, all Company employees covered under this program will be re-examined by competent medical authorities to ensure their continued fitness to wear respiratory protection on the job.

Each department director will specify which job classifications require the employee to use respirators. A medical doctor will conduct preemployment and five-year follow-up medical evaluation. The safety division will conduct spirometric evaluation. The safety division will forward the employee's spirometry results to the human resources manager for inclusion in the employee's personnel file.

B. Annual Spirometric Evaluation

Annual spirometric evaluations will be conducted by the safety division on all employees filling job classifications requiring the use of respirators in the performance of their duties. Spirometry testing will be used to measure Forced Vital Capacity (FVC) and Forced Expiratory Volume-1 second (FEV1). If FVC is less than 75% and/or FEV1 is less than 70%, the employee will not be allowed to wear a respirator unless a written waiver is obtained from a medical doctor. The supervisor determines whether the employee can be exempted from work functions that require wearing a respirator.

Note: Company will make reasonable accommodations to allow employees to retain their current positions with specified medical restrictions on respirator use.

The safety division will route annual results of spirometric testing to human resources manager for inclusion in each employee's personnel file and will notify appropriate supervisors of any employee who fails the test.

IX. TRAINING

No worker may wear a respirator before spirometric evaluation, medical evaluation, fit testing, and training have all been completed and documented.

A. The safety division holds the responsibility for providing employee respirator training.
B. Supervisors are the day-to-day monitors of the program and have the responsibility to perform refresher training and to ensure self-fit testing is accomplished by their employees as needed.

Available dates for Safety Division administered training sessions will be published on a routine basis. Supervisors are responsible for scheduling their new employees for the next available session. Training on respiratory protection is also conducted at New Employee Safety Orientation sessions.

This respiratory protection program is subject to changes and improvements as new regulations and technologies emerge. The safety division will train supervisors and employees as applicable on any new information.

X. DOCUMENTATION PROCEDURES

Documentation of safety training is very important. OSHA insists that certain records be maintained on all employees. All safety-training records should be considered legal records; the likelihood of having to use safety-training records in a court of law is real.

A. The following information will be maintained by the Safety Division:
  a. Date and location of initial employee training;
  b. Inventory records of all Company respirators.
B. The following information will be processed by the human resources manager for inclusion in the employee's personnel file.
  a. Results of annual employee fit testing;
  b. Results of new employee medical evaluation and annual spirometric testing (to remain on file for five years).
C. Supervisors will maintain:
  a. A file of respirator inspection records;
  b. Respirator inventory records

Note: The maintenance and accuracy of all records specified in this will be inspected quarterly by the safety division.

XI. PROCEDURE FOR SAFE USE OF SCBA/SUPPLIED AIR RESPIRATORS

To be in compliance with 1910.134 (e)(3), Company is providing these written procedures covering the safe use of respirators (SCBA & Supplied Air Respirators only).

Note: Air-purifying/chemical cartridge respirators are to be used only for coatings and sand blasting operations, and **NEVER** for confined space entry or any other activity where oxygen deficiency or atmospheric contaminants are present.

SCBAs and/or supplied air (with emergency escape bottles) are to be used in all situations that involve chemical handling, confined space entry during normal operations, and in emergencies.

1. Safe Use Procedure in Dangerous Atmospheres

This written procedure is prepared for safe respirator use in IDLH atmospheres that may occur in normal operations or emergencies.

All Company personnel covered under this program are to be familiar with these procedures and respirators.

a. Inspect all respirator equipment prior to use to ensure that it is complete and in good repair.
b. Ensure respirator face piece is correct size for your face; perform a self-fit test.
c. Ensure that available air is adequate for the expected time to be used.

Note: No Company employee should use an SCBA that is not 100% full.

d. Test all alarms on the respirator to ensure that they work.
e. At least two fully trained and certified standby/rescue persons, equipped with proper rescue equipment (including an SCBA) will be present in the nearest safe area for emergency rescue of those wearing respirators in an IDLH atmosphere.
f. Communications (visual, voice, signal line, telephone, radio, or other suitable type) will be maintained among all persons present (those in the IDLH atmosphere and the standby person or persons). The respirator wearers are to be equipped with safety harnesses and safety lines to permit their removal from the IDLH atmosphere if they are overcome.
g. The atmospheres in a confined space may be immediately dangerous to life or health (IDLH) because of toxic air contaminants or lack of oxygen. Before any Company employee enters a confined space, tests must be performed to determine the presence and concentration of any flammable vapor or gas, or any toxic airborne particulate, vapor, or gas, and to determine the oxygen concentration (follow all procedures as outlined in Company's Confined Space Program).
h. No one is to enter if a flammable substance exceeds the lower explosive limit (LEL). No one should enter without wearing the proper type of respirator if any air contaminant exceeds the established permissible exposure limit (PEL), or if there is an oxygen deficiency. Ensure that the confined space is force-ventilated to keep the flammable substance at a safe level.

Note: Even if the contaminant concentration is below the established breathing time-weighted average (TWA) limit and there is enough oxygen, the safest procedure is to ventilate the entire space continuously and to monitor the contaminant and oxygen concentrations continuously if people are to work in the confined space without respirators.

i. If the atmosphere in a confined space is IDLH owing to a high concentration of an air contaminant or oxygen deficiency, those who must enter the space to perform work must wear a pressure-demand SCBA or a combination pressure-demand airline and self-contained breathing apparatus that always maintains positive air pressure inside the respiratory inlet covering. Fully trained and equipped rescue must be on-site and ready to respond if needed. This is the best safety practice for confined space entry and **is required** at Company.

***

RESPIRATORY PROGRAM EVALUATION

The safety and environmental manager must not only ensure that his or her organization's respiratory protection program complies with the ten elements covered in the sample program above, but must also ensure that the eleventh element, respirator program evaluation is also accomplished. Why? Because the OSHA standard (29 CFR

1910.134) requires regular inspection and evaluation of the respirator program to determine its continued effectiveness in protecting employees. Remember that periodic air monitoring is also required, to determine if the workers are adequately protected. The overall program should be evaluated at least annually, and the written program or standard operating procedure (SOP) modified if necessary.

Do you have questions about how to evaluate your respiratory protection program? Good. You should. The NIOSH guidelines in *NIOSH Guide to Industrial Respiratory Protection,* publication No. 87-116 (1987), probably provide the best answer: an evaluation checklist. Here is a sample from the NIOSH Guide.

SAMPLE EVALUATION CHECKLIST

Respirator Program Evaluation Checklist

In general, the respirator program should be evaluated for each job or at least annually, with program adjustments, as appropriate, made to reflect the evaluation results. Program function can be separated into administration and operation.

A. Program Administration

_____ (1) Is there a written policy that acknowledges employer responsibility for providing a safe and healthful workplace and assigns program responsibility, accountability, and authority?

_____ (2) Is program responsibility vested in one individual who is knowledgeable and who can coordinate all aspects of the program at the jobsite?

_____ (3) Can feasible engineering controls or work practices eliminate the need for respirators?

_____ (4) Are there written procedures/statements covering the various aspects of the respirator program, including:

_____ designation of an administrator;

_____ respirator selection;

_____ purchase of OSHA/NIOSH-certified equipment;

_____ medical aspects of respirator usage;

_____ issuance of equipment;

_____ fitting;

_____ training;

_____ maintenance, storage, and repair;

_____ inspection;

_____ use under special conditions; and

_____ work area surveillance?

B. Program Operation

(1) Respiratory protective equipment selection

_____ Are work area conditions and worker exposures properly surveyed?

_____ Are respirators selected on the basis of hazards to which the worker is exposed?

_____ Are selections made by individuals knowledgeable of proper selection procedures?

_____ (2) Are only certified respirators purchased and used? Do they provide adequate protection for the specific hazard and concentration of the contaminant?

_____ (3) Has a medical evaluation of the prospective user been made to determine physical and psychological ability to wear the selected respiratory protective equipment?

_____ (4) Where practical, have respirators been issued to the users for their exclusive use, and are there records covering issuance?

(5) Respiratory protective equipment fitting

_____ Are the users given the opportunity to try on several respirators to determine whether the respirator they will subsequently be wearing is the best fitting one?

_____ Is the fit tested at appropriate intervals?

_____ Are those users who require corrective lenses properly fitted?

_____ Are users prohibited from wearing contact lenses when using respirators?

_____ Is the facepiece-to-face seal tested in a test atmosphere?

_____ Are workers prohibited from wearing respirators in contaminated work areas when they have facial hair or other characteristics may cause face seal leakage?

(6) Respirator use in the work area

_____ Are respirators being worn correctly (i.e., head covering over respirator straps)?

_____ Are workers keeping respirators on all the time while in the work area?

(7) Maintenance of respiratory protective equipment

Cleaning and Disinfecting

_____ Are respirators cleaned and disinfected after each use when different people use the same device, or as frequently as necessary for devices issued to individual users?

_____ Are proper methods of cleaning and disinfecting utilized?

Storage

_____ Are respirators stored in a manner so as to protect them from dust, sunlight, heat, excessive cold or moisture, or damaging chemicals?

_____ Are respirators stored properly in a storage facility so as to prevent them from deforming?

_____ Is storage in lockers and toolboxes permitted only if the respirator is in a carrying case or carton?

Inspection

_____ Are respirators inspected before and after each use and during cleaning?

_____ Are qualified individuals/users instructed in inspection techniques?
_____ Is respiratory protective equipment designated as "emergency use" inspected at least monthly (in addition to after each use)?
_____ Are SCBA incorporating breathing gas containers inspected weekly for breathing gas pressure?
_____ Is a record kept of the inspection of "emergency use" respiratory protective equipment?

Repair

_____ Are replacement parts used in repair those of the manufacturer of the respirator?
_____ Are repairs made by manufacturers or manufacturer-trained individuals?

Special Use Conditions

_____ Is a procedure developed for respiratory protective equipment usage in atmospheres immediately dangerous to life or health?
_____ Is a procedure developed for equipment usage for entry into confined spaces?

(8) Training
_____ Are users trained in proper respirator use, cleaning, and inspection?
_____ Are users trained in the basis for selection of respirators?
_____ Are users evaluated, using competency-based evaluation, before and after training?

THOUGHT-PROVOKING QUESTIONS

- What are the two basic types of respirators?
- When is respiratory protection needed?
- What are the inherent problems in not providing training and a comprehensive respiratory protection program?
- Discuss the eleven elements of a respiratory protection program.
- What type of respirator should be used for confined space entry?
- How does the chain of responsibility fall in a respiratory protection program?
- What are the safety and environmental manager's responsibilities for a respiratory protection program?
- What are the supervisor's responsibilities in a respiratory protection program?
- What are the employee's responsibilities in a respiratory protection program?
- Define vapor, fume, smoke, mist, gas, dust, and aerosol.
- Describe three types of respirators.
- How should respirators be stored, cleaned, maintained, and inspected?
- What is the purpose of fit testing? When should it be done?
- What is the use of positive and negative pressure testing? When should it be done?
- What records should be kept for respiratory protection program training and procedures?
- When should air-purifying chemical cartridge respirators be used?

- What do safe work practices for respiratory protection in confined space entry (dangerous atmosphere) entail?
- When and how often should you evaluate your respiratory protection program? Why?
- What is the importance of a respiratory protection program for confined space entry?

REFERENCES AND RECOMMENDED READING

American National Standards Institute, Inc. *American National Standard for Respirator Protection-Respirator Use-Physical Qualifications for Personnel*, ANSI Z88.6. New York: ANSI, Inc., 1984.

Code of Federal Regulations. Occupational Safety and Health Administration, Department of Labor, title 29, sec. 1900-1910.134.

DaRoza, R.A., and W. Weaver. "Is it safe to Wear Contact Lenses with a Full-Facepiece Respirator?" Lawrence Livermore National Laboratory manuscript, UCRL-53653. 1985.

Janpuntich, D.A. "Respiratory Particulate Filtration." *Journal of the International Society for Respiratory Protection* 2, no. 1 (1984): 137–169.

NIOSH Guide to Industrial Respiratory Protection, NIOSH Publication No. 87–116. Cincinnati, OH: National Institute for Safety and Health, 1987.

Spellman, F.R. *Confined Space Entry*. Lancaster, PA: Technomic Publishing Company, 1999.

Chapter 12

Hearing Safety

Prevention of noise-induced hearing loss is the primary and ultimate goal of all occupational hearing conservation efforts. Although this goal is simple to state, it is not easy to achieve. In spite of stringent and vigorous efforts to control potentially hazardous noise exposure among workers, many employees continue to acquire noise-induced losses, losses that should have been prevented. In effect, any noise-induced hearing loss among those included in an occupational hearing conservation program indicates failure of elements that were designed to prevent such occurrences. Why is it that the most vigorous and comprehensive programs are still unable to achieve greater success? The reasons are many. All programs of hearing conservation, the large and the small, contain essentially the same ingredients. Yet, in one setting little noise-induced hearing loss is observed, while in other settings the occurrences are considerable. What are the differences between the successful and the not-so-successful programs? Anyone who has carefully studied programs that are apparently more effective than others will soon recognize the key factors. The elemental differences are usually easy to identify, but the real task consists of using these insights to establish procedures that avoid less effective approaches and that ensure success.

—D.C. Gasaway, 1985

INTRODUCTION

David C. Gasaway, a hearing conservation expert, makes a couple of interesting points in the chapter's opening statement. The first point of significance to us is, "any noise-induced hearing loss among those included in a hearing conservation program indicates failure of elements that were designed to prevent such occurrences." Failure of elements is the important point addressed throughout this chapter (Figure 12.1 shows the elements we are referring to). The second important point is, "Why is it that the most vigorous and comprehensive programs are still unable to achieve success?"

We can explain both these key points by pointing out that even if you have all the elements shown in Figure 12.1 in place, and even if you have the most vigorous and

Figure 12.1　Elements of a Hearing Conservation Program

comprehensive program possible, without proper program management and follow-through, you will achieve less than stellar results with any safety program. The key is leadership and management. Leadership is important simply by the example that it makes or fails to make. When employees see company leadership buying into and observing the elements of the company's safety program, they generally follow this positive leadership example. Unfortunately, the same can be said when the example is not positive.

Follow-through or follow-up is an important factor in any management task. Anyone can lay out a set of directions and say that they must be followed. But giving directions does not guarantee that the directions will be followed. This is where follow-up comes into play. Even a well-written safety program is powerless without leadership, direction, and management. They are empty words unless someone does manage, unless someone takes charge of the assignment from start to finish. In our experience, most safety programs fail because they were not properly managed; the failure we see most often is in follow-through.

In this chapter, we discuss the elements for which the safety and environmental manager is responsible in ensuring that the organization's hearing conservation program is in full compliance with OSHA. The safety and environmental manager who does not understand that safety and health must be managed and that he or she is the key manager of the program may want to seek career employment in some other field. Devising a concept that is designed to protect workers and not implementing and managing it is a wasted effort.

OSHA REQUIREMENTS

In 1983, OSHA adopted a Hearing Conservation Amendment to OSHA 29 CFR 1910.95 that requires employers to implement hearing conservation programs in any work setting where employees are exposed to an eight-hour time-weighted average of 85 dBA and above (LaBar, 1989). Employers are required to designate areas as noise "hazard areas" in settings where the noise level exceeds a time-weighted average of 90 dBA. They are also required to provide personal protective equipment for any employee who shows evidence of hearing loss regardless of the noise level at his or her worksite.

In addition to concerns over noise levels, the OSHA standard also addresses the issue of duration of exposure. G. LaBar, a safety and health expert, explains the duration aspects of the regulation as follows:

> Duration is another key factor in determining the safety of workplace noise. The regulation has a 50 percent 5 dBA logarithmic tradeoff. That is, for every 5-decibel increase in the noise level, the length of exposure must be reduced by 50 percent. For example, at 90 decibels (the sound level of a lawnmower or shop tools), the limit on 'safe' exposure is 8 hours. At 95 dBA, the limit on exposure is 4 hours, and so on. For any sound that is 106 dBA and above—this would include such things as a sandblaster, rock concert, or jet engine—exposure without protection should be less than 1 hour, according to OSHA's rule.

The basic requirements of OSHA's Hearing Conservation Standard are explained here:

- *Monitoring noise levels*. Noise levels should be monitored on a regular basis. Whenever a new process is added, an existing process is altered, or new equipment is purchased, special monitoring should be undertaken immediately.
- *Medical surveillance*. The medical surveillance component of the regulation specifies that employees who will be exposed to high noise levels be tested upon being hired and again at least annually.
- *Noise controls*. The regulation requires that steps be taken to control noise at the source. Noise controls are required in situations where the noise level exceeds 90 dBA. Administrative controls are sufficient until noise levels exceed 100 dBA. Beyond 100 dBA engineering controls must be used.
- *Personal protection*. Personal protective devices are specified as the next level of protection when administrative and engineering controls do not reduce noise hazards to acceptable levels. They are to be used in addition to, rather than instead of, administrative and engineering controls.
- *Education and training*. The regulation requires the provision of education and training to do the following: ensure that employees understand (1) how the ear works, (2) how to interpret the results of audiometric tests, (3) how to select personal protective devices that will protect them against the types of noise hazards to which they will be exposed, and (4) how to properly use personal protective devices (LaBar, 1989).

OCCUPATIONAL NOISE EXPOSURE

Noise is commonly defined as any unwanted sound. Noise literally surrounds us every day, and is with us just about everywhere we go. However, the noise we are concerned with here is that produced by industrial processes. Excessive amounts of noise in the work environment (and outside it) cause many problems for workers, including increased stress levels, interference with communication, disrupted concentration, and most importantly, varying degrees of hearing loss. Exposure to high noise levels also adversely affects job performance and increases accident rates.

One of the major problems with attempting to protect workers' hearing acuity is the tendency of many workers to ignore the dangers of noise. Because hearing loss, like cancer, is insidious, it's easy to ignore. It sort of sneaks up slowly and is not apparent (in many cases) until after the damage is done. Alarmingly, hearing loss from occupational noise exposure has been well documented since the eighteenth century, yet since the advent of the industrial revolution, the number of exposed workers has greatly increased (Mansdorf, 1993). However, today the picture of hearing loss is not as bleak as it has been in the past, as a direct result of OSHA's requirements. Now that noise exposure must be controlled in all industrial environments, that well-written and well-managed hearing conservation programs must be put in place, and that employees are made aware of the dangers of exposure to excessive levels of noise, job-related hearing loss is coming under control.

HEARING CONSERVATION: THE WRITTEN PROGRAM

As with all other industrial safety requirements, the safety and environmental manager must ensure that the specifics of any safety and health requirement be itemized and spelled out in a well-written program. Not only is it an OSHA requirement that the Hearing Conservation Program be in writing, the safety and environmental manager who finds him or herself without a written program soon discovers implementing any program is virtually impossible.

What information and guidelines should be included in the organization's Hearing Conservation Program? This question is best answered by referring to OSHA's 29 CFR 1910.95 Standard, *Occupational Noise Exposure.*

The introduction to the written Hearing Conservation Program should include a purpose statement, one that clearly declares that protection against the effects of noise exposure will be provided when the sound levels exceed those shown in Table 12.1 (when measured on the A scale of a standard sound level meter at slow response).

In addition to stating the purpose of the Hearing Conservation Program, the written program should contain a statement about the Hearing Conservation program itself and define terms pertinent to the written program. For example, a statement declaring that the Hearing Conservation Program is designed to comply with OSHA requirements, and that a continuing, effective Hearing Conservation Program will be administered whenever employee noise exposures equal or exceed an 8-hour time-weighted average sound level (TWA) of 85 decibels measured on the A scale (slow response) or, equivalently, a dose of 50 percent, clearly defines the perimeters of the program. For the purposes of this program, an 8-hour time-weighted average of 85 decibels or a dose of 50 percent will also be referred to as the *action level.* At this point, the written program, along with action level, should list and define the other pertinent terms.

Table 12.1 Permissible Noise Exposures*

Duration per Day, Hours	Sound Level dBA Slow Response
8	90
6	92
4	95
3	97
2	100
1.5	102
1	105
1/2	110
1/4 or less	115

Exposure to impulsive or impact noise should not exceed 140-dB peak sound pressure level.

* When the daily noise exposure is composed of two or more periods of noise exposure of different levels, their combined effect should be considered, rather than the individual effect of each. If the sum of the following fractions $C_1/T_1 + C_2/T_2 + C_n/T_n$ exceeds unity, then, the mixed exposure should be considered to exceed the limit value. C_n indicates the total time of exposure at a specified noise level, and T_n indicates the total time of exposure permitted at that level.
Source: 29 CFR 1910.95, OSHA.

Hearing Safety: Definitions

Attenuate: To reduce the amplitude of sound pressure (noise).

Audible range: The frequency range over which normal ears hear: approximately 20 Hz through 20,000 Hz.

Audiogram: A chart, graph, or table resulting from an audiometric test showing an individual's hearing threshold levels as a function of frequency.

Audiologist: A professional specializing in the study and rehabilitation of hearing, who is certified by the American Speech-Language-Hearing Association or licensed by a state board of examiners.

Background noise: Noise coming from sources other than the particular noise sources being monitored.

Baseline audiogram: The audiogram against which future audiograms are compared.

Criterion sound level: A sound level of 90 decibels.

Decibel (dB): Unit of measurement of sound level.

Double hearing protection: A combination of both ear plug and ear muff type hearing protection devices is required for employees who have demonstrated temporary threshold shift during audiometric examination and for those who have been advised to wear double protection by a medical doctor in work areas that exceed 104 dBA.

Frequency: Rate in which pressure oscillations are produced. Measured in hertz (Hz).

Hearing conservation record: Employee's audiometric record. Includes name, age, job classification, TWA exposure, date of audiogram, and name of audiometric technician. To be retained for duration of employment for OSHA. Kept indefinitely for Workers' Compensation.

Hertz (Hz): Unit of measurement of frequency, numerically equal to cycles per second.

Medical pathology: A disorder or disease. For purposes of this program, a condition or disease affecting the ear, which a physician specialist should treat.

NIOSH: National Institute of Occupational Safety & Health.

Noise dose: The ratio, expressed as a percentage, of (1) the time integral, over a state time or event, of the 0.6 power of the measured SLOW exponential time-averaged, squared A-weighted sound pressure, and (2) the product of the criterion duration (8 hours) and the 0.6 power of the squared sound pressure corresponding to the criterion sound level (90 dB).

Noise dosimeter: An instrument that integrates a function of sound pressure over a period of time to directly indicate a noise dose.

Noise hazard area: Any area where noise levels are equal to or exceed 85 dBA. OSHA requires employers to designate work areas as "noise hazard areas" when work practices exceed 90 dBA. They must post warning signs and warn employees and hearing protection must be worn whenever 90 dBA is reached or exceeded.

Noise hazard work practice: Performing or observing work where 90 dBA is equaled or exceeded. Some work practices will be specified. However, as a "Rule of Thumb," if a normal conversation with someone who is one foot away requires shouting to be heard, one can assume that a 90 dBA noise level or greater exists and hearing protection is required. Typical examples of work practices where hearing protection is required are jackhammering, heavy grinding, heavy equipment operations, and similar activities.

Noise level measurement: Total sound level within an area. Includes workplace measurements indicating the combined sound levels of tool noise (from ventilation systems, cooling compressors, circulation pumps, etc.).

Noise reduction ratio: The number of decibels of sound reduction actually achieved by a particular hearing protection device.

Otoscopic examination: Inspection of external ear canal and tympanic membrane.

Permanent threshold shift (PTS): Hearing loss with less than normal recovery.

Personal protective device: Items such as earplugs or earmuffs used as protection against hazardous noise.

Presbycusis: Hearing loss due to age.

Sensorineural: Type of hearing loss characterized as having been induced by industrial noise exposure. This hearing loss type is permanent.

Temporary threshold shift (TTS): Temporary loss of normal hearing level brought on by brief exposure to high-level sound. TTS is greatest immediately after exposure to excessive noise and progressively diminishes with increasing rest time.

Otolaryngologist: A physician specializing in diagnosis and treatment of disorders of the ear, nose, and throat.

Representative exposure: Measurements of an employee's noise dose or 8-hour time-weighted average sound level that the employers deem to be representative of the exposures of other employees in the workplace.

Sound level: Ten times the common logarithm of the ratio of the square of the measured A-weighted sound pressure to the square of the standard reference pressure of 20 micropascals. Unit: decibels (dB).

Sound level meter: An instrument for the measurement of sound level.

Time-weighted average sound level: That sound level, which if constant over an 8-hour exposure, would result in the same noise dose as is measured.

In the written Hearing Conservation Program, also list or designate who is responsible for managing and enforcing the various components in effecting compliance with the program. Let's take a look a sample Designation of Responsibilities so that you can understand what is required.

HEARING CONSERVATION PROGRAM (SAMPLE)

A. Designation of Responsibilities

The Hearing Conservation Program requires good direction, management, supervision, and conduct at all levels within the company. Assigned duties and responsibilities of company personnel should not be delegated to subordinates.

1. Each company director will be responsible for implementing and ensuring compliance with The Company's Hearing Conservation Program for their department. Each director shall establish the responsibilities for managing this program and for designating the employee classification that will perform the following:
   a. Supervise program within departmental work centers.
   b. Report potential noise hazards to safety and environmental manager for further evaluation.
   c. Provide hearing protection devices as required.
   d. Maintain work center employee training records.
2. The company safety and environmental manager is the Hearing Conservation Program manager. The safety and environmental manager will be responsible for the following:
   a. Writing and modifying, as necessary, the Company's Hearing Conservation Program.
   b. Conducting noise level measurements and maintaining a current and accurate noise level measurement summary of all company workplaces.
   c. Providing training on the Hearing Conservation Program as required, including training in how to wear hearing protection devices.
   d. Ensuring that otoscopic and audiometric examinations of all Company personnel who come under the Hearing Conservation Program are conducted.
   e. Ensuring Hearing Conservation Records are forwarded to Company Human Resources Manager.

   f. Providing work center supervisors an ongoing, current approved listing of hearing protection equipment.

3. The company human resource manager will be responsible for maintaining employee Hearing Conservation Records and will facilitate referral of employees requiring further medical examination by a physician.

4. Assigned supervisors have the following responsibilities under the Company's Hearing Conservation Program:
   a. Reporting to safety and environmental manager any installation or removal of equipment that might alter the noise level within designated work centers
   b. Ensuring that hearing protection devices, as prescribed by safety and environmental manager, are available to employees.
   c. Ensuring hearing protection devices are utilized as required.
   d. Maintaining current work center training records on Hearing Conservation Program training of employees.
   e. Ensuring that all new employees receive a baseline audiogram within six months of hire date.
   f. Ensuring proper hazard labeling practices for all noise hazard areas.
   g. Ensuring that suitable hearing protection is provided to work center visitors.

5. Company personnel will be responsible for:
   a. Familiarizing themselves with the Hearing Conservation Program.
   b. Wearing hearing protection devices as required.

Monitoring: Sound Level Survey

The Hearing Conservation Program begins with noise monitoring and sound-level surveys. Common sense dictates that if a workplace noise hazard is not identified, it will probably be ignored and no attempt at protecting workers' hearing will be made. According to OSHA, when information indicates that any employee's exposure equals or exceeds an 8-hour time-weighted average of 85 decibels, the employer must develop and implement a *monitoring program*. The responsibility for noise monitoring is typically assigned to the organization safety engineer.

    Additional OSHA monitoring procedural requirements include:

1. The noise monitoring protocol that is to be followed, which includes fashioning a sampling strategy designed to (a) identify employees for inclusion in the hearing conservation program and (b) to enable the proper selection of hearing protectors.

2. If circumstances (such as high worker mobility, significant variations in sound level, or a significant component of impulse noise) make area monitoring generally inappropriate, the employer is required to use representative personal sampling to comply with the monitoring requirements, unless the employer can show that area sampling produces equivalent results.

3. All continuous intermittent and impulsive sound levels from 80 decibels to 130 decibels must be integrated into the noise measurements.

4. Instruments used to measure employee noise exposure must be calibrated to ensure measurement accuracy.

5. Monitoring must be repeated whenever a change in production, process, equipment, or controls increases noise exposures to the extent that:
    a. additional employees may be exposed at or above the action level; or
    b. the attenuation provided by hearing protectors being used by employees might be rendered inadequate.
6. The employer is required to notify each employee exposed at or above an 8-hour time-weighted average of 85 decibels of the results of the monitoring.
7. The employer is required to provide affected employees or their representatives with an opportunity to observe any noise measurements conducted.

Audiometric Testing

Audiometric testing is an important element of the Hearing Conservation Program for two reasons: It helps to determine the effectiveness of hearing protection and administrative and/or engineering controls and it helps to detect hearing loss before it noticeably affects the employee and before the loss becomes legally compensable under workers' compensation. Audiometric examinations are usually done by an outside contractor but can be done in-house with the proper equipment. Wherever they are done, they require properly calibrated equipment used by a trained and certified audiometric technician.

The importance of audiometric evaluations cannot be overstated. Not only do they satisfy the regulatory requirement, but they also work to tie the whole program together. One thing is certain: if the Hearing Conservation Program is working, employees' audiometric results will not show changes associated with on-the-job noise-induced hearing damage. If suspicious hearing changes are found, the audiometric technician and the audiologist who reviews the record can counsel the employee to wear hearing protection devices more carefully, can assess whether better hearing protection devices are needed, and can use the test results to point out to the employee the need to be more careful in protecting his or her hearing both on and off the job.

The organizational safety and environmental manager needs to ensure that designation of audiometric evaluation procedures is included in the written Hearing Conservation Program. A sample written procedure is included below.

Designation of Audiometric Evaluation Procedures

The Hearing Conservation Program requires audiometric evaluation for all company employees who come under this program.

1. The safety and environmental manager will ensure that otoscopic and audiometric examinations are conducted.
2. A baseline audiogram will be required of all employees within six months of hire date.
    a. Testing will be conducted after 14 hours of nonexposure to workplace noise.
    b. Testing will be performed by a certified audiometric technician, using a calibrated (annual requirement) audiometer in an environment of less than 50 dBA background noise.

3. Follow-up audiograms are required yearly or within 60 days of temporary or permanent threshold shift.
4. If follow-up audiograms suggest permanent threshold shift rather than temporary threshold shift, employee will be referred to a physician for evaluation as to whether damage is Presbycusis or Sensorineural.
5. All cases of occupationally related hearing loss must be recorded on the OSHA 200 form.
6. Safety and environmental manager will review results of temporary or permanent threshold shifts to determine if cause was work related or other than work related.
7. Safety and environmental manager will forward the employee Hearing Conservation Record to the human resource manager for inclusion in the employee's medical record.

Hearing Protection

The *hearing protection* element of the Hearing Conservation Program provides hearing protection devices for employees and training on how to wear them effectively, as long as hazardous noise levels exist in the workplace. Hearing protection comes in various sizes, shapes, and materials, and the cost of this equipment can vary dramatically. Two general types of hearing protection are used widely in industry: the cup muff (commonly called Mickey Mouse Ears) and the plug insert type. Because feasible engineering noise controls have not been developed for many types of industrial equipment, hearing protection devices are the best option for preventing noise-induced hearing loss in these situations.

As with the other elements of the Hearing Conservation Program, the hearing protective device element must be in writing and included in the Hearing Conservation Program. A sample designation of hearing protection devices is presented below.

Designation of Hearing Protection Devices

The Hearing Conservation Program requires providing hearing protection devices for employees in designated noise hazard areas.

1. After audiometric examination, the safety and environmental manager will fit each employee with the proper hearing protection devices. Two types of hearing protection devices will be made available to the employees: Earmuff and/or plug types. If the employee has demonstrated a temporary threshold shift (TTS), the safety and environmental manager will issue double hearing protection for those employees who work in areas that exceed 104 dBA. A complete listing of each employee's hearing protection requirements will be provided in writing to the work center supervisor.
2. The work center supervisor is responsible for ensuring that employees wear approved hearing protection devices in designated areas.

Training

For a Hearing Conservation Program (or any other safety program) to be effective, the participants in the program must be *trained*. OSHA requires that the employer include this important element in the written Hearing Conservation Program. The training program must be repeated annually for each employee included in the Hearing Conservation Program. The safety and environmental manager needs to ensure that the information included in the training program is current and informs the employees of the effects of noise on hearing, the purpose of hearing protectors, the advantages, disadvantages, and attenuation of various types, and instructions on selection, fitting, use, and care. The purpose of audiometric testing and an explanation of the test procedures must also be included.

To facilitate compliance with all regulatory standards and the company's safety and health requirements (including the Hearing Conservation Program), organizational management and the safety and environmental manager should ensure that emphasis for compliance is made a condition of employment. Remember: ensure and document employee participation.

Safe Work Practices

Safe work practices are an important element in the Hearing Conservation Program. Written safe work practices for hearing conservation should focus on relaying noise hazard information to the employee. For instance, if an employee is required to perform some kind of maintenance function in a high noise hazard area, the written procedure for doing the maintenance should include a statement that warns the employee about the noise hazard and lists the personal protective devices that he or she should use to protect themselves from the noise. Our experience has shown that when such warnings (safe work practices) are placed in preventive maintenance procedures (e.g., noise hazard area, confined space, lockout/tagout required), not only is the program much more efficient, the repeated reminder also helps workers to maintain compliance with regulatory standards.

Recordkeeping

Under OSHA's 29 CFR 1910.95 (Hearing Conservation Standard), the employer is required to keep and maintain certain records. Along with an accurate record of all employee exposure measurements, the employer is required to retain all employee audiometric test records. Audiometric test records must include:

- Name and job classification of the employee;
- Date of the audiogram;
- The examiner's name;
- Date of the last acoustic or exhaustive calibration of the audiometer; and
- Employee's most recent noise exposure assessment.

The employer must maintain accurate records of the measurements of the background sound pressure levels in audiometric test rooms.

The employer is required to retain records of noise exposure measurement for two years. Audiometric test records must be retained for the duration of the affected employee's employment. Employee noise exposure records must be made available to employees whenever they request them. Note that whenever an employee is transferred, the employer is required to transfer the records to the employee's successor employer.

Administrative and Engineering Controls

Two more important element that must be included in any Hearing Conservation Program are *administrative and engineering controls.* Administrative controls, simply stated, involve controlling the employee's exposure to noise. If a certain work area has a noise source that exceeds safe exposure levels, the employee is allowed within such a space only up to the time in which he or she has reached their maximum allowed time-weighted exposure limit. For example, if the noise hazard area consistently produces noise at the 100-dBA level, the employee would only be allowed in such an area up to 2 hours per 8-hour shift. NOTE: A word of caution is advised here: keep in mind that we are referring to an employee who has no recorded hearing loss. If an employee has suffered permanent hearing loss, then his or her time exposure at such high noise levels should be significantly reduced. Under no circumstance should the employee with documented hearing loss be exposed to high noise hazards without proper hearing protection.

We have said all along that the preferred hazard control method is the employment of engineering controls to "engineer out" the hazard. In hearing conservation, engineering controls play a vital role in providing the level of protection employees need. Not only should existing equipment be evaluated for possible engineering control applications, new equipment should be evaluated for noise emissions before purchase.

Engineering controls used in controlling hazardous noise levels can be accomplished at the source of the noise through preventive maintenance, speed reduction, vibration isolation, mufflers, enclosures, and substitution of machines. In the air path, engineering controls such as absorptive material, sound barriers, and increasing the distance between the source and the receiver can be employed. At the receiver, the best engineering control is to enclose and isolate the employee from the noise hazard.

THOUGHT-PROVOKING QUESTIONS

- What does the success of this (or any other safety program) depend on?
- Define noise.
- What can the safety engineer and management do to ensure workers follow directives and safe work practices?
- Why is hearing loss so often ignored?
- Define action level.
- Define noise dose.
- How do you determine a noise hazard area?
- What is the rule of thumb for a needed Noise Hazard Safe Work Practice?
- How do you take a noise level measurement?
- How do you determine representative exposure levels?
- Define decibels. How are they measured?

- What is the company director responsible for in implementing a hearing conservation program?
- What are the safety and environmental manager's Hearing Conservation Program responsibilities?
- What do supervision responsibilities involve for Hearing Conservation Programs?
- What do the human resource manager's responsibilities include?
- Why is monitoring important?
- What does a sound survey level entail?
- How do you determine a time-weighted average?
- Define audiometric surveillance.
- Why are audiometric evaluations important?
- How should audiometric evaluations be conducted?
- What are the common types of PPE for hearing protection?
- What should hearing protection training include?
- What safe work practices should be developed for hearing conservation?
- What should audiometric test records include?
- How do administrative controls work to eliminate noise hazards?
- How can engineering controls eliminate noise hazards?

REFERENCES AND RECOMMENDED READING

Code of Federal Regulations. Occupational noise exposure, title 29, sec. 1910.95.

Gasaway, D.C. *Hearing Conservation: A Practical Manual and Guide.* Englewood Cliffs, NJ: Prentice-Hall, 1985.

LaBar, G. "Sound Policies for Protecting Workers' Hearing." *Occupational Hazards* July 1989: 46.

Royster, J.D., and L.H. Royster. *Hearing Conservation Programs: Practical Guidelines for Success.* Chelsea, MI: Lewis Publishers, 1990.

Chapter 13

Electrical Safety

The common use of electricity and electrical equipment and appliances has resulted in failure of most persons to appreciate the hazards involved. These hazards can be divided into five principal categories: (1) shock to personnel, (2) ignition of combustible (or explosive) materials; (3) overheating and damage to equipment; (4) electrical explosions, and (5) inadvertent activation of equipment.

—W. Hammer, 1989

INTRODUCTION

If you were to take a look at the annual on-the-job injury statistics for all employers in the United States, you would quickly notice that many of these injuries are typically the result of electrical shock, injuries received during electrical fires, and/or injuries received when some electrical component failed due to faulty installation, faulty maintenance conducted on electrical equipment, or equipment malfunction caused by manufacturers' errors.

While normally true that most workers fear electricity and its power, or at least have a healthy respect for electricity, it is also true that on-the-job electrocutions do occur and that the number one cause of fire in the workplace is from electrical causes.

For the organization safety and environmental manager, electrical safety in the workplace is not only an important priority, but also requires constant vigilance on his or her part, and on the part of all supervisors and workers, to ensure that safe work practices are followed when working with or around electrical circuits and components. The company employees must maintain the integrity of all electrical equipment and systems that require constant vigilance. This includes an organization standing order that any discovered electrical discrepancy is to be reported to responsible parties immediately.

Another important element in any electrical safety program is employee awareness. This is accomplished through training and written safe work practices and policies (see Figure 13.1). Employees should be routinely trained on the hazards of electricity and on what to look for and what to do if electrical discrepancies are discovered. Safe

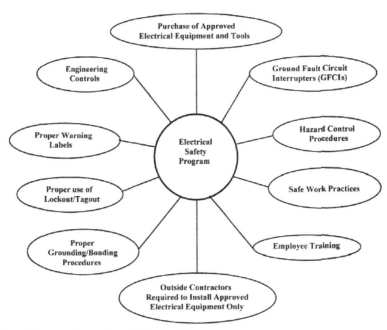

Figure 13.1   Elements in an Electrical Safety Program

work practices are required for those employees required to work with or on electrical circuits and components. The safety and environmental manager must include a close look at all electrical installations during his or her organizational safety inspection (audit).

The facility safety and environmental manager must also insist that outside contractors hired to install new equipment, renovations, and upgrades accomplish their construction projects in accordance with OSHA, National Electrical Code (NEC), and all local code requirements. The safety and environmental manager must also ensure that any planned electrical equipment is suitable for installation in the proposed installation areas. For example, if a new electrical motor and controller is to be installed in an area that contains explosive vapors, the proper class of electrical motor and control equipment (see Table 13.1) must be installed in such a space to prevent the possibility of explosion, based on NFPA recommendations.

OSHA's standards relating to electricity are found in 29 CFR 1910 (Subpart S). They are extracted from the National Electrical Code (NEC). Subpart S is divided into the following two categories of standards: (1) Design of Electrical Systems; and (2) Safety-Related Work Practices. The standards in each of these categories are as follows:

Design of Electrical Systems

1910.302 Electric utilization systems
1910.303 General requirements

Table 13.1 Classification of Areas for Electric Installations

Class	Presence of
I	Flammable gases or vapors
II	Combustible dust
III	Ignitable fibers or filings

Group	Atmosphere containing
A	Acetylene
B	Butadiene, ethylene oxide, propylene oxide, acrolein, or hydrogen
C	Cyclopropane, ethyl ether, or ethylene
D	Acetone, alcohol, ammonia, benzene, benzol, butane, gasoline, Hexane, lacquer solvent vapors, naphtha, natural gas, propane
E	Combustible metal dusts having resistivity of less than 105 ohms-cm
G	Combustible dusts having resistivity of 105 ohm-cm or greater

Division
1. Flammable atmosphere under normal conditions
2. Flammable atmosphere only under abnormal conditions

CONTROL OF ELECTRICAL HAZARDS

We have stated consistently throughout this text that when the object is to control hazards, the goal should first be to engineer out any hazard whenever possible. This (of course) is also the case with electrical hazards. For example, a company policy that insists that only intrinsically safe electrical equipment and tools (i.e., double- and triple-insulated handtools) will be purchased and used within the organization is a type of engineering control. Another type of electrical engineering control is the installation of low-voltage systems. Other types of controls can reduce or eliminate electrical hazards, including switching devices, grounding and bonding, ground fault circuit interrupters and procedures, and lessening the hazardous effects of static electricity.

The facility safety and environmental manager must be fully aware of the hazards of electricity, electrical circuits and components and must also be familiar with the common means of electrical hazard controls. This includes knowledge of applicable codes, regulations, and standards that provide detailed specifications and procedures for safeguarding electrical equipment and systems.

Because safety and environmental managers need to have some knowledge of electricity, electrical equipment and systems, and electrical hazard control methodologies, they need to also have some basic understanding of electricity itself, its uses, and the potential hazards it presents to all who might come into contact with it. In hazard control, the facility safety engineer must have fundamental knowledge of the electrical materials used, design of components, and placement of electrical equipment. An understanding of shielding methods and the enclosing and positioning of electrical devices can reduce contact by employees.

Note: This is not to say that the safety and environmental manager must be an electrical engineer. Instead, we recommend some training in the fundamentals of electricity.

This training should be included in the safety and environmental manager's formal college or advanced short-school training.

The minimum electrical system and component operation knowledge we recommend the safety and environmental manager have include the following electrical hazard controls:

(1) The facility safety and environmental manager should understand that *overcurrent devices,* which limit the current that can flow through a circuit or electrical device, should be included in any electrical system design. Such a device cuts off power if current exceeds a given limit. The two most common overcurrent devices in use at present are fuses and circuit breakers.

Fuses are composed of materials (usually lead or a lead alloy) that are designed to limit the current flow in the circuit. When current in the circuit exceeds some limiting value, the lead or lead alloy material heats above its melting point and separates, opening the circuit, thereby stopping the flow of current. Safety and environmental managers must understand that fuses are rated at certain design levels. In other words, not every fuse is suited for every electrical circuit. In fact, the danger with fused circuits is when the fuses are replaced with fuses that are too large for the circuit they are designed to protect. When this occurs, so does the danger that too much current will be allowed to flow in a circuit not designed to handle the high level of current flow, which could lead to electrical fires and other problems.

Circuit breakers are actually a form of switch designed to open when current passing through them exceeds a designed limit. Circuit breakers are designed to limit current flow in two different ways. One type is designed to open when the temperature of the breaker reaches a predetermined level. A common problem with this type of breaker is that the temperature of the environment around it can affect its operation. The second type is magnetic, and it opens when a predetermined current level is reached. The advantage of this type breaker is that environmental conditions have little impact on its operation.

(2) In addition to overcurrent devices, certain switching devices can reduce or eliminate electrical hazards. These include interlocks, lockouts, and thermal or overspeed switches.

Interlocks are switches that prevent access to an energized or dangerous location. Often attached to access doors, panels, and gates, interlocks act to shut off power to the equipment whenever these devices are opened. Probably the most commonly used and most familiar interlock device is the one installed in most washing machine lids, which shuts down the machine when the lid is opened. See Chapter 8 for more about lockout/tagout procedures.

You should recall that a lockout procedure involves placing a lock on a switch, circuit breaker, or other device to prevent the switch, circuit breaker, or equipment from being turned on or energized.

Thermal and overspeed cutout devices are commonly used to protect electrical equipment (and thus the operator). A thermal cutout is simply a temperature sensitive switch with a preset limit designed to interrupt power when the temperature exceeds a certain value. As its name implies, an overspeed switch operates when it senses that a motor or other device operates too fast. Excessive speed may create dangerous conditions and/or indicate failure of equipment. The overspeed switch operates to shut down an overspeeding device by interrupting power to it.

(3) *Grounding* and *bonding* control the electrical potential between two bodies. If there is a difference of potential between two bodies, a conductor between them will allow charge or current to flow. That flow may be dangerous, particularly as a source of ignition.

R. H. Lee in *Electrical Grounding: Safe or Hazardous* (1969) and W. Hammer in *Occupational Safety Management and Engineering* (1989) provide important information on grounding and bonding.

Note: The information provided in the following assumes that the reader has some fundamental knowledge of electrical terms and their meaning.

The earth acts as an infinite store from which electrons (current flow) can be drawn, or to which they can return. Providing a path from where it exists to earth can eliminate any undesirable excess or deficiency. Gaining electrons can then neutralize positive ions in a system; electrons can be conducted to earth (called "earthing" in some countries). In the United States the term "grounding" is preferred, and the path to earth or the earth itself is a "ground." In some instances (such as in electronic equipment), a massive metallic body acts as the reservoir of electrons and ions (the ground) in place of the earth.

Grounds can be designed and installed into a system or they can be accidental. Unless noted otherwise, the word "ground" used here indicates one of design. Installed grounds are basically safety mechanisms to prevent (1) overloading of circuits and equipment which would destroy them or shorten their lives, (2) shock to personnel, and (3) arcing or sparking that might act as an ignition source.

Grounds may protect a system, equipment, or personnel. Certain designs used on high-voltage transmission lines are sophisticated types that follow the standards set by the American Institute of Electrical Engineers or other codes. The ground systems and standards of the National Electrical Code (NEC), which apply to buildings and related facilities, are more common.

Safety and environmental managers should know several terms used in the NEC that are related to grounding and bonding. System ground refers to an electric circuit and is designed to protect conductors (wires/wiring) for a transmission, distribution, or wiring system.

The term voltage to ground is often used in electrical codes. It indicates the maximum voltage in a grounded circuit measured between the ground wire and a wire that is not grounded. Where a ground is not used, voltage to ground indicates the maximum voltage between any two wires. The wire that connects the circuit to earth is the grounding wire or ground; the wire to which it is connected is the grounded wire.

Probably the simplest way in which to illustrate the principles of grounding is to use a typical three-wire system as an example. In a three-wire system, current generally flows along two wires—the third is neutral. In distribution systems for building and related facilities, the neutral wire is always the one grounded when grounding is installed. High-voltage transmission lines sometimes ground all three wires, but this is less common. The types of grounding systems that have been used on transmission lines include:

1. Solid grounds: The neutral wire is grounded without any impedance, which might restrict current flow.
2. Resistance grounds: The neutral wire is connected to ground through a high resistance at a transformer.
3. Reactance grounds: The neutral wire is connected to ground through impedance which is principally reactance.
4. Capacitance grounds: Each line of a circuit is connected to a capacitor; the other side of each capacitor is grounded.
5. Resonant grounds: This is a tuned, parallel system that uses capacitance grounds and a ground from a transformer neutral through an induction coil.

Solid grounds are the most commonly used, especially in interior electrical systems of buildings. Resistance and capacitance grounds are designed into most electronic equipment. These types of grounds involve circuitry comparable to two-wire systems in which it is necessary to maintain potentials within prescribed limits.

One purpose of grounding the neutral in a three-wire system is to activate overcurrent protection devices before damage is done when a fault occurs. Should one of the two wires that normally carries current be broken or accidentally grounded, current will flow through the neutral, through the installed ground, and back to the power source. This short circuit will open the protection devices and de-energize the affected portion of the system.

Where the neutral is not grounded, accidental grounding of one of the other wires will cause an increase in voltage to ground of the remaining system. The definitions of voltage to ground for grounded and ungrounded systems will illustrate this point. According to these, a 220-volt three-wire grounded neutral system will have a voltage between any two wires. The excessively high voltages may cause burnout of equipment, burning or breakdown of insulation, arcing and sparking, and shock to personnel in contact with metal energized through the breaks.

Other possibilities exist by which an excessively high voltage can be produced, which would create similar hazards if the system is not grounded. A fault in a stepdown transformer could result in the distribution system potential, or part of it greater than normal, being applied to a building wiring system. An accidental connection between the two systems would produce the same result. Where grounds existed, the overcurrent protection devices would de-energize and safeguard the system.

Equipment grounds: may be used on the metal parts of a wiring system, such as the conduit, armor, switch boxes, and connected apparatus other than the wire, cable, or other circuit components. They may also be provided for equipment such as metal tables and cabinets that might come in contact with an energized circuit or source of electrical charges. Equipment on which undesirable charges may be induced or generated should also be grounded.

Metal of electrical equipment may come in contact with an energized circuit whose insulation is deteriorated or cut, or through which arcing can take place. A person may then touch the metal surface inadvertently, receiving a shock. The degree of shock would depend on whether the equipment was grounded. If it was not, the person in contact with the metal would act as a ground, the current passing through his or her body. If the equipment was grounded, the person might or might not receive a shock at all. If current did pass through this body, the amount would be inversely proportional to the resistance of his body compared to that of the equipment ground. If the resistance of his body were high enough, no current would pass.

Bonding: ensures that all major parts of a piece of equipment are linked to provide a continuous path to ground. A bond is a mechanical connection, which provides a low-resistance path to current flow between two surfaces that are physically separated or may become separated. A bond can be permanent, such as one in which the connection is welded or brazed to the two surfaces, or it may be semi-permanent, bolted or clamped where required.

Where permanent types are used, the parts themselves can be joined and narrow gaps filled with weld or brazing metal. Where separation is wider, a strip of metal can be welded or brazed at both ends across the gap. Bonds connecting one vibrating part to another part that may or may not vibrate should be of a flexible material that will not fail under vibration. Corrosion because of the joining of dissimilar metals may cause the electrical resistance across the bond to increase. This is especially noticeable in humid or corrosive atmospheres. The types of metal for the bond and its fastenings must therefore be selected with care.

Grounding and bonding requirements
Grounds and bonds should:

- be permanent wherever possible
- have ample capacity to conduct any possible current flow (Note: a ground should not normally be designed to be part of a current-carrying circuit.)
- have as low impedance as possible
- be continuous, and wherever possible, be made directly to the basic structure rather than through other bonded parts
- be secured so that vibration, expansion, contraction, or other movement will not break the connection or loosen so that the resistance varies
- have connections located in protected areas and where accessible for inspection or replacement
- not impede movement of movable components
- not be compression-fastened through nonmetallic materials
- not have dissimilar metals in contact
- have metals selected to minimize corrosion

Grounding is not always advantageous in all cases; some electrical systems are safer ungrounded. R. H. Lee has pointed out:

Some electrical systems (necessarily of limited extent), must be left ungrounded for safety reasons. For example, the electrical system of a hospital operating room is purposely ungrounded because a spark from an insulation failure would otherwise ignite the anesthesia-permeated atmosphere. When ungrounded, an insulation failure "to ground" produces no current low and hence no spark, no ignition, and no explosion. Electric blasting caps present a similar condition; a short-circuit current returning through the earth could fire the caps if their two connecting wires touched the earth more than a few inches apart (p. 162).

(4) *Ground Fault Circuit Interrupters* (GFCIs) are designed to open the circuit before a fault path through the operator can cause harm, at levels as low as five milliamps (5 Ma). A GFCI compares current normally flowing through the power distribution wire and the grounded neutral wire of a circuit. The current flowing through one must pass through the other for the circuit to work. If current is not equal, some electrical energy is flowing to the ground through other than the normal route, perhaps through a person. When the current is not equal, the GFCI detects this current differential and shuts off the current.

Though GFCIs protect normal 115 volt circuits where users can form a ground with energized equipment, they do not work on line-to-line connections found in distribution of 220 volt and higher. GFCIs are required by the NEC for outdoor receptacles or circuits and for bathrooms and other locations.

(5) *Static electricity* is a workplace hazard because of its potential to ignite (by arc) certain vapor or dust mixtures in air. Various controls are available in minimizing the effects of static charges, dependent on the individual case.

1. Selection of suitable materials (i.e., avoiding the use of materials such as clothing composed of synthetic fabrics that generate static electricity) is often the simplest method.
2. Modifying a material by spraying its surface to make it conductive can frequently reduce or eliminate the static electricity problem.
3. Bonding and grounding can be utilized to provide a path by which various surfaces on which charges could accumulate can be neutralized.
4. Electrostatic neutralizers can be used to neutralize charges on materials.
5. Humidification (raising the relative humidity above 65%) permits static charges to leak off and dissipate.

THOUGHT-PROVOKING QUESTIONS

- Why does electrical safety take special attention?
- How should electrical discrepancies be handled?
- What are the classes of electrical motors and control equipment, and why is classification important?
- What are three ways to use engineering controls to reduce or eliminate electrical hazards?
- As safety and environmental manager, to what degree should you understand electrical systems and component operation?
- What's the difference between a circuit breaker and a fuse?

- What are common switching devices?
- Define grounding.
- Define bonding.
- What is meant by "voltage to ground" and why is it important?
- Name five types of grounds.
- What are the consequences of improper grounding?
- What are grounding and bonding requirements?
- When shouldn't an electrical system be grounded?
- What's GFCI, and when and where should it be used?
- What hazards are associated with static electricity? How can you reduce or eliminate them?

REFERENCES AND RECOMMENDED READING

American National Standards Institute, National Fire Protection Association Standards:

70 *National Electrical Code*
76A *Essential Electrical Systems for Health Care Facilities*
76C *High-Frequency Electricity in Health Care Facilities*
77 *Static Electricity*

ANSI/AAMI ES1, *Safe Current Limits for Electromechanical Apparatus*
ANSI Z244.1, *Minimum Safety Requirements for A Lockout/Tagout of Energy Sources*
ANSI/UL 817, *Cord Sets and Power-Supply Cords*
ANSI/UL 859, *Electrical Personal Grooming Appliances*
Fordham-Cooper, W. *Electrical Safety Engineering*, 2nd ed. London: Butterworths, 1986.
Hammer, W. *Occupational Safety Management and Engineering*. Englewood Cliffs, NJ: Prentice-Hall, Inc., 1989.
Hermack, F.L. *Static Electricity in Fibrous Materials*. National Bureau of Standards Report 4455, December 1955.
Lee, R.H. "Electrical Grounding: Safe or Hazardous?" *Chemical Engineering* July 1969.

Chapter 14

Ergonomics

Private sector employers spend about $60 billion each year on workers' compensation claims associated with musculoskeletal disorders, which involve illnesses and injuries linked to repetitive stress or sustained exertion on the body. The Occupational Safety and Health Administration (OSHA) has tried to develop a workplace standard that would require employers to reduce ergonomic hazards in the workplace. A draft of the standard that OSHA circulated for comment in 1995 generated stiff opposition from many employers because they believed that it required an unreasonable level of effort to address ergonomic issues. Since then, Congress has limited OSHA's ability to issue a proposed or final ergonomic standard. GAO (General Accounting Office) found that employers can reduce the costs and injuries associated with ergonomic hazards, thereby improving employees' health and morale as well as productivity and product quality, through simple, flexible approaches that are neither costly nor complicated. Effective ergonomics programs share certain core elements: Management commitment, employee involvement, identification of problem jobs, development of solutions, training and education of employees, and appropriate medical management. OSHA may wish to consider a framework for a worksite ergonomics program that gives employers the flexibility to introduce site-specific efforts and the discretion to determine the appropriate level of effort to make, as long as the effort effectively addresses the hazards.

—HEHS-97-163, 1997

INTRODUCTION

What is ergonomics? We state this question right up front (and subsequently answer it) because we have found that few people understand the meaning of the word, and even fewer understand what ergonomics is all about. Until recently, the term *ergonomics* was used primarily in Europe and the rest of the world to describe *human factors engineering* (a synonymous term most commonly used in the United States).

At present, common practice in the United States and in general safety practice now uses the term ergonomics. Thus we use the term ergonomics as the appropriate and accepted substitute for human factors engineering, human engineering, and engineering psychology in this chapter.

So, again, what is ergonomics? Let's break the word down a bit and see if the Greek it is derived from (*ergon* and *nomos*) will help us in defining the term. *Ergon* means work, and *nomos* means law. Thus, ergonomics means the laws of work? What does that mean? Let's further define ergonomics by pointing out that it relates to the interface between people and a variety of elements: equipment, environments, facilities, vehicles, printed materials, and so forth (Brauer, 2005). Grimaldi & Simonds (1989) define ergonomics as "the measurement of work" (p. 512). Ergonomics could be defined as how human physical considerations affect work.

So, we end up with ergonomics meaning the laws of work, the interface between people and a variety of elements, and/or the measurement of work. However, the definition does not seem complete. To solve this problem, let's sum it all up by using the best definition we have been able to find to date. This term is actually derived, though slightly modified for our purposes, from Chapanis (1985): Ergonomics discovers and applies information about human behavior, abilities, limitations, and other characteristics to the design of tools, machines, systems, tasks, jobs, and environments for productive, safe, comfortable, and effective human use.

Chapanis' definition for ergonomics seems logical (and it is the definition we use to describe ergonomics in this text), but what is the goal of ergonomics? Stated simply, the goal of ergonomics is to protect the worker, to minimize worker error, and to maximize worker efficiency.

To have an effective Ergonomics Program, certain elements must be included:

- Hazard identification
- Program evaluation
- Training
- Medical Management
- Management commitment and employee participation
- Hazard prevention and control

However, OSHA has had trouble convincing Congress to institute ergonomics legislation. Consider the following from *Safety Compliance Alert*, June 18, 1998:

OSHA spelled out the new proposed ergo rule

> Soon everyone will need an ergonomics program. You don't have to guess any longer what OSHA's ergonomics rule will mean to you.

> Soon after October—when a congressional restriction on issuing the rule runs out—OSHA will release the formal draft. And OSHA officials made it clear at the recent American Industrial Hygiene Conference & Exposition that the agency won't back down on a few sticking points (p. 1).

OSHA was moving forward with its Ergonomics Standard, which would have been promulgated as early as late 1998.

However, when George W. Bush became president in 2001, the new Ergonomics Standard was put on hold for further review. Thus, at the present time, OSHA's ergonomic standard is not applicable as an enforcement tool. However, it should be pointed out that under OSHA's General Duty Clause, it has the authority to ensure that affected workplaces comply with many of the original tenets of the federal standard.

Was it a good decision to put the original Ergonomics Standard on hold? It depends on your point of view. For example, according to the United Steelworkers of America (USWA, 2001), OSHA's Ergonomics Standard was assassinated. After a decade of struggle, workers in this country finally won protections to prevent crippling repetitive strain injuries, the nation's biggest job safety problem. OSHA's ergonomics standard was issued in November 2000 and went into effect in January 2001. This important worker safeguard would have prevented hundreds of thousands of injuries a year by requiring employers to implement ergonomics programs and fix jobs where musculoskeletal disorders (MSDs) occur. The standard would have provided worker protection in a large majority of USWA workplaces in the United States.

But it was not to be. On March 7 the U.S. House of Representatives voted 223 to 206 to kill OSHA's ergonomics standard. The Senate voted the previous evening 56 to 44 to repeal the standard. The 10 years of work intended to protect workers was debated in the Senate for 10 hours, while the House gave the issue only one hour of consideration. President George W. Bush signed the Resolution of Disapproval on March 20. This was the final move to kill the OSHA ergonomics standard.

A differing view is provided by the National Coalition on Ergonomics (NEC, 2003). It says the requirements of the rule are subjective and ambiguous, assuring that an employer could never achieve compliance because, in its more than 1,000 pages, not a single proven solution is provided. Even the best-intentioned employer will be unable to understand what this proposal requires, no matter how many lawyers or experts he or she hires.

Although the authors support the need for regulated and standardized ergonomics guidelines, we agree with the NEC that the original proposal, passed and later rejected, needed to be fixed. Based on our experience, the original proposal was not doable. It was simply unmanageable, for a variety of reasons.

In this chapter, we discuss each of the elements of the ergonomics proposal that are "doable."

ELEMENTS OF ERGONOMICS PROGRAM

Hazard Identification

The first and most obvious step in devising an organizational Ergonomics Program is to conduct a worksite hazard analysis to identify all hazards. The best way to conduct a hazard identification procedure is to use the checklist approach.

Hazard Prevention and Control

Once the worksite has been thoroughly surveyed and the hazards identified, hazard prevention and control measures need to be put in place. As with all hazards, the safety and environmental manager either attempts to eliminate the hazard or to engineer it out. When this is not possible, other prevention and control measures must be adopted. For example, when evaluating a particular workstation, the safety and environmental manager may determine that an employee who operates a video display terminal (VDT) complains about persistent neck pain. When the safety and environmental manager evaluates that particular VDT station, he or she discovers that the VDT monitor is not adjusted to the proper height. If this is the case, then the remedial action (the prevention and control phase) is easily put into place by adjusting the monitor correctly to fit the individual.

Management Commitment & Employee Participation

We have stated throughout this text that without proper management commitment and employee participation, any effort to include safety in any workplace is an empty effort. You must have top management support and include employee participation in the organization's safety and environmental health program. The same holds true for the incorporation of an ergonomics program into the organization.

Medical Management

When an employee who operates a VDT complains that he or she begins to notice a loss of feeling in his or her hands, then a tingling in his or her arms, and complains that at night he or she can't sleep because of a burning sensation in his or her wrists, this may be symptomatic of *repetitive strain injury* (RSI). The question is, "What is the employer to do about it? What is the employer "required" to do about it?" The employer and the safety and environmental manager are going to be confronted with many such medical questions.

The question remains, "What is to be done about it?" At the present, this question is difficult to answer. Many insurance companies do not recognize RSI as a compensable injury. Even if the insurance company does recognize RSI as compensable, in some states (e.g., Virginia), RSI injuries are not recognized as compensable injuries under workers' compensation laws.

Program Evaluation

Any active safety and environmental health program must periodically undergo evaluation. This should be accomplished not only to verify the effectiveness of the program and remedial actions taken, but also to verify the currency of the program. The effectiveness of any safety and environmental management program can be measured by the results. For example, suppose an employee operates a VDT all day and each day during the workweek complains about wrist pain. An ergonomic evaluation indicates that the employee's arms and wrists are not properly supported. The remediation was to adjust the employee's position by whatever means determined. Continued employee complaints would indicate, obviously, that the remedial actions taken were not correct. A different approach might need to be taken. The results in this particular case illustrate the importance of evaluation.

The program should also be evaluated to ensure that it is current with applicable regulatory requirements. Regulations are often dynamic (changing constantly). The only way to ensure compliance with the latest requirements is to evaluate the program regularly.

Training

To aid in reducing ergonomically related hazards in the workplace, employee participation is critical. Employee participation is normally increased whenever employees are properly trained on both program requirements and on those elements that make up the program. As with almost all safety and health provisions, training is an essential, required ingredient. Employees need to be aware of the organization's efforts, not only to reduce, eliminate, evaluate, and control ergonomic hazards, but also to be aware of the types of workplace situations and practices that lead to ergonomic problems. Your training program should enable the employee to answer the question, "What do I do if I experience eyestrain, wrist pain, back pain, neck pain, and other discomforts from the performance of my day-to-day work activities?" Your training program should enable each employee to know how to go about (and to whom) to report suspected ergonomic problems in the workplace.

THOUGHT-PROVOKING QUESTIONS

- Define ergonomics.
- What is a medical surveillance program, and what function does it serve?
- How should safety and environmental managers monitor and track results of their ergonomics programs?
- What questions should your training enable workers to answer?

REFERENCES AND RECOMMENDED READING

Brauer, R.L. *Safety and Health for Engineers.* New York: Van Nostrand Reinhold, 2005.

Chapanis, A. "Some reflections on progress." Proceedings of the Human Factors Society 29th Annual Meeting. Santa Monica, CA: Human Factors Society, 1985.

Ergonomics Guides, including *Ergonomics of VDT Workplaces* by K.H.E. Kroemer. Akron, OH: American Industrial Hygiene Association, 1984.

Grimaldi, J.V., and R.H. Simonds. *Safety Management,* 5th ed. Homewood, IL: Irwin, 1989.

Morris, B.K., ed. "Ergonomic Problems at Paper Mill Prompt Fine," *Occupational Health & Safety Letter* 20, no. 16 (1990): 127.

National Coalition on Ergonomics. "Home Page." Accessed August 11, 2003. http://www. necergor.org.

National Safety Council. *Fundamentals of Industrial Hygiene*, 3rd ed. Chicago: National Safety Council, 1988.

Safety Compliance Alert. Malvern, PA: Progressive Business Publications, 1998.

Sanders, M.S., and E.J. McCormick. *Human Factors In Engineering and Design*, 7th ed. New York: McGraw-Hill, 1993.

United States General Accounting Office. HEHS-97-163: *Worker Protection: Private Sector Ergonomics Programs Yield Positive Results*, 1997.

United Steelworkers of America. "OSHA Ergonomics Standard Assassinated." Accessed August 11, 2003. http://www.uswa.or/services/erog040301a.html.

Chapter 15

Machine Guarding

The basic objective of machine guarding is to prevent personnel from coming in contact with revolving or moving parts such as belts, chains, pulleys, gears, flywheels, shafts, spindles, and any working part that creates a shearing or crushing action or that may entangle the worker.

Machine guarding is visible evidence of management's interest in the worker and its commitment to a safe work environment. It is also to management's benefit, as unguarded machinery is a principal source of costly accidents, waste, compensation claims, and lost time.

—Ted Ferry, 1990

INTRODUCTION

As Ted Ferry points out, the basic purpose of machine guarding is to prevent contact of the human body with dangerous parts of machines. Moving machine parts have the potential for causing severe workplace injuries, such as crushed fingers or hands, amputations, burns, and blindness, just to name a few. Machine guards are essential for protecting workers from these needless and preventable injuries. Any machine part, function, or process that may cause injury must be safeguarded. When the operation of a machine or accidental contact with it can injure the operator or others in the vicinity, the hazards must be either eliminated or controlled (OSHA, 2003). Our experience has clearly (and much too frequently) demonstrated that when arms, fingers, hair, or any body part enters into or makes contact with moving machinery, the results can be not only gory, bloody, and disastrous, but also sometimes fatal.

Depending on the machine and the types of hazards it presents, methods of machine guarding vary greatly. The intent of this chapter is to familiarize safety and environmental managers and other safety professionals with the hazards of unguarded machines, common safeguarding methods, and the safeguarding of machines. All of these, if followed, combine to ensure that Ferry's main point, "Machine guarding is

Figure 15.1   Elements of a Machine Guarding Safety Program

visible evidence of management's interest in the worker and its commitment to a safe work environment," becomes a reality. As for the second part of Ferry's statement (on the benefits to the employer if correct machine guarding practices are followed) it logically follows that if the employer provides a safe work place, then all sides benefit from the results. Incorporating the elements of Figure 15.1 into your facility safety and environmental health program pays huge dividends in making your facility a safer place to work.

BASICS OF SAFEGUARDING MACHINES

OSHA points out that any mechanical motion that threatens a worker's safety should not remain unguarded. The reasoning behind this point is quite clear and is reinforced anytime the safety and environmental manager investigates on-the-job injuries involving crushed hands and arms, severed fingers, blindness or other horrifying machinery-related injuries. For the safety and environmental manager, the goal is quite clear; when the operation of a machine or accidental contact with it can injure the operator or others in the vicinity, the hazards must be either controlled or eliminated.

Safeguarding Defined

Application of appropriate safeguards keeps people and their clothing from coming into contact with hazardous parts of machines and equipment. They also prevent flying particles from an operation, and/or broken machine parts from striking or injuring people. Guards may also serve to enclose noise or dust hazards.

The National Safety Council (1987) defines *safeguarding* as follows:

". . . machine safeguarding is to minimize the risk of accidents of machine-operator contact. The contact can be:

1. An individual making the contact with the machine—usually the moving part—because of inattention caused by fatigue, distraction, curiosity, or deliberate chance taking;
2. From the machine via flying metal chips, chemical and hot metal splashes, and circular saw kickbacks, to name a few;
3. Caused by the direct result of a machine malfunction, including mechanical and electrical failure."

According to Brauer (2005), guards should have certain characteristics. They should be a permanent part of the machine or equipment, must prevent access to the danger zone during operation and must be durable and constructed strongly enough to resist the wear and abuse expected in the environment where machines are used. Guards should not interfere with the operation of the machine—that is, guards must not create hazards. Finally, machine guards should be designed to allow the more frequently performed maintenance tasks to be accomplished without the removal of the guards.

Types of Machine Safeguards

Safeguards can be broadly categorized as

1. **point-of-operation guards:** that point where work is performed on the material, such as cutting, shaping, boring, or forming of stock.
2. **point-of-operation devices (power transmission apparatus):** all components of the mechanical system that transmit energy to the part of the machine performing the work. These components include flywheels, pulleys, belts, connecting rods, couplings, cams, spindles, chains, cranks, and gears.
3. **feeding/ejection devices and other moving parts:** all parts of the machine that move while the machine is working. These can include reciprocating, rotating, and transverse moving parts, as well as feed mechanisms and auxiliary parts of the machine.

Mechanical Hazards: Motions and Actions

Three types of machine motion and four types of actions may present hazards to the worker. These can include the movement of rotating members, reciprocating arms, moving belts, meshing gears, cutting teeth, and any parts that impact or shear. These different types of hazardous mechanical motions and actions are basic (in varying combinations) to nearly all machines, and recognizing them is the first step toward protecting workers from the danger they present.

The basic types of hazardous mechanical motions and actions are:

Motions

- rotating (including in-running nip points)
- reciprocating
- transversing

Actions

- cutting
- punching
- shearing
- bending

COMMON SAFEGUARDING METHODS

The safety and environmental manager has several safeguarding methods to consider when he or she has determined that machine guarding is needed. The type of operation, the size or shape of stock, the method of handling, the physical layout of the work area, the type of material, as well as production requirements or limitations will help to determine the appropriate safeguarding method for the individual machine.

OSHA points out that as a general rule, power transmission apparatus is best protected by fixed guards that enclose the danger areas. For hazards at the point of operation, where moving parts actually perform work on stock, several kinds of safeguarding may be possible. The safety and environmental manager must always choose the most effective and practical means available.

Safeguards include guards, devices, automatic and semiautomatic feeding and ejecting methods, location and distance, and miscellaneous safeguarding accessories.

Guards

Guards are barriers that prevent access to danger areas. Guards can be of several types. These include fixed, interlocked, adjustable, and self-adjusting.

Fixed guards, as its name implies, are a permanent part of the machine. Unlike other types of guards, these do not move to accommodate the work being performed. They are not dependent upon moving parts to perform their intended function. They may be constructed of sheet metal, screen, wire cloth, bars, plastic, or any other material that is substantial enough to withstand the impact they may receive, and to endure prolonged use. If feasible, these guards are usually preferable to all other types, because of their relative simplicity and permanence. Limitations include interference with visibility; that they are limited to specific operations; and machine adjustment and repair may require removal, thereby necessitating other means of protection for maintenance personnel.

Interlocked guards shut off or disengage power and prevent starting of the machine when the guard is open. An interlocked guard may use electrical, mechanical, hydraulic, or pneumatic power, or any combination of these. Interlocked guards have the advantage of providing the maximum protection and allow access to the machine for setup, adjustment, or maintenance purposes. However, this type of guard requires careful adjustment/maintenance and can be made inoperable.

Adjustable guards provide a barrier that may be adjusted to facilitate a wide variety of production operations. Advantages include their ability to be constructed to suit many specific applications and that they can be adjusted to admit varying sizes of stock. However, protection may not be complete at all times because hands may enter the danger area. They often require frequent adjustment and maintenance.

Self-adjusting guards also accommodate different sizes of stock, but the movement of the stock determines the openings of these barriers. As the operator moves the stock into the danger area, the guard is pushed away, providing an opening that is only large enough to admit the stock. After the stock is removed, the guard returns to the rest position. This guard protects the operator by placing a barrier between the danger area and the operator. The guards may be constructed of plastic, metal, or other substantial material. Self-adjusting guards offer different degrees of protection, and are often easier to purchase and fit to machine. However, this type of guard does not always provide maximum protection, can limit visibility, and requires frequent adjustment and maintenance.

Devices

Devices can also be used to safeguard machinery. A safety device may perform many functions. It may stop the machine if a hand or any part of the body is inadvertently placed in the danger area; restrain or withdraw the operator's hands from the danger area during operation; require the operator to use both hands on machine controls, thus keeping both hands and body out of danger; or provide a barrier synchronized with the operating cycle of the machine to prevent entry to the danger area during the hazardous part of the cycle. This category includes presence-sensing devices, pullback mechanisms, restraints, safety controls, and gates.

Presence-Sensing Devices commonly operate on photoelectric, radiofrequency, or electromagnetic principles to disengage the machine when something is detected in the zone of concern. The photoelectric (optical) presence-sensing device uses a system of light sources and controls to interrupt the machine's operating cycle. If the light field is broken the machine stops and will not cycle. This device must be used only on machines which can be stopped before the worker can reach the danger area. The design and placement of the guard depends upon the time it takes to stop the mechanism and the speed at which the employee's hand can reach across the distance from the guard to the danger zone. This type of device allows freer movement for the operator, is simple to use, can be used by multiple operators, provides passerby protection, and requires no adjustment. However, it does not protect against mechanical failure and is limited to machines that can be stopped.

The radio frequency (capacitance) presence-sensing device uses a radio beam as part of the machine control circuit. When the capacitance field is broken the machine stops or will not activate. Like the photoelectric device, this device is only to be used on machines that can be stopped before the worker can reach the danger area. This requires the machine to have a friction clutch or other reliable means for stopping. This device allows freer movement for the operator but does not protect against mechanical failure. In addition, antennae sensitivity must be properly adjusted and maintained.

The electromechanical sensing device has a probe or contact bar that descends to a predetermined distance when the operator initiates the machine cycle. If an obstruction prevents it from descending its full predetermined distance, the control circuit does not actuate the machine cycle. This device allows for access at point of operation, but the contact bar or probe must be properly adjusted for each application. This adjustment must be maintained properly.

Pullback devices use cables attached to the operator's hands, wrists, and/or arms to prevent hands from entering the point of operation. This type of device is primarily used on machines with stroking action. When the slide/ram is up between cycles, the operator is allowed access to the point of operation. When the slide/ram begins to cycle by starting its descent, a mechanical linkage automatically assures withdrawal of the hands from the point of operation. This type of device eliminates the need for auxiliary barriers or other interference at the danger area. However, it limits movement of the operator and may obstruct workspace around the operator.

The *restraint* (holdback) device uses cables or straps that are attached to the operator's hands at a fixed point. The cables or straps must be adjusted to let the operator's hands travel within a predetermined safe area, with no extending or retracting action involved. Consequently, hand-feeding tools are often necessary if the operation involves placing material into the danger area. Because restraints prevent the operator from reaching into the danger area there is little risk of danger. However, adjustments must be made for specific operations and for each individual; frequent inspections and regular maintenance are required; close supervision of the operator's use of the equipment is required; movement of operator is limited; work space may be obstructed; and adjustments must be made for specific operations and each individual.

Safety controls use involvement of the operator as a safeguarding method and include safety trip controls, two-hand control, and two-hand trips. Safety trip controls use a machine. If the operator or anyone trips, loses balance, or is drawn toward the machine, applying pressure to the bar will stop the operation. The positioning of the bar, therefore, is critical. It must stop the machine before a part of the employee's body reaches the danger area.

While safety trip controls offer simplicity of use, they must still be manually activated, which may be difficult because of their location. Other limitations include that safety trip controls work to protect only the operator. They may require special fixtures to hold work and often require a machine-braking mechanism.

Another type of safety control is the two-hand control, which requires constant concurrent pressure by the operator to activate the machine. This kind of control requires a part-revolution clutch, brake, and a brake monitor if used on a power press. With this type of device, the operator's hands must be at a safe location (on control buttons), and at a safe distance from the danger area while the machine completes its closing cycle. The advantages of this type of safety control is that the operator's hands are at a predetermined location and that the operator's hands are free to pick up a new part after first half of cycle is completed. However, some two-handed controls can be rendered unsafe by holding with an arm or blocking, thereby permitting one-hand operation. The safety control only protects the operator.

The two-hand trip requires concurrent application of both the operator's control buttons to activate the machine cycle, after which the hands are free. This device is usually used with machines equipped with full-revolution clutches. The trips must be placed far enough from the point of operation to make it impossible for operators to move their hands from the trip buttons or handles into the point of operation before the first half of the cycle is completed. The distance from the trip button depends upon the speed of the cycle and the band speed constant, so that the operator's hands are kept far enough away to prevent them from being placed in the danger area prior to the slide/ram or blade reaching the full "down" position.

The two-hand trip has the advantage of keeping the operator's hands away from the danger area; it can be adapted to multiple operations; presents no obstruction to hand feeding; and does not require adjustment for each operation. However, the operator may try to reach into danger area after tripping machine. Some trips can be rendered unsafe by holding with arm or blocking, thereby permitting one-hand operation.

Note that to be effective, both two-hand controls and trips must be located so that the operator cannot use two hands or one hand and another part of his/her body to trip the machine.

Gates can also provide a high degree of protection to both the operator and other workers in the area. A gate is a movable barrier that protects the operator at the point of operation before the machine cycle can be started. Gates, in many instances, are designed to be operated with each machine cycle. To be effective, the gate must be interlocked so that the machine will not begin a cycle unless the gate guard is in place. It must be in the closed position before the machine can function. The main advantage of using gates is that they prevent reaching into or walking into the danger area. However, gates may require frequent inspection and regular maintenance and may interfere with the operator's ability to see the work.

Feeding and Ejection Methods

Automatic and semiautomatic feeding and ejection of parts are other ways of safeguarding machine processes. These methods eliminate the need for the operator to work at the point of operation. In some situations, no operator involvement is necessary after the machine is set up. In other cases, operators can manually feed the stock with the assistance of a feeding mechanism. Properly designed ejection methods do not require any operator involvement after the machine starts to function. Note that using these feeding and ejection methods does not eliminate the need for guards and devices. Guards and devices must be used wherever they are necessary and possible to provide protection from exposure to hazards.

Safeguarding by Location/Distance

Location and distance can also be used to safeguard machinery. A thorough hazard analysis of each machine and particular situation is absolutely essential before attempting this safeguarding technique.

To consider a part of a machine to be safeguarded by location, the dangerous moving part of a machine must be so positioned that those areas are not accessible, or do not present a hazard to a worker during the normal operation of the machine. This may be accomplished by locating a machine so that the hazardous parts of the machine are located away from operator workstations or areas where employees walk or work. For example, a machine could be positioned with its power transmission apparatus against a wall, leaving all routine operations to be conducted on the other side of the machine. Enclosure walls or fences could restrict access to machines. Another possible solution is to have dangerous parts located high enough to be out of the normal reach of any worker.

The feeding process can be safeguarded by location if a safe distance can be maintained to protect the worker's hands. The dimensions of the stock being worked on may provide adequate safety.

For instance, if the stock is several feet long and only one end of the stock is being worked on, the operator may be able to hold the opposite end while the work is being performed. An example would be a single-end-punching machine. However, depending upon the machine, protection might still be required for other personnel.

The positioning of the operator's control station provides another potential approach to safeguarding by location. Operator controls may be located at a safe distance from the machine if there is no reason for the operator to tend it.

Miscellaneous Safeguarding Accessories

A variety of methods and tools can be used to help lower the hazard potential created by certain machines, even though they do not provide full or complete machine safeguarding. Note that sound judgment is needed in their application and usage.

Awareness barriers may be used. Though the barrier does not physically prevent a person from entering the danger area, it calls attention to it. For an employee to enter the danger area an overt act must take place; the employee must either reach or step over, under, or through the barrier.

Shields may be used to provide protection from flying particles, splashing cutting oils, or coolants.

Special devices or hand tools for placing objects in power presses allow the operator's hands and arms to remain away from the point of operation.

Push sticks/blocks and jigs allow employees to keep their hands at a safe location when guiding wood or other materials during joiner and shaper operations.

Spreaders and nonkickback devices help prevent work from being thrown back at the operator, particularly with woodworking machines such as circular and radial saws.

SAFE WORK PRACTICES

Complying with OSHA's 19 CFR 1910.212-.219 standards regarding safeguarding machines is an important step the safety and environmental manager takes in ensuring control of workplace hazards and protecting the safety of employees. However, ensuring machines are safeguarded with the types of guards and devices discussed in the previous section are only part of the compliance effort. *Safe work practices* are another important element of any machine guarding safety program (and most other specialized safety programs). Our experience has clearly demonstrated that if written safe work practices are not in place, giving employees a written protocol to follow in safeguarding themselves from the hazards presented by many machines, the machine safeguarding safety program is incomplete and less than fully effective.

Consider the following safe work practices provided by Hoover et al (1989), which are designed to be employed in addition to the machine safeguarding guards and devices, as well as other practices:

Safe Work Practices: Machine Guarding

1. Guards should not be removed unless:
   a. permission is given by a supervisor,
   b. the person concerned is trained, and
   c. machine adjustment is a normal part of his/her job.
2. Do not start machinery unless guards are in place and in good condition.
3. Report missing or defective guards to your supervisor immediately.
4. When removing safeguards for repair, adjustment, or service, turn off power and lock and tag the main switch.
5. Do not permit employees to work on or around equipment while wearing ties, loose clothing, watches, rings, etc.
6. Inspect and conduct a maintenance program of guards on a regularly scheduled basis.
7. Instruct operators of mechanical equipment in all safe practices for operation of that machine.

Training, Enforcement, and Inspections

As with all other safety and environmental health programs, *training* is at the heart of the safety effort, because even the most elaborate safeguarding system and precise step-by-step safe work practices cannot offer effective protection unless the worker knows how to use it and why. Specific and detailed training is therefore a crucial element of any effort to provide safeguarding against machine-related hazards. Thorough operator training should involve instructions and/or hands-on training in the following:

1. a description and identification of the hazards associated with particular machines;
2. the safeguards themselves, how they provide protection, and the hazards for which they are intended;
3. how to use the safeguards and why;
4. how and under what circumstances safeguards can be removed, and by whom (in most cases, repair or maintenance personnel only); and
5. what to do (e.g., contact the supervisor) if a safeguard is damaged, missing, or unable to provide adequate protection.

This kind of safety training is necessary for new operators and maintenance or setup personnel, when any new or altered safeguards are put in service, or when workers are assigned to a new machine or operation.

Properly installed machinery safeguards, well-written safe work practices, and a strong training program are all important elements of the Machine Guarding Safety Program. However, if employees are allowed to overtly disregard company safe work practices and rules, the Machine Guarding Safety Program is worthless. *Enforcement* of safety rules and safe work practices is required. Though the safety engineer is not normally associated with disciplinary action, he or she must take an active role in enforcing company safety policies; likewise, the safety engineer must ensure that supervisors and workers alike understand the importance of company safety polices, rules, regulations, and safe work practices, and more importantly, that they will be strictly enforced.

Machinery safety guards must be periodically *inspected* and maintained to ensure their integrity: to ensure that they are in place and working as designed, to ensure they are continually effective, and to ensure that they have not been tampered with or bypassed in any way. Generally, machinery safety guards are inspected through the company's preventive maintenance program checks. Whether discovered through scheduled maintenance or while in operation, broken or inoperable parts must be replaced. However, good engineering practice dictates that machine safety guards should be inspected before and after each use to ensure their operability.

To aid the safety and environmental manager in inspecting his or her workplace machinery to determine the safeguarding needs of his or her own workplace, OSHA, in its 3067 booklet, provides the following Machine Guarding Checklist:

MACHINE GUARDING CHECKLIST

Answers to the following questions should help the interested reader determine the safeguarding need of his or her own workplace by drawing attention to hazardous conditions or practices requiring correction.

	Yes	No

Requirements for all Safeguards

1. Do the safeguards provided meet the minimum requirements? ____ ____

2. Do the safeguards prevent workers' hands, arms, and other body parts from making contact with dangerous moving parts? ____ ____

3. Are the safeguards firmly secured and not easily removable? ____ ____

4. Do the safeguards ensure that no object will fall into the moving parts? ____ ____

5. Do the safeguards permit safe, comfortable, and relatively easy operation of the machine? ____ ____

6. Can the machine be oiled without removing the safeguard? ____ ____

7. Is there a system for shutting down the machinery before safeguards are removed? ____ ____

8. Can the existing safeguards be improved? ____ ____

MECHANICAL HAZARDS

The Point of Operation

1. Is there a point-of-operation safeguard provided for the machine? ____ ____

2. Does it keep the operator's hands, fingers, and body out of the danger area? ____ ____

3. Is there evidence that the safeguards have been tampered with or removed? ____ ____

4. Could you suggest a more practical, effective safeguard? ____ ____

5. Could changes be made on the machine to eliminate the point-of-operation hazard entirely? ____ ____

Power Transmission Apparatus

1. Are there any unguarded gears, sprockets, pulleys, or flywheels on the apparatus? ____ ____

2. Are there any exposed belts or chain drives? ____ ____

3. Are there any exposed set screws, key ways, collars, etc.? ____ ____

4. Are starting and stopping controls within easy reach of the operator? ____ ____

5. If there is more than one operator, are separate controls provided? ____ ____

Other moving parts

1. Are safeguards provided for all hazardous moving parts of the machine, including auxiliary parts? ____ ____

	Yes	No

Nonmechanical Hazards

1. Have appropriate measures been taken to safeguard workers against noise hazards? ____ ____

2. Have special guards, enclosures, or personal protective equipment been provided where necessary to protect workers from exposure to harmful substances used in machine operation? ____ ____

Electrical Hazards

1. Is the machine installed in accordance with National Fire Protection Association and National Electrical Code requirements? ____ ____

2. Are there loose conduit fittings? ____ ____

3. Is the machine properly grounded? ____ ____

4. Is the power supply correctly fused and protected? ____ ____

5. Do workers occasionally receive minor shocks while operating any of the machines? ____ ____

Training

1. Do operators and maintenance workers have the necessary training in how to use the safeguards and why? ____ ____

2. Have operators and maintenance workers been trained in where the safeguards are located, how they provide protection, and what hazards they protect against? ____ ____

3. Have operators and maintenance workers been trained in how and under what circumstances guards can be removed? ____ ____

4. Have workers been trained in the procedures to follow if they notice guards that are damaged, missing, or inadequate? ____ ____

Protective Equipment and Proper Clothing

1. Is protective equipment required? ____ ____

2. If protective equipment is required, is it appropriate for the job, in good condition, kept clean and sanitary, and stored carefully when not in use? ____ ____

3. Where several maintenance persons work on the same machine, are multiple lockout devices used? ____ ____

4. Do maintenance persons use appropriate and safe equipment in their repair work? ____ ____

5. Is the maintenance equipment itself properly guarded? ____ ____

6. Are maintenance and servicing workers trained in the requirements of 29 CFR 1910. 147, lockout/tagout hazard, and do the procedures for lockout/tagout exist **before** they attempt their tasks? ____ ____

Source: OSHA 3067, *Concepts and Techniques of Machine Safeguarding*, 1992 (Revised).

MACHINE HAZARD WARNINGS

One or more warnings are needed on a machine to communicate hazards that may be present. *Machine hazard warnings* are of several different types. Hazard signs and/ or labels use **signal words** such as DANGER, WARNING, or CAUTION. *Danger signs* indicate an imminently hazardous situation, which if not avoided could result in death or serious injury. *Warning signs* indicate a potentially hazardous situation, which if not avoided could result in death or serious injury. *Caution signs* indicate that a hazard may result in moderate or minor injury. For example, warning signs with the appropriate signal word are often used to indicate dangerous or hazardous conditions. Examples are signs such as "KEEP HANDS OUT OF MACHINERY," "EYE PROTECTION REQUIRED IN THIS AREA," and Danger signs such as "DANGER: PINCH POINTS! WATCH YOUR HANDS," "DANGER: THIS MACHINE HAS NO BRAIN, USE YOUR OWN," "DANGER: THIS MACHINE CYCLES," or "DANGER: THIS MACHINE STARTS AUTOMATICALLY."

Note that another kind of sign is often used—the NOTICE sign. However, *notice signs* are used to state a company policy and should not be associated directly with a hazard or hazardous situation. They must not be used in place of "Danger," "Warning," or "Caution." For example, if a machine has guards, the warnings should include a "notice" to keep guards in place, and not to operate the machine without them. On guard devices, the warnings should state the hazards or "danger," any limitations the device may have, and protective actions the operator must take.

In addition to mechanical guards and warning or notice signs, *color-coding* may be used to alert workers of hazards. Typically, standard colors that workers can learn to recognize are used. In many cases, individual industries (such as wastewater treatment) have their own color coding systems.

EMPLOYEE CLOTHING AND JEWELRY

Engineering controls that eliminate the hazard at the source and do not rely on the worker's behavior for their effectiveness offer the best and most reliable means of safeguarding workers. Therefore, engineering controls must be the employer's first choice for eliminating machine hazards. But whenever engineering controls are not available or are not fully capable of protecting the employee, an extra measure of protection is necessary. Operators must wear protective clothing or personal protective equipment.

Note that it is management's responsibility to assure employees wear appropriate clothing when operating or working around hazardous machines. If it is to provide adequate protection, the protective clothing and equipment selected must always be:

1. appropriate for the particular hazards
2. maintained in good condition
3. properly stored to prevent damage or loss when not in use
4. kept clean, fully functional, and sanitary.

Note also that protective clothing and equipment can create hazards. Protective gloves which can become caught between rotating parts, or respirator facepieces which hinder the wearer's vision, for example, require alertness and continued attentiveness whenever they are used.

Other parts of the worker's clothing may present additional safety hazards. For example, loose-fitting, oversized clothing might possibly become entangled in rotating spindles or other kinds of moving machinery. Rings, bracelets, or watchbands can catch on machine parts or stock and lead to serious injury by pulling a hand into the danger area. Employees with long hair may need to wear hats or hair nets if the long hair represents a hazard because of the proximity of moving machinery.

LOCKOUT/TAGOUT

Setup maintenance and servicing of machinery often requires that existing safeguarding be removed or disengaged to provide access to machine parts. At such times, the machine should be *locked out* and *tagged out* of service, to prevent anyone from activating it while someone else expects it to be de-energized.

THOUGHT-PROVOKING QUESTIONS

- What are the common sources of mechanical hazards in the workplace?
- What actions is machine guarding intended to protect against?
- Why is attention to guarding against physical danger presented to workers by machinery so important for the success of the whole safety program?
- Describe three types of machine-operator controls that could lead to injury.
- What are the three most common general safeguard types?
- What three types of machine motions present hazards to workers?
- What four types of actions present hazards to workers?
- What devices and means can safeguards include?
- What are the common guard types? Describe how they protect.
- What are the common safety device types? How do they protect workers?
- How are feeding and ejection methods used? Do they replace guarding? Why or why not?
- What advantages do location and distance as safeguards offer? What are its disadvantages?
- What benefits do awareness barriers and warning signs provide?
- Several handheld devices and some mechanical elements can help safeguard workers from machinery. What are their specific types and purposes?
- What place do written safe work practices hold in such a physical process as machine guarding?
- What role should supervisors play in enforcing and promoting machine guarding safety?
- What specifics and specific areas should training cover?
- When is training necessary?
- How important is enforcement? Why?

- When should inspection of machine guarding elements occur?
- What purpose do warning signs have? What are the general categories?
- What special purpose do notice signs hold?
- How can color-coding be used to alert workers?
- What should employees do for protecting themselves in terms of how they prepare for work?
- Who is responsible for choosing and providing protective clothing?
- What special considerations for machine guarding should be present for lockout/tagout?

REFERENCES AND RECOMMENDED READING

ANSI A13.1. New York: American National Standards Institute, 1981.

ANSI Z535.1. New York: American National Standards Institute, 1991.

Baumeister, T., ed. *Standard Handbook for Mechanical Engineers*, 9th ed. New York: McGraw-Hill, 1987.

Blundell, J.K. *Safety Engineering—Machine Guarding Accidents*. Del Mar, CA: Hanrow Press, 1987.

Brauer, R.L. *Safety and Health for Engineers*. New York: Van Nostrand Reinhold, 2005.

Code of Federal Regulations. Safety color code for marking physical hazards, title 29, sec. 1910.144.

Code of Federal Regulations. Machinery and Machine Guarding, title 29, sec. 1910.212-219.

Ferry, T. *Safety and Health Management Planning*. New York: Van Nostrand Reinhold, 1990.

Goetsch, D.L. *Occupational Safety and Health*, 2nd ed. Englewood Cliffs, NJ: Prentice Hall, 1996.

Hoover, R.L., R.L. Hancock, K.L. Hylton, O.B. Dickerson, and G.E. Harris. *Health, Safety and Environmental Control*. New York: Van Nostrand Reinhold, 1989.

National Safety Council. *Guards: Safeguarding Concepts Illustrated*, 5th ed. Chicago: National Safety Council, 1987.

National Safety Council. *Power Press Safety Manual*, 3rd ed. Chicago: National Safety Council, 1988.

Occupational Safety and Health Administration. *Concepts and Techniques of Machine Safeguarding*, OSHA 3067. 1992.

Occupational Safety and Health Administration. "Machine Guarding." Accessed August 7, 2003. http://www.osha-sle.gov/SLTC/machineguarding/.

Roberts, V.L. *Machine Guarding—Historical Perspective*. Durham, NC: Institute for Product Safety, 1980.

Chapter 16

Workplace Environmental Concerns

Most safety and environmental managers are already involved in some aspects of industrial hygiene. They study work operations, look for potential hazards, and make recommendations to minimize these hazards. The safety and environmental manager, through specialized study and training, has the expertise to deal with many complex industrial hygiene problems. If the safety and environmental manager carries on the day-to-day safety and environmental health functions involving immediate decisions beyond the scope of his or her expertise, he or she must know when and where to get help on industrial hygiene problems.

After the workplace environmental specialist or industrial hygienist surveys the plant, makes recommendations, and suggests certain control measures, it may become the safety and environmental manager's responsibility to see that the control measures are being applied and followed. Or such responsibility may be vested in an individual whose education and training is in the combined disciplines of safety and health.

—J.B. Olishifski, 1988

INTRODUCTION

While it is true that safety and environmental managers typically deal with the subject areas described and discussed to this point in the text, it is important to point out that environmental issues beyond those already mentioned and covered or not covered by standard OSHA requirements are also important to maintaining a safe and healthy workplace. For example, workplace stressors are real; they present a challenging area to investigate and resolve for safety and environmental managers. In addition to environmental stressors, controls for fall protection, office safety, indoor air quality, and on-the-job thermal stress issues are important. In this chapter, we describe and discuss these areas of concern to provide awareness and possible mitigation measures to assist safety and environmental managers in the proper performance of their responsibilities to ensure a safe and healthy work environment.

WHAT IS INDUSTRIAL HYGIENE?

The American Industrial Hygiene Association (AIHA) defines industrial hygiene as "that science and art devoted to the anticipation, recognition, evaluation, and control of those environmental factors or stresses, arising in or from the workplace, which may cause sickness, impaired health and well-being, or significant discomfort and inefficiency among workers or among citizens of the community."

What is an industrial hygienist?

A well-trained, well-prepared industrial hygienist (IH) is equipped to deal with virtually any situation that arises (Ogle, 2003).

Is the industrial hygienist also a safety and environmental manager?

Maybe. It depends. Safety and environmental management and industrial hygiene have commonly been thought to be separate entities (especially by safety professionals and industrial hygienists). In fact, over the years, a considerable amount of debate and argument has risen between those in the safety and environmental management and industrial hygiene professions on many areas concerning safety and health issues in the workplace and on exactly who is best qualified to administer a workplace safety and health program.

Historically, the safety and environmental manager had the upper hand in this argument; that is, prior to the enactment of the OSH Act. Until OSHA went into effect, industrial hygiene was not a topic that many thought about, cared about, or had any understanding of. Safety was safety—and safety included health—and that was that.

After the OSH Act, however, things changed. In particular, people began to look at work injuries and work diseases differently. In the past, they were regarded as separate problems. Why? The primary reason for this view was obvious, and not so obvious. The obvious was work-related injury. Work injuries occurred suddenly and their agents (i.e., the electrical source, chemical, machine, tool, work or walking surface, or whatever unsafe element caused the injury) usually was readily obvious.

Not so readily obvious were the workplace agents (occupational diseases) that caused illnesses. Why? Because most occupational diseases develop rather slowly, over time. In asbestos exposure, for example, workers who abate (remove) asbestos-containing materials without the proper training (awareness) and personal protective equipment are subject to exposure. Typically, asbestos exposure may be a one-time exposure event (the silver bullet syndrome) or the exposure may go on for years. No matter the length of exposure, one thing is certain, with asbestos contamination, pathological change occurs slowly. Some time will pass before the worker notices a difference in his or her pulmonary function. Disease from asbestos exposure has a latency period that may be as long as 20–30 years before the effects are realized. The point? Any exposure to asbestos, short term or long term, may eventually lead to a chronic disease (in this case, restrictive lung disease) that is irreversible (e.g., asbestosis). Of course, many other types of workplace toxic exposures can affect workers' health. The prevention, evaluation, and control of such occurrences is the role of the industrial hygienist.

Thus, because of the OSH Act, and also because of increasing public awareness and involvement by unions in industrial health matters, the role of the industrial hygienist has continued to grow over the years. Certain colleges and universities have incorporated industrial hygiene majors into environmental health programs.

The other offshoot of the OSH Act has been, in effect (though many practitioners in the field will disagree with this point of view), a continuing tendency toward uniting safety and industrial hygiene into one entity.

This presents a problem with definition. When we combine the two entities, do we combine them into a "safety" or an "industrial hygiene" title or profession? The debate on this issue continues. What is the solution to this problem? How do we end the debate? To attempt an answer we need to look at a couple of factors and at the actual experience gained from practice in the real world of safety and environmental management and industrial hygiene.

We have stated throughout this text that the safety and environmental manager must be a generalist, a Jack or Jill of all trades. You should already have a pretty good feel for what the safety and environmental manager is required to do and what he or she is required to know to be effective in the workplace.

How about the industrial hygienist? What are industrial hygienists required to do and know to be effective in the workplace? Let's take a look at what a typical industrial hygienist does (you should be able to determine the level of knowledge they should have).

The primary mission of the industrial hygienist is to examine the workplace environment and its environs by studying work operations and processes. From these studies, he or she is able to obtain details related to the nature of the work, materials and equipment used, products and by-products, number of employees, hours of work, and so on. At the same time, appropriate measurements are made to determine the magnitude of exposure or nuisance (if any) to workers and the public. The hygienist's next step is to interpret the results of the examination of the workplace environment and environs, in terms of ability to impair worker health (i.e., Is there a health hazard in the workplace that must be mitigated?). With examination results in hand, the industrial hygienist then presents specific conclusions to the appropriate managerial authorities.

Is the process described above completed? Yes and no. In many organizations, the industrial hygienist's involvement stops there. But remember, discovering a problem is only half the safety battle. Knowing that a problem exists, but not taking any steps to mitigate it is leaving the job half-done. In this light, the industrial hygienist normally will make specific recommendations for control measures—an important part of industrial hygiene's anticipate-recognize-evaluate-control paradigm.

Any further involvement in working toward permanently changing the hazard depends on the industrial hygienist's role in a particular organization. Organization size is one of the two most important factors that determine the industrial hygienist's role within an organization. Obviously, an organization that consists of several hundred (or thousands) of workers will increase the work requirements for the industrial hygienist.

The second factor has to do with what the organization produces. Does it produce computerized accounting records? Does it perform a sales or telemarketing function? Does it provide office supplies to those businesses requiring such service? If the organization accomplishes any of these functions (and many similar functions) the organization probably does not require the services of a full-time staff industrial hygienist.

On the other hand, if the organization handles, stores, or produces hazardous materials, if the organization is a petroleum refinery, or if the organization is a large environmental laboratory, then there might be a real need for the services performed by a full-time staff industrial hygienist.

Let's look at these factors in combination and assume the organization employs 5,000 full-time workers in the production of chemical products. In this situation, not only are the services normally performed by an industrial hygienist required, but the company probably needs more than just one full-time industrial hygienist—perhaps they need several. In this case, each industrial hygienist might be assigned duties that are narrow in scope, with limited responsibility. In short, each hygienist would specialize.

What this all means, of course, is that the size of the organization and the type of work performed dictates need for an industrial hygienist, and each organization's needs sets the extent of their work requirements.

An important area of responsibility for the industrial hygienist that we have not mentioned here (one often overlooked in the real work world) is the industrial hygienist's responsibility to conduct training. If the industrial hygienist examines a workplace environment and environs for occupational hazards and discovers one (or more), he or she will perform the anticipate-recognize-evaluate-control actions. However, an important part of the "control" area of the industrial hygiene paradigm is informing those exposed to the hazards about the hazards, to train them. In the field of industrial hygiene in general, preparation in this area and emphasis on its importance on the job has been weak.

What does all this mean? Good question—one answered most simply by pointing out that safety and environmental managers are (or should be) generalists. Industrial hygienists are specialists. This simple statement sums up the main difference between the two professions: one is a generalist and the other a specialist.

Which one is best? Another good question. Actually, which one is "best" is not the issue—rather, which one is better suited to perform the functions of the safety engineer. Based on our own personal experience, we profess the need for generalization (we have stated this throughout the text and continue to do so to the end). The safety engineer needs to be well versed in all aspects of industrial hygiene—no question about this. We also hold no doubt about the industrial hygienist being efficacious to safety.

The problem lies in present perception and use. The safety engineer views his role as all-encompassing (as he or she rightly should), and the industrial hygienist views him or herself as a step above safety and environmental management occupying that high pinnacle position known as "specialist." Which one is right and which one is wrong? Neither. Again, it is a matter of perception. On the use side of the issue, the safety and environmental management is usually employed to run or manage an organization's safety and health program (as safety and environmental manager, safety engineer, safety professional, safety director, safety coordinator, or another similar title), while the industrial hygienist is usually hired to fill only a position of organizational industrial hygienist.

Can you see the difference? Believe us when we say that practicing safety and environmental managers and industrial hygienists not only see the difference, they feel the difference—and learn to live (work) with the difference—sometimes as colleagues and sometimes as competitors.

So what is the solution to this dilemma (yes; indeed it is a dilemma—especially for those practitioners active in the field)? Based on our experience we feel the solution lies in merging the two disciplines into one: safety professionals. The safety and environmental manager must be well grounded in industrial hygiene, but must also have knowledge of many other fields; again, he or she must be a generalist.

Will this scenario work? To a degree, it already has. Of course, pundits out there still argue against such recommendations, but in the work world the argument really comes down to one thing and one thing only: personal economics. In some cases in the past, safety and environmental managers have had the opportunity to earn a better living than industrial hygienists. Why? Because safety and environmental managers seem to be better suited to become managers. In the work world, management is where the money is. Why is this the case? Simply put, a specialist has little exposure to managing, while the safety and environmental management generalist is often involved in and with management. The bottom line: If you owned a company that employed generalists and specialists, who would you most likely promote to managerial positions? Notwithstanding the tendency to promote the "best person" (based on ability and performance) for the job, consider that the organization's safety and environmental manager (and not the organization's industrial hygienist) is more likely to be promoted to the safety and environmental director's position.

What do we base this on? First we refer you back to the chapter opening and the statement by Olishifski and Plog: . . . "most safety professionals are already involved in some aspects of industrial hygiene . . . after the industrial hygienist surveys the plant, makes recommendations, and suggests certain control measures, it may become the safety professional's responsibility to see that the control measures are being applied and followed." Secondly, we also base our opinion on one simple fact that for management, a well-rounded generalist accustomed to authority is more useful in management than a narrowly focused specialist. How do we know? We've been there.

Industrial Hygiene: Stressors

The industrial hygienist focuses on evaluating the healthfulness of the workplace environment, either for short periods or for a work life of exposure. When required, the industrial hygienist recommends corrective procedures to protect health, based on solid quantitative data, experience, and knowledge. The control measures he or she often recommends include: isolation of a work process, substitution of a less harmful chemical or material, and/or other measures designed solely to increase the healthfulness of the work environment.

To ensure a healthy workplace environment and environs, the industrial hygienist focuses on the recognition, evaluation, and control of chemical, physical, or biological and ergonomic stressors that can cause sickness, impaired health, or significant discomfort to workers.

The key word was *stressors*, or simply, stress—the stress caused by the workplace external environmental demands placed upon a worker. Increases in external stressors beyond a worker's tolerance level affect his or her on-the-job performance.

The industrial hygienist must not only understand that workplace stressors exist but also that they are sometimes cumulative (additive). For example, studies have shown that some assembly line processes are little affected by neither low illumination nor vibration; however, when these two stressors are combined, assembly-line performance deteriorates.

Other cases have shown just the opposite effect. For example, the worker who has had little sleep and then is exposed to a work area where noise levels are high actually benefits (to a degree, depending on the intensity of the noise level and the worker's exhaustion level) from increased arousal level; a lack of sleep combined with a high noise level is compensatory (Ferry, 1990).

To recognize environmental stressors and other factors that influence worker health, the industrial hygienist must be familiar with work operations and processes. An essential part of the new industrial hygienist's employee orientation process should include orientation on all pertinent company work operations and processes. Obviously, the newly hired industrial hygienist who has not been fully indoctrinated on company work operations and processes not only is not qualified to study the environmental effects of such processes, but also suffers from another disability—lack of credibility with supervisors and workers. This point cannot be emphasized strongly enough.

Note: Woe be it to the rookie industrial hygienist or safety engineer who has the audacity (and downright stupidity) to walk up to any supervisor (or any worker with experience at task) and announce that he or she is going to find out everything that is unhealthy (and thus injurious to workers) about the work process—without having any idea how the process operates, what it does, or what it is all about. We do not recommend this scenario. As an industrial hygienist (or as a safety engineer), you must understand the work operations and processes to the point that you could almost operate the system efficiently and safely yourself.

What are the workplace stressors the industrial hygienist should be concerned with? According to Pierce (1984), the industrial hygienist should be concerned with those workplace stressors that are likely to accelerate the aging process, cause significant discomfort and inefficiency, or may be immediately dangerous to life and health. Several stressors fall into these categories; the most important ones include:

1. **Chemical stressors**—gases, dusts, fumes, mists, liquids, or vapors.
2. **Physical stressors**—noise, vibration, extremes of pressure and temperature, and electromagnetic and ionizing radiation.
3. **Biological stressors**—bacteria, fungi, molds, yeasts, insects, mites, and viruses.
4. **Ergonomic stressors**—repetitive motion, work pressure, fatigue, body position in relation to work activity, monotony/boredom, and worry.

Industrial Hygiene: Areas of Concern

From the list of stressors above, you can see that the industrial hygienist has many areas of concern related to protecting the health of workers on the job. In this section, we focus on the major areas that the industrial hygienist typically is concerned with in the workplace. We also discuss the important areas of industrial toxicology, industrial health hazards, industrial noise, vibration, and environmental control. All of these areas are important to the industrial hygienist (and to the worker, of course), but they are not all-inclusive; the industrial hygienist also is concerned with other areas: ionizing and nonionizing radiation, for example, and many others.

Industrial Toxicology

Normally, we give little thought to the materials (chemical substances, for example) we are exposed to on a daily, almost constant basis, unless they interfere with our lifestyles, irritate us, or noticeably physically affect us. Most of these chemical substances do not present a hazard under ordinary conditions. However, keep in mind that all chemical substances have the potential for being injurious at some sufficiently high concentration and level of exposure. The industrial hygienist understands this, and to prevent the lethal effects of overexposure for workers, they must have an adequate understanding and knowledge of general *toxicology*.

What is toxicology? Toxicology is a very broad science that studies the adverse effects of chemicals on living organisms. It deals with chemicals used in industry, drugs, food, and cosmetics, as well as those occurring naturally in the environment. Toxicology is the science that deals with the poisonous or toxic properties of substances. The primary objective of industrial toxicology is the prevention of adverse health effects in workers exposed to chemicals in the workplace. The industrial hygienist's responsibility is to consider all types of exposure and the subsequent effects on living organisms. Following the prescribed precautionary measures and limitations placed on exposure to certain chemical substances by the industrial toxicologist is the worker's responsibility. The industrial hygienist uses toxicity information to prescribe safety measures for protecting workers.

To gain a better appreciation for what industrial toxicology is all about, you must understand some basic terms and factors, many of which contribute to determining the degree of hazard particular chemicals present. You must also differentiate between *toxicity* and *hazard*. *Toxicity* is the intrinsic ability of a substance to produce an unwanted effect on humans and other living organisms when the chemical has reached a sufficient concentration at a certain site in the body. *Hazard* is the probability that a substance will produce harm under specific conditions. The industrial hygienist and other safety professionals employ the opposite of hazard: *safety*, that is, the probability that harm will not occur under specific conditions. A toxic chemical used under safe conditions may not be hazardous.

Basically, all toxicological considerations are based on the *dose-response relationship*, another toxicological concept important to the industrial hygienist. In its simplest terms, the dose of a chemical to the body resulting from exposure is directly related to the degree of harm. This relationship means that the toxicologist is able to determine a *threshold level* of exposure for a given chemical, the highest amount of a chemical substance to which one can be exposed with no resulting adverse health effect. Stated another way, chemicals present a threshold of effect, or a no-effect level.

Threshold levels are critically important parameters. For instance, under the OSH Act, threshold limits have been established for the air contaminants most frequently found in the workplace. The contaminants are listed in three tables in 29 CFR 1910

Subpart Z—Toxic and Hazardous Substances. The threshold limit values listed in these tables are drawn from values published by the American Conference of Governmental Industrial Hygienists (ACGIH) and from the "Standards of Acceptable Concentrations of Toxic Dusts and Gases," issued by the American National Standards Institute (ANSI).

An important and necessary consideration when determining levels of safety for exposure to contaminants is their effect over a period of time. For example, during an 8-hour work shift, a worker may be exposed to a concentration of Substance A [with a 10 ppm (parts per million—analogous to a full shot glass in a swimming pool) TWA (time-weighted average), 25 ppm ceiling and 50 ppm peak] above 25 ppm (but never above 50 ppm) only for a maximum period of 10 minutes. Such exposure must be compensated by exposures to concentrations less than 10 ppm, so that the cumulative exposure for the entire 8-hour work shift does not exceed a weighted average of 10 ppm. Formulas are provided in the regulations for computing the cumulative effects of exposures in such instances. Note that the computed cumulative exposure to a contaminant may not exceed the limit value specified for it.

Although air contaminant values are useful as a guide for determining conditions that may be hazardous and may demand improved control measures, the industrial hygienist must recognize that the susceptibility of workers varies.

Even though it is essential not to permit exposures to exceed the stated values for substances, note that even careful adherence to the suggested values for any substance will not assure an absolutely harmless exposure. Thus, the air contaminant concentration values should only serve as a tool for indicating harmful exposures, rather than the absolute reference on which to base control measures.

For a chemical substance to cause or produce a harmful effect, it must reach the appropriate site in the body (usually via the bloodstream) at a concentration (and for a length of time) sufficient to produce an adverse effect. Toxic injury can occur at the first point of contact between the toxicant and the body, or in later, systemic injuries to various organs deep in the body (Hammer, 1989). Common routes of entry are ingestion, injection, skin absorption, and inhalation. However, entry into the body can occur by more than one route (e.g., inhalation of a substance that can be absorbed through the skin).

Ingestion of toxic substances is not a common problem in industry. Most workers do not deliberately swallow substances they handle in the workplace. However, ingestion does sometimes occur either directly or indirectly. Industrial exposure to harmful substance through ingestion may occur when workers eat lunch, drink coffee, chew tobacco, apply cosmetics, or smoke in a contaminated work area. The substances may exert their toxic effect on the intestinal tract or at specific organ sites.

Injection of toxic substances may occur just about anywhere in the body where a needle can be inserted but is a rare event in the industrial workplace.

Skin absorption or contact is an important route of entry in terms of occupational exposure. While the skin may act as a barrier to some harmful agents, other materials may irritate or sensitize the skin and eyes, or travel through the skin into the bloodstream, thereby impacting specific organs.

Inhalation is the most common route of entry for harmful substances in industrial exposures. Nearly all substances that are airborne can be inhaled. Dusts, fumes, mists, gases, vapors, and other airborne substances may enter the body via the lungs and

Table 16.1 Comparison of Selected Chemical Agents and their Harmful Effects Resulting from Overexposure

Agent Type	Major Route of Entry	Acute/Chronic Effects
Asbestos	Inhalation	Chronic: asbestosis, mesothelioma, lung cancer
Arsenic	Skin absorption	Acute: skin irritation, conjunctivitis, sensitization dermatitis Chronic: possible epidermal cancer
Cadmium	Inhalation	Acute: chest pain, shortness of breath, pulmonary edema, digestive effects
Lead	Inhalation	Chronic: gastrointestinal disturbance, anemia due to red blood cell effects, kidney disease and reproductive effects
Aromatic Solvents	Inhalation	Acute: central nervous system effects, depression, narcotic effects Chronic: liver, blood system disorders
	Skin absorption	Dermatitis (chronic or acute)
Sulfur dioxide	Inhalation	Acute: eye and respiratory irritation Chronic: bronchitis

may produce local effects on the lungs, or may be transported by the blood to specific organs in the body.

Upon finding a route of entry into the body, chemicals and other substances may exert their harmful effects on specific organs of the body, such as the lungs, liver, kidneys, central nervous system, and skin. These specific organs are termed *target organs* and will vary with the chemical of concern.

The toxic action of a substance can be divided into *short-term (acute)* and *long-term (chronic)* effects. Short-term adverse (acute) effects are usually related to an accident where exposure symptoms (effects) may occur within a short time period following either a single exposure or multiple exposures to a chemical. Long-term adverse (chronic) effects usually occur slowly after a long period of time, following exposures to small quantities of a substance (as lung disease may follow cigarette smoking). Chronic effects may sometimes occur following short-term exposures to certain substances.

Table 16.1 shows the harmful effects that can result from overexposure to some chemical agents.

Industrial Health Hazards

NIOSH and OSHA's *Occupational Health Guidelines for Chemical Hazards*, DHHS (NIOSH) Publication No. 81-123 (Washington, DC: Superintendent of Documents, U.S. Government Printing Office, current edition) illustrates quite clearly that the number of known industrial poisons is quite large, and also that their effects and means of control are generally understood. Generally, determining if a substance is hazardous or not is simple, if the following is known: (1) what the agent is and what form it is in; (2) the concentration; and (3) the duration and form of exposure.

However, practicing safety and environmental managers and industrial hygienists come face-to-face with one problem rather quickly. Many new compounds of somewhat uncertain toxicity are introduced into the workplace each year. Another related problem occurs when manufacturers develop chemical products with unfamiliar trade names and do not properly labeled them to indicate the chemical constituents of the compounds (of course, under OSHA's Hazard Communication Standard this practice is illegal.)

One of the primary categories of industrial health hazards that the safety and environmental manager must deal with is airborne contaminants. Two main forms of airborne contaminants are of chief concern: particulates, and gases or vapors. Particulates include dusts, fumes, smoke, aerosols, and mists, classified additionally by size and chemical makeup and sometimes by shape.

Dusts are solid particles of matter produced by grinding, crushing, handling, detonation, rapid impact, etc. Size may range from 0.5 to 100 mm (micron: 1 mm = 1/25000 inch), with most (over 90%) airborne dust in the 0.5–5-mm range. Dusts do not tend to flocculate except under electrostatic forces.

Fumes are solid particles of matter formed by condensation of vapors. Heating or volatilizing metals (welding) or other solids usually produces them. Size usually ranges from 0.01 to 0.5 microns. Fumes flocculate and sometimes coalesce.

Gases are normally formless fluids (a state of matter separate from solids and liquids) that occupy the space of an enclosure and that can change to liquid or solid states only by the combined effects of increased pressure and decreased temperatures. Gases diffuse.

Mists are fine liquid droplets suspended in or falling through air. Mist is generated by condensation from the gaseous to liquid state, or by breaking up liquid into fine particles through atomizing, spraying, mechanized agitation, splashing, or foaming.

Smoke is the visible carbon or soot particles (generally less than 0.1 micron in size) resulting from the incomplete combustion of carbonaceous materials such as oil, tobacco, coal, and tar.

Vapor is the gaseous phase of a substance that is liquid or solid at normal temperature and pressure. Vapors diffuse.

Industrial atmospheric contaminants exist in virtually every workplace. Sometimes they are readily apparent to workers because of their odor or because they can actually be seen. Safety and environmental managers, however, cannot rely on odor or vision to detect or measure airborne contaminants. They must rely on measurements taken by detection and sampling devices. Many different commercially available instruments permit the detection and concentration evaluation of many different contaminants. Some of these instruments are so simple that nearly any worker can learn to properly operate them. A note of caution, however; the untrained worker may receive an instrument reading that seems to indicate a higher degree of safety than may actually exist. Thus, the qualitative and quantitative measurement of atmospheric contaminants generally is the job of the safety and environmental manager. Any samples taken should also be representative. Samples should be taken of the actual air the workers breathe, at the point they breathe them, in their breathing zone: between the top of the head and the shoulders.

ENVIRONMENTAL CONTROLS

Workplace exposure to toxic materials can be reduced or controlled by a variety of individual control methods or by a combination of methods. Various control methods available to safety and environmental managers are broken down into three categories: Engineering controls, administrative controls, and personal protective equipment.

Engineering Controls

Engineering controls are methods of environmental control whereby the hazard is "engineered out," either by initial design specifications or by applying methods of substitution (e.g., replacing toxic chlorine used in disinfection processes with relatively nontoxic sodium hypochlorite). Engineering control may entail utilization of isolation methods. For example, a diesel generator that, when operating, produces noise levels in excess of 120 decibels (dBA) could be controlled by enclosing it inside a soundproofed enclosure—effectively isolating the noise hazard. Another example of hazard isolation can be seen in the use of tightly closed enclosures that isolate an abrasive blasting operation. This method of isolation is typically used in conjunction with local exhaust ventilation. Ventilation is one of the most widely used and effective engineering controls (because it is so crucial in controlling workplace atmospheric hazards) and is discussed in detail in the following section.

Ventilation

Simply put, *ventilation* is "the" classic method and the most powerful tool of control used in safety engineering to control environmental airborne hazards. Experience (much experience) has shown that the proper use of ventilation as a control mechanism can assure that workplace air remains free of potentially hazardous levels of airborne contaminants. In accomplishing this, ventilation works in two ways: (1) by physically removing the contaminated air from the workplace, or (2) by diluting the workplace atmospheric environment to a safe level by the addition of fresh air.

A ventilation system is all very well and good (virtually essential, actually), but an improperly designed ventilation system can make the hazard worse. This essential point cannot be over-emphasized. At the heart of a proper ventilation system are proper design, proper maintenance, and proper monitoring. The safety and environmental manager plays a critical role in ensuring that installed ventilation systems are operating at their optimum level.

Because of the importance of ventilation in the workplace, the safety engineer must be well versed in the general concepts of ventilation, the principles of air movement, and monitoring practices. The safety and environmental manager must be properly prepared (through training and experience) to evaluate existing systems and design new systems for control of the workplace environment. In the next few sections, we present the general principles and concepts of ventilation system design and evaluation. This material should provide the basic concepts and principles necessary for the proper application of industrial ventilation systems. This material also serves to refresh the knowledge of the practitioner in the field. Probably the best source of information on ventilation is the ACGIH's *Industrial Ventilation: A Manual of Recommended Practice* (current edition). This text is a must-have reference for every safety engineer.

Concepts of Ventilation

The purpose of industrial ventilation is essentially to (under control) recreate what occurs in natural ventilation. Natural ventilation results from differences in pressure. Air moves from high-pressure areas to low-pressure areas. This difference in pressure is the result of thermal conditions. We know that hot air rises, which (for example) allows smoke to escape from the smokestack in an industrial process, rather than disperse into areas where workers operate the process. Hot air rises because air expands as it is heated, becoming lighter. The same principle is in effect when air in the atmosphere becomes heated. The air rises and is replaced by air from a higher-pressure area. Thus, convection currents cause a natural ventilation effect through the resulting winds.

What does all of this have to do with industrial ventilation? Actually, quite a lot. Simply put, industrial ventilation is installed in a workplace to circulate the air within, to provide a supply of fresh air to replace air with undesirable characteristics.

Could this be accomplished simply by natural workplace ventilation? That is, couldn't we just heat the air in the workplace so that it will rise and escape through natural ports—windows, doors, cracks in walls, or mechanical ventilators in the roof (installed wind-powered turbines, for example)? Yes, we could design a natural system like this, but in such a system, air does not circulate fast enough to remove contaminants before a hazardous level is reached, which defeats our purpose in providing a ventilation system in the first place. Thus, we use fans to provide an artificial, mechanical means of moving the air.

Along with controlling or removing toxic airborne contaminants from the air, installed ventilation systems perform several other functions within the workplace. These functions include:

1. Ventilation is often used to maintain an adequate oxygen supply in an area. In most workplaces, this is not a problem because natural ventilation usually provides an adequate volume of oxygen; however, in some work environments (deep mining and thermal processes that use copious amounts of oxygen for combustion) the need for oxygen is the major reason for an installed ventilation system.
2. An installed ventilation system can remove odors from a given area. This type of system (as you might guess) has applications in such places as athletic locker rooms, rest rooms, and kitchens. In performing this function, the noxious air may be replaced with fresh air, or odors may be masked with a chemical masking agent.
3. One of the primary uses of installed ventilation is one that we are familiar with: providing heat, cooling, and humidity control.
4. A ventilation system can remove undesirable contaminants at their source, before they enter the workplace air (e.g., from a chemical dipping or stripping tank). Obviously, this technique is an effective way to ensure that certain contaminants never enter the breathing zone of the worker—exactly the kind of function safety engineering is intended to accomplish.

Earlier, we stated that installed ventilation is able to perform its designed function via the use of a mechanical fan. Actually, a mechanical fan is the heart of any ventilation system, but like the human heart, certain ancillaries are required to make it function as a system. Ventilation is no different. Four major components make up a ventilation system: (1) The fan forces the air to move; (2) an inlet or some type of opening allows air to enter the system; (3) an outlet must be provided for air to leave the system; and (4) a conduit or pathway (ducting) not only directs the air in the right direction, but also limits the amount of flow to a predetermined level.

An important concept regarding ventilation systems is the difference between exhaust and supply ventilation. An *exhaust ventilation system* removes air and air-borne contaminants from the workplace. Such a system may be designed to exhaust an entire work area, or it may be placed at the source to remove the contaminant prior to its release into the workplace air. The second type of ventilation system is the *supply system,* which (as the name implies) adds air to the work area, usually to dilute work area contaminants to lower the concentration of these contaminants. However, a supplied air system does much more; it also provides movement to air within the space (especially when an area is equipped with both an exhaust and supply system—a usual practice because it allows movement of air from inlet to outlet and is important in replenishing exhausted air with fresh air).

Air movement in a ventilation system is a result of differences in pressure. Note that pressures in a ventilation system are measured in relation to atmospheric pressure. In the workplace, the existing atmospheric pressure is assumed to be the zero point. In the supply system, the pressure created by the system is *in addition* to the atmospheric pressure that exists in the workplace (i.e., a positive pressure). In an exhaust system, the objective is to lower the pressure in the system below the atmospheric pressure (i.e., a negative pressure).

When we speak of increasing and decreasing pressure levels within a ventilation system, what we are really talking about is creating small differences in pressure— small when compared to the atmospheric pressure of the work area. For this reason, these differences are measured in terms of *inches of water* or *water gauge*, which results in the desired sensitivity of measurement. Air can be assumed to be incompressible, because of the small-scale differences in pressure.

Let's get back to the water gauge or inches of water. Since one pound per square inch of pressure is equal to 27 inches of water, one inch of water is equal to 0.036 pounds pressure, or 0.24% of standard atmospheric pressure. Remember the potential for error introduced by considering air to be incompressible is very small at the pressure that exists with a ventilation system.

The safety and environmental manager must be familiar with the three pressures important in ventilation: velocity pressure, static pressure, and total pressure. To understand these three pressures and their functions in ventilation systems, you must first be familiar with pressure itself. In fluid mechanics, the energy of a fluid (air) that is flowing is termed *head.* Head is measured in terms of unit weight of the fluid or in foot pounds/pound of fluid flowing. *NOTE*: The usual convention is to describe head in terms of feet of fluid that is flowing.

So what is pressure? *Pressure* is the force per unit area exerted by the fluid. In the English system of measurement, this force is measured in lbs/ft^2. Since we have stated that the fluid in a ventilation system is incompressible, the pressure of the fluid is equal to the head.

Velocity pressure (VP) is created as air travels at a given velocity through a ventilation system. Velocity pressure is only exerted in the direction of airflow and is *always* positive (i.e., above atmospheric pressure). When you think about it, velocity pressure has to be positive, and obviously the force or pressure that causes it also must be positive.

Note that the velocity of the air moving within a ventilation system is directly related to the velocity pressure of the system. This relationship can be derived into the standard equation for determining velocity (and clearly demonstrates the relationship between velocity of moving air and the velocity pressure):

$$v = 4005/\sqrt{VP} \tag{16.1}$$

Static pressure (SP) is the pressure that is exerted in all directions by the air within the system, which tends to burst or collapse the duct. It is expressed in inches of water gauge (wg). A simple example may help you grasp the concept of static pressure. Consider the balloon that is inflated at a given pressure. The pressure within the balloon is exerted equally on all sides of the balloon. No air velocity exits within the balloon itself. The pressure in the balloon is totally the result of static pressure. Note that static pressure can be both negative and positive with respect to the local atmospheric pressure.

Total pressure (TP) is defined as the algebraic sum of the static and velocity pressures or

$$TP = SP + VP \tag{16.2}$$

The total pressure of a ventilation system can be either positive or negative (i.e., above or below atmospheric pressure). Generally, the total pressure is positive for a supply system and negative for an exhaust system.

For the safety engineer to evaluate the performance of any installed ventilation system, he or she must make measurements of pressures in the ventilation system. Measurements are normally made using instruments such as a manometer or a Pitot tube.

The *manometer* is often used to measure the static pressure in the ventilation system. The manometer is a simple, U-shaped tube, open at both ends, and usually constructed of clear glass or plastic so that the fluid level within can be observed. To facilitate measurement, a graduated scale is usually present on the surface of the manometer. The manometer is filled with a liquid (water, oil, or mercury). When pressure is exerted on the liquid within the manometer, the pressure causes the level of liquid to change as it relates to the atmospheric pressure external to the ventilation system. The pressure measured, therefore, is relative to atmospheric pressure as the zero point.

When manometer measurements are used to obtain positive pressure readings in a ventilation system, the leg of the manometer that opens to the atmosphere will contain the higher level of fluid. When a negative pressure is being read, the leg of the tube that opens to the atmosphere will be lower, thus indicating the difference between the atmospheric pressure and the pressure within the system.

The *Pitot tube* is another device used to measure static pressure in ventilation systems. The Pitot tube is constructed of two concentric tubes. The inner tube forms the impact portion, while the outer tube is closed at the end and has static pressure holes normal to the surface of the tube. When the inner and outer tubes are connected to opposite legs of a single manometer, the velocity pressure is obtained directly. If the engineer wishes to measure static pressure separately, two manometers can be used. Positive and negative pressure measurements are indicated on the manometer as above.

Local Exhaust Ventilation

Local exhaust ventilation (the most predominant method of controlling workplace air) is used to control air contaminants by trapping and removing them near the source. In contrast to dilution ventilation (which lets the contamination spread throughout the workplace, later to be diluted by exhausting quantities of air from the workspace), local exhaust ventilation surrounds the point of emission with an enclosure, and attempts to capture and remove the emissions before they are released into the worker's breathing zone. The contaminated air is usually drawn through a system of ducting to a collector, where it is cleaned and delivered to the outside through the discharge end of the exhauster. Figure 16.1 shows a typical local exhaust system, which consists of a hood, ducting, an air-cleaning device, fan, and a stack. W.G. Hazard (1988) points out that a local exhaust system is usually the proper method of contaminant control if:

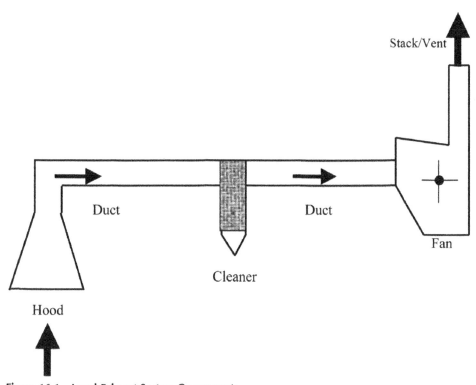

Figure 16.1   Local Exhaust System Components

- the contaminant in the workplace atmosphere constitutes a health, fire, or explosion hazard
- national or local codes require local exhaust ventilation at a particular process
- maintenance of production machinery would otherwise be difficult
- housekeeping or employee comfort will be improved
- emission sources are large, few, fixed and/or widely dispersed
- emission rates vary widely by time
- emission sources are near the worker breathing zone

The safety and environmental manager must remember that determining beforehand the precise effectiveness of a particular system is often difficult. Thus, measuring exposures and evaluating how much control has been achieved after a system is installed is essential. A good system may collect 80 to 90+ percent, but a poor system may capture only 50 percent or less. Without total enclosure of the contaminant sources (where capture is obviously very much greater), the safety and environmental manager must be familiar with handling problems like these.

Once the system is installed and has demonstrated that it is suitable for the task at hand, the system must be well maintained. Careful maintenance is a must. In dealing with ventilation problems, the safety and environmental manager soon finds out that his or her worst headache in maintaining the system is poor—or no—maintenance.

A phenomenon that many practitioners in the safety field forget (or never knew in the first place) is that ventilation, when properly designed, installed, and maintained, can go a long way to ensure a healthy working environment. However, ventilation does have limitations. For example, the effects of blowing air from a supply system and removing air through an exhaust system are different. To better understand the difference and its significance, let's take an example of air supplied through a standard exhaust duct.

When air is exhausted through an opening, it is gathered equally from all directions around the opening. This includes the area behind the opening itself. Thus, the cross-sectional area of airflow approximates a spherical form, rather than the conical form that is typical when air is blown out of a supply system. To correct this problem, a flange is usually placed around the exhaust opening, which reduces the air contour, from the large spherical contour to that of a hemisphere. As a result, this increases the velocity of air at a given distance from the opening. This basic principle is used in designing exhaust hoods. Remember that the closer the exhaust hood is to the source, and the less uncontaminated air it gathers, the more efficient the hood's percentage of capture will be. Simply put, it is easier for a ventilation system to blow air than it is for one to exhaust it. Keep this in mind whenever you are dealing with ventilation systems and/or problems.

General and Dilution Ventilation

Along with local exhaust ventilation are two other major categories of ventilation systems: general and dilution ventilation. Each of these systems has a specific purpose, and finding all three types of systems present in a given workplace location is not uncommon.

General ventilation systems (sometimes referred to as heat control ventilation systems) are used to control indoor atmospheric conditions associated with hot industrial environments (such as those found in foundries, laundries, bakeries, and other workplaces that generate excess heat) for the purpose of preventing acute discomfort or injury. General ventilation also functions to control the comfort level of the worker in just about any indoor working environment. Along with the removal of air that has become process-heated beyond a desired temperature a general ventilation system supplies air to the work area to condition (by heating or cooling) the air or to make up for the air that has been exhausted by dilution ventilation in a local exhaust ventilation system.

A *dilution ventilation system* dilutes contaminated air with uncontaminated air to reduce the concentration below a given level (usually the threshold limit value of the contaminant) to control potential airborne health hazards, fire and explosive conditions, odors, and nuisance-type contaminants. This is accomplished by removing or supplying air, to cause the air in the workplace to move, and as a result, mix the contaminated with incoming uncontaminated air.

This mixing operation is essential. To mix the air there must be, of course, air movement. Air movement can be accomplished by natural draft caused by prevailing winds moving through open doors and windows of the work area.

Thermal draft can also move air. Whether the thermal draft is the result of natural causes or is generated from process heat, the heated air rises, carrying any contaminant present upward with it. Vents in the roof allow this air to escape into the atmosphere. Makeup air is supplied to the work area through doors and windows.

A mechanical air-moving device provides the most reliable source for air movement in a dilution ventilation system. Such a system is rather simple. It requires a source of exhaust for contaminated air, a source of air supply to replace the air mixture that has been removed with uncontaminated air, and a duct system to supply or remove air throughout the workplace. Dilution ventilation systems often are equipped with filtering systems to clean and temper the incoming air.

Industrial Noise Controls

We discussed hearing safety in Chapter 12 of this text. We present industrial noise control here again because of its importance to maintaining a healthy workplace and also because the safety and environmental manager's duties in this particular area of safety and health compliance is important and ongoing; in many instances virtually never ending.

Only recently has noise been recognized as a significant industrial health problem. In fact, now, workers' compensation laws in all states recognize hearing losses due to industrial noise as an occupational disease.

The obvious question is, "What is noise?" Simply put, *noise* is any unwanted sound. The safety engineer is concerned about noise (or any workplace sound) that exceeds OSHA regulated levels and may be injurious to workers, causing hearing damage.

Hearing damage risk criteria for exposure to noise are found in OSHA's 29 CFR 1910.95 (Hearing Conservation Standard) and are stated in Table 16.2.

Table 16.2 Permissible Noise Exposures*

Duration per Day, Hours	Sound Level dBA** Slow Response
8	90
6	92
4	95
3	97
2	100
1.5	102
1	105
0.5	110
0.25	115

* When the daily noise exposure is composed of two or more periods of noise exposure of different levels, their combined effect should be considered, rather than the individual effect of each. If the sum of the following fractions $C_1/T_1 + C_2/T_2 + C_n/T_n$ exceeds unity, then the mixed exposure should be considered to exceed the limit value. C_n indicates the total time of exposure at a specified noise level, and T_n indicates the total time of exposure permitted at that level.
Exposure to impulsive or impact noise should not exceed 140-dB peak sound pressure level.
** Measured on the A-weighting scale of a standard sound level meter is slow response mode.
Source: U.S. Department of Labor, Part 1910. *Occupational Safety and Health Standards* Subpart G., section 1910.95. Washington, D.C.: Occupational Safety and Health Administration, 1995.

The safety and environmental manager's primary concern when starting a noise reduction or control program is first to determine if any "noise-makers" in the facility exceed the OSHA limits for worker exposure and discover exactly which machines or processes produce noise at unacceptable levels. Making this determination is accomplished by conducting a noise level survey of the plant or facility.

When conducting the noise level survey, the safety engineer should use an ANSI-approved *sound level meter* (a device used most commonly to measure sound pressure). Sound is measured in decibels. One decibel is one-tenth of a bel (a unit of measure in electrical communication engineering) and is the minimum difference in loudness that is usually perceptible.

The sound level meter consists of a microphone, an amplifier and an indicating meter, which responds to noise in the audible frequency range of about 20–20,000 Hz. Sound level meters usually contain "weighting" networks designated "A," "B," or "C." Some meters have only one weighting network; others are equipped with all three. The A-network approximates the equal loudness curves at low sound pressure levels, the B-network is used for medium sound pressure levels, and the C-network is used for high levels.

In conducting a routine workplace sound level survey, using the A-weighted network (referenced dBA) in the assessment of the overall noise hazard has become common practice. The A-weighted network is the preferred choice because it is thought to provide a rating of industrial noises that indicates the injurious effects such noise has on the human ear (gives a frequency response similar to that of the human ear at relatively low sound pressure levels).

With an approved and freshly calibrated (always calibrate test equipment prior to use) sound level meter in hand, the safety engineer is ready to begin the sound level survey. In doing so, the safety and environmental manager is primarily interested in answering the following questions: (1) What is the noise level in each work area, (2) what equipment or process is generating the noise, (3) which employees are exposed to the noise, and (4) how long are they exposed to the noise?

In answering these questions, safety and environmental managers record their findings as they move from workstation to workstation, following a logical step-by-step procedure. The first step involves using the sound level meter set for A-scale slow response mode to measure an entire work area. When making such measurements, restrict the size of the space being measured to under 1,000 square feet. If the maximum sound level does not exceed 80 dBA, it can be assumed that all workers in this work area are working in an environment with a satisfactory noise level. However, a note of caution is advised here: The key words in the preceding statement are "maximum sound level." To assure an accurate measurement, the safety and environmental manager must ensure that all "noise-makers" are actually in operation when measurements are taken. Measuring an entire work area does little good when only a small percentage of the noise-makers are actually in operation.

The next step depends on the readings recorded when the entire work area was measured. For example, if the measurements indicate sound levels greater than 80 dBA, then another set of measurements needs to be taken at each worker's workstation. The purpose here, of course, is to determine two things: which machine or process is making noise above acceptable levels (i.e., >80 dBA), and which workers are exposed to these levels. Remember that the worker who operates the machine or process might not be the only worker exposed to the noise-maker. You need to inquire about other workers who might, from time to time, spend time working in or around the machine or process. Our experience in conducting workstation measurements has shown us noise levels usually fluctuate. If this is the case, you must record the minimum and maximum noise levels. If you discover that the noise level is above 90 dBA (and it remains above this level), you have found a noisemaker that exceeds the legal limit (90 dBA). However, if your measurements indicate that the noise level is never greater than 85 dBA (OSHA's action level), the noise exposure can be regarded as satisfactory.

If workstation measurements indicate readings that exceed the 85 dBA level (the level OSHA's Hearing Conservation Requirements kick in), you must perform another step. This step involves determining the length of time of exposure for workers. The easiest, most practical way to make this determination is to have the worker wear a noise dosimeter, which records the noise energy to which the worker was exposed during the workshift.

What happens next?

You must then determine if the worker is exposed to noise levels that exceed the permissible noise exposure levels listed in Table 16.2. The key point to remember is that your findings must be based on a time-weighted average (TWA). For example, from Table 16.2 you will notice that a noise level of 95 dBA is allowed up to 4 hours per day.

Note: This parameter assumes that the worker has good hearing acuity with no loss. If the worker has documented hearing loss, then exposure to 95 dBA or higher may be unacceptable under any circumstances without proper hearing protection.

So exactly what does 4-hour maximum exposure per day mean? It means that, cumulatively, a worker cannot be exposed to noise at the 95-dBA level for more than 4 hours. Cumulative maximum exposures are used because all noise-makers are not necessarily continuous; instead, they may be intermittent or impact-type noise-makers. Consider this: a worker who runs a machine operates the machine 8 hours each day. When the machine is running, it continuously produces 95 dBA. Obviously, this worker must be protected from the 95-dBA noisemaker, because his or her exposure will be over an 8-hour period, which is not allowed under OSHA. Another worker operates a machine that produces 95-dBA noise, but the operator only operates it for a few minutes at a time, with several minutes without the machine running in between operations. The worker is also exposed to noise from other workstations, at varying levels. This is considered intermittent operation, with intermittent noise generation—and possibly intermittent exposure (depending upon the level of noise). Is this worker exposed to noise levels above the permissible exposure limit of 4 hours maximum (i.e., without hearing protection)?

It depends. To make this determination we must calculate the daily noise dose. We can accomplish by using equation 16.3.

$$Em = C_1/T_1 + C_2/T_2 + C_3/T_3 + \cdots + C_n/T_n \qquad (16.3)$$

where

E_m = mixed exposure
C = total time of exposure at a specified noise level
T = total time of exposure permitted at that level

For purposes of illustration let's assume that the worker's intermittent noise levels expose him or her to the following noise levels during the workday:

85 dBA for 2.75 hours
90 dBA for 1 hour
95 dBA for 2.25 hours
100 dBA for 2 hours

The question is has the worker received an excessive exposure during the workday.

To answer this question we use equation 16.3 and plug in the parameters. From our calculation, if we find that the sum of the fractions equals or exceeds 1, then the mixed exposure is considered to exceed the limit value. Daily noise dose (D) is expressed as a percentage of E_m = 1, the mixed exposure is equivalent to a noise dose of 100 percent. Keep in mind that noise levels below 90 dBA are not considered in the calculation of daily noise.

So, again, has our worker received an excessive exposure during her workday? Let's find out:

$$Dose = \frac{1}{8} + \frac{2.25}{4} + \frac{2}{2} = 1.69$$

The sum exceeds 1, therefore, the results indicate that the employee has received an excessive exposure during the workday.

Note: A final word on noise exposure. From Table 16.2, you can see that the highest sound level listed is 115 dBA. Any exposure above this level is not permissible for any length of time.

Engineering Controls for Industrial Noise

When the safety and environmental manager investigates the possibility of using engineering controls to control noise, the first thing he or she recognizes is that reducing and/or eliminating all noise is virtually impossible. And this should not be the focus in the first place . . . eliminating or reducing the "hazard" is the goal. While the primary hazard may be the possibility of hearing loss, the distractive effect (or its interference with communication) must also be considered. The distractive effect of excessive noise can certainly be classified as hazardous whenever the distraction might affect the attention of the worker. The obvious implication of noise levels that interfere with communications is emergency response. If ambient noise is at such a high level that workers can't hear fire or other emergency alarms, this is obviously a hazardous situation.

So what does all this mean? The safety and environmental manager must determine the "acceptable" level of noise. Then he or she can look into applying the various noise control measures. These include making alterations in engineering design (obviously this can only be accomplished in the design phase) or making modifications after installation. Unfortunately, this latter method is the one the safety and environmental manager is usually forced to apply and it is also the most difficult, depending upon circumstances.

Let's assume that the safety engineer is trying to reduce noise levels generated by an installed air compressor to a safe level. Obviously, the first place to start is at the source: the air compressor. Several options are available for the safety engineer to employ at the source. First, the safety and environmental manager would look at the possibility of modifying the air compressor to reduce its noise output. One option might be to install resilient vibration mounting devices. Another might be to change the coupling between the motor and the compressor.

If the options described for use at the source of the noise are not feasible or are only partially effective, the next component the safety and environmental manager would look at is the path along which the sound energy travels. Increasing the distance between the air compressor and the workers could be a possibility. (*Note:* Sound levels decrease with distance). Another option might be to install acoustical treatments on ceilings, floors, and walls. The best option available (in this case) probably is to enclose the air compressor, so that the dangerous noise levels are contained within the enclosure, and the sound leaving the space is attenuated to a lower, safer level. If total enclosure of the air compressor is not practicable, then erecting a barrier or baffle system between the compressor and the open work area might be an option.

The final engineering control component the safety and environmental manager might incorporate to reduce the air compressor's noise problem is to consider the receiver (the worker). An attempt should be made to isolate the operator by providing a noise reduction or soundproof enclosure or booth for the operator.

Industrial Vibration Control

Vibration is often closely associated with noise, but it is frequently overlooked as a potential occupational health hazard. Vibration is defined as the oscillatory motion of a system around an equilibrium position. The system can be in a solid, liquid, or gaseous state, and the oscillation of the system can be periodic or random, steady state or transient, continuous or intermittent (NIOSH, 1973). Vibrations of the human body (or parts of the human body) are not only annoying, they also affect worker performance, sometimes causing blurred vision and loss of motor control. Excessive vibration can cause trauma, which results when external vibrating forces accelerate the body or some part so that amplitudes and restraining capacities by tissues are exceeded.

Vibration results in the mechanical shaking of the body or parts of the body. These two types of vibration are called *whole-body vibration* (affects vehicle operators) and *segmental vibration* (occurs in foundry operations, mining, stonecutting, and a variety of assembly operations, for example). Vibration originates from mechanical motion, generally occurring at some machine or series of machines. This mechanical vibration can be transmitted directly to the body or body part or it may be transmitted through solid objects to a worker located at some distance away from the actual vibration.

The effect of vibration on the human body is not totally understood; however, we do know that vibration of the chest may create breathing difficulties, and that an inhibition of tendon reflexes is a result of vibration. Excessive vibration can cause reduced ability on the part of the worker to perform complex tasks, and indications of potential damage to other systems of the body also exist.

More is known about the results of segmental vibration (typically transmitted through hand to arm), and a common example is the vibration received when using a pneumatic hammer—jackhammer. One recognized indication of the effect of segmental vibration is impaired circulation to the appendage, a condition known as *Raynaud's Syndrome*, also known as "dead fingers" or "white fingers." Segmental vibration can also result in the loss of the sense of touch in the affected area. Some indications exist that decalcification of the bones in the hand can result from vibration transmitted to that part of the body. In addition, muscle atrophy has been identified as a result of segmental vibration.

As with noise, the human body can withstand short-term vibration, even though this vibration might be extreme. The dangers of vibration are related to certain frequencies that are resonant with various parts of the body. Vibration outside these frequencies is not nearly so dangerous as vibration that results in resonance.

Control measures for vibration include substituting some other device (one that does not cause vibration) for the mechanical device that causes the vibration. An important corrective measure (often overlooked) that helps in reducing vibration is proper maintenance of tools, or support mechanisms for tools, including coating the tools with materials that attenuate vibrations. Another engineering control often employed to reduce vibration is the application of balancers, isolators, and damping devices/materials that help to reduce vibration.

Administrative Controls

After the design, construction, and installation phase, installing engineering controls to control a workplace hazard or hazards often becomes difficult. Some exceptions were mentioned in the previous section. A question safety and environmental managers face on almost a continuous basis is, "If I can't engineer out the hazard, what can I do?"

This question would not arise, of course, if the safety and environmental manager had been allowed to participate in the design, construction, and installation phases. However, our experience has shown us that more often than not the safety and environmental manager is excluded from such preliminary construction phases, and this certainly is not "good engineering practice," but it happens. And thus the questions arise on how best to reduce or remove hazards after they have been installed. The safety and environmental manager is tasked with finding the answers.

As a second line of defense, after engineering controls are determined to be impossible, not practicable, not feasible, or cannot be accomplished for technological reasons—or for any reasons—*administrative controls* might be an alternative.

What are administrative controls?

Administrative controls attempt to limit the worker's exposure to the hazard. Normally accomplished by arranging work schedules and related duration of exposures so that employees are minimally exposed to health hazards, another procedure transfers workers who have reached their upper permissible limits of exposure to an environment where no additional exposure will be experienced. Both control procedures are often used to limit worker exposure to air contaminants or noise. For example, a worker who is required to work in an extremely high noise area where engineering controls are not possible would be rotated from the high noise area to a quiet area when the daily permissible noise exposure is reached.

Reducing exposures by *limiting the duration of exposure* (basically by modifying the work schedule) must be carefully managed (most managers soon find that attempting to properly manage this procedure takes a considerable amount of time, effort, and "imagination"). When practiced, reducing worker exposure is based on limiting the amount of time a worker is exposed, ensuring that OSHA permissible exposure limits (PELs) are not exceeded.

Let's pause right here and talk about Permissible Exposure Limits (PELs) and threshold limit values (TLVs). You should know what they are and what significance they play in the safety engineer's daily activities. Let's begin with TLVs.

Threshold limit values (TLVs) are published by the American Conference of Governmental Industrial Hygienists (ACGIH) (an organization made up of physicians, toxicologists, chemists, epidemiologists, and industrial hygienists) in its *Threshold Limit Values for Chemical Substances and Physical Agents in the Work Environment*. These values are useful in assessing the risk of a worker exposed to a hazardous chemical vapor; concentrations in the workplace can often be maintained below these levels with proper controls. The substances listed by ACGIH are evaluated annually, limits are revised as needed, and new substances are added to the list, as information becomes available. The values are established from the experience of many groups in industry, academia, and medicine, and from laboratory research.

The chemical substance exposure limits listed under both ACGIH and OSHA are based strictly on airborne concentrations of chemical substances in terms of milligrams per cubic meter (mg/m^3), parts per million (ppm), and fibers per cubic centimeters ($fibers/cm^3$). Allowable limits are based on three different time periods of average exposure: (1) 8-hour work shifts known as TWA (time-weighted average), (2) short terms of 15 minutes or STEL (short-term exposure limit), and (3) instantaneous exposure of "C" (ceiling). Unlike OSHA's PELs, TLVs are recommended levels only and do not have the force of regulation to back them up.

OSHA has promulgated limits for personnel exposure in workplace air for approximately 400 chemicals listed in Tables Z1, Z2, and Z3 in Part 1910.1000 of the Federal Occupational Safety and Health Standard. These limits are defined as permissible exposure limits (PEL), and like TLVs are based on 8-hour time-weighted averages (or ceiling limits when preceded by a "C"). Keeping within the limits in the Z Tables is the only requirement specified by OSHA for these chemicals. The significance of OSHA's PELs is that they have the force of regulatory law behind them to back them up—compliance with OSHA's PELs is the law.

Evaluation of personnel exposure to physical and chemical stresses in the industrial workplace requires the use of the guidelines provided by TLVs and the regulatory guidelines of PELs. For the safety and environmental manager to carry out the goals of recognizing, measuring, and effecting controls (of any type) of workplace stresses, such limits are a necessity and have become the ultimate guidelines in the science of safety and environmental management. A word of caution is advised, however. These values are set only as guides for the best practice and are not to be considered absolute values. What are we saying here? These values provide reasonable assurance that occupational disease will not occur if exposures are kept below these levels. On the other hand, occupational disease is likely to develop in some people if the recommended levels are exceeded on a consistent basis.

Let's get back to administrative controls.

We stated that one option available to the safety and environmental manager in controlling workplace hazards is the use of an administrative control that involves modifying workers' work schedules to limit the time of their exposure so that the PEL/TLV is not exceeded. We also said (or at least implied) that this procedure is a manager's nightmare to implement and manage. Practicing safety and environmental managers don't particularly like it, either; they feel that such a strategy merely spreads the exposure out and does nothing to control the source. Experience has shown that in many instances this statement is correct. Nevertheless, work schedule modification is commonly used for exposures to such stressors as noise and lead.

Another method of reducing worker exposure to hazards is by ensuring good *housekeeping practices*. Housekeeping practices? Absolutely. Think about it. If dust and spilled chemicals are allowed to accumulate in the work area, workers will be exposed to these substances. This is of particular importance for flammable and toxic materials. Ensuring that housekeeping practices do not allow toxic or hazardous materials to disperse into the air is also an important concern.

Administrative controls can also reach beyond the workplace. For example, if workers work to abate asbestos eight hours a day, they should only wear approved tyvek protective suits and other required personal protective equipment (PPE). After the work assignment is completed, these workers must decontaminate following standard protocol. The last thing these workers should be allowed to do is to wear their personal clothing for such work, avoid decontamination procedures, and then take their contaminated clothing with them when they leave for home. The idea is to leave any contaminated clothing at work (safely stored or properly disposed of).

Implementation of standardized *materials handling* or *transferring procedures* are another administrative control often used to protect workers. In handling chemicals, any transfer operations taken should be closed system or should have adequate exhaust systems to prevent worker exposure or contamination of the workplace air. This practice should also include the use of spill trays to collect overfill spills or leaking materials between transfer points.

Programs that involve visual inspection and automatic sensor devices (*leak detection programs*) allow not only for quick detection but also for quick repair and minimal exposure. When automatic system sensors and alarms are deployed as administrative controls, tying the alarm system into an automatic shutdown system (close a valve, open a circuit, etc.) allows the sensor to detect a leak, sound the alarm, and initiate corrective action (for example, immediate shutdown of the system).

Finally, two other administrative control practices that go hand-in-hand are *training* and *personal hygiene*. For workers to best protect themselves from workplace hazards (to reduce the risk of injury or illness), they must be made aware of the hazards. OSHA puts great emphasis on the worker-training requirement. This emphasis is well placed. No worker can be expected to know the entire workplace, process, or equipment hazards, unless he or she has been properly trained on the hazards.

An important part of the training process is worker awareness. Legally (and morally) workers have the right to know what they are working with, what they are exposed to while on the job; they must be made aware of the hazards. They must also be trained on what actions to take when they are exposed to specific hazards. Personal hygiene practices are an important part of worker protection. The safety and environmental manager must ensure that appropriate cleaning agents and facilities such as emergency eyewashes and showers, and changing rooms are available and conveniently located for worker use.

Personal Protective Equipment

As a hazard control method, *personal protective equipment* (PPE) should only be used when other methods fail to reduce or eliminate the hazard; PPE is the safety and environmental manager's last line (last resort) of defense against hazard exposure in the workplace. A detailed discussion of PPE can be found in Chapter 10. Briefly, the types commonly used to control materials-related hazards include:

• *Respiratory Protection*—when engineering controls are not feasible or are in the process of being instituted, appropriate respirators should be used to control exposures to airborne hazardous materials. Note that under OSHA regulations (specifically 29 CFR 1910.134), the employer is required to provide such equipment whenever it is necessary. Use of such equipment automatically requires you to implement a Respiratory Protection Program.
• *Protective clothing*—includes chemical, thermal, and/or electrical clothing such as gloves, aprons, coveralls, suits, etc. Many materials and types are available to suit different applications and needs (see Chapter 10).

- *Head, eye, hand, foot protection*—This type of PPE is required in any situation that presents a reasonable probability of injury. These items are worn for protection from physical injury and include hard hats, safety glasses, goggles, leather gloves, laboratory gloves, and steel-toed safety shoes.

A final word on PPE as a method of environmental control: PPE has one serious drawback—it does nothing to reduce or eliminate the hazard. This critical point is often ignored or overlooked. What PPE really does is afford the wearer a barrier between him/herself and the hazard—and that is all. Sometimes workers gain a false sense of security when they don PPE, thinking that somehow the PPE is the element that makes them safe, not working safely. An electrician wears the proper type of electrical insulating gloves and stands on a rubber mat while she works on high-voltage electrical switchgear. If she performs her work in a haphazard manner, will she be safe? Will the gloves and rubber mat protect her from electrocution? Maybe. Maybe not. PPE provides some personal protection, but does not substitute for safe work practices.

Another problem with PPE is that often PPE offers the temptation to employ its use without first attempting to investigate thoroughly the possible methods of correcting the unsafe physical conditions. This results in substituting PPE in place of safety engineering methods to correct the hazardous environment (Grimaldi & Simonds, p. 428, 1989).

The safety and environmental manager also learns (rather quickly) that employees often resist using PPE. We see a constant struggle between the safety professional and the worker about ensuring that the worker wears his or her PPE. We hear their excuses. "Those safety glasses get in my way." "That hard hat is too heavy for my head. "I can't do my work properly with those clumsy gloves on my hands." "Gee, I forgot my safety shoes. I think I left them at my girlfriend/boyfriend's house." Like homework assignment excuses, these statements are common, frequent, often irritating, and never-ending (Though some of the more original ones are even quite entertaining). But one thing is certain—workers who do not wear PPE when required are leaving themselves wide open to injury or death. For the novice safety and health practitioner, we can only add: "Welcome to the challenging field of safety and environmental management."

INDOOR AIR QUALITY

The quality of the air we breathe and the attendant consequences for human health are influenced by a variety of factors. These include hazardous material discharges indoors and outdoors, meteorological and ventilation conditions, and pollutant decay and removal processes. Over 80% of our time is spent in indoor environments so that the influence of building structures, surfaces, and ventilation are important considerations when evaluating air pollution exposures (Wadden & Scheff, p. 1, 1983).

For those familiar with *Star Trek*, Trekees (and for those who are not), consider a quotable quote: "The air is the air." However, in regard to the air we breathe, according to USEPA (2001), few of us realize that we all face a variety of risks to our health as we go about our day-to-day lives. Driving our cars, flying in planes, engaging in recreational activities, and being exposed to environmental pollutants all pose varying degrees of risk. Some risks are simply unavoidable. Some we choose to accept because to do otherwise would restrict out ability to lead our lives the way we want. And some are risks we might decide to avoid if we had the opportunity to make informed choices. Indoor air pollution is one risk that we can do something about.

Between 1972 and 2001, a growing body of scientific evidence has indicated that the air within homes and other buildings can be more seriously polluted than the outdoor air in even the largest and most industrialized cities. Other research indicates that people spend approximately 90 percent of their time indoors. A type of microclimate we don't often think about (if at all) is the microclimate we spend 80% of our time in: the office and/or the home (indoors) (Wadden and Scheff 1983). Thus, for many people, the risks to health may be greater due to exposure to air pollution indoors than outdoors (USEPA, 2001).

Not much attention was given to indoor microclimates until after two events took place a few decades ago. The first event had to do with Legionnaires' disease and the second with Sick Building Syndrome. In addition, people who may be exposed to indoor air pollutants for the longest periods of time are often those most susceptible to the effects of indoor air pollution. Such groups include the young, the elderly, and the chronically ill, especially those suffering from respiratory or cardiovascular disease.

The impact of energy conservation on inside environments may be substantial, particularly with respect to decreases in ventilation rates ("Impact of Infiltration...," 1979) and "tight" buildings constructed to minimize infiltration of outdoor air (Woods, 1980; "Impact of Energy Conservation...," 1979). The purpose of constructing "tight buildings" is to save energy—to keep the heat or air conditioning inside the structure. The problem is indoor air-contaminants within these tight structures are not only trapped within but also can be concentrated, exposing inhabitants to even more exposure.

These topics and others along with causal factors leading to indoor air pollution are covered in this section. What about indoor air quality problems in the workplace? In this section we also discuss this pervasive but often overlooked problem. In this regard, we discuss the basics of IAQ (as related to the workplace environment) and the major contaminants that currently contribute to this problem. Moreover, mold and mold remediation, although not new to the workplace, are the new buzzwords attracting attention these days. Contaminants such as asbestos, silica, lead, and formaldehyde contamination are also discussed. Various related remediation practices are also discussed.

Legionnaire's Disease

Since that infamous event that occurred in Philadelphia in 1976 at the Belleview Stratford Hotel during a convention of American Legion members, which included 182 cases and 29 deaths, *Legionella pneumophila* (the deadly bacterium) has become synonymous with the term *Legionnaires' disease.* The deaths were attributed to colonized bacteria in the air-conditioning system cooling tower.

Let's take a look at this deadly killer—a killer that inhabits the microclimates we call offices, hotels, and other indoor spaces.

Organisms of the genus *Legionella* are ubiquitous in the environment and are found in natural fresh water, potable water, as well as in closed-circuit systems, such as evaporative condensers, humidifiers, recreational whirlpools, air handling systems, and, of course, in cooling tower water.

The potential for the presence of *Legionella* bacteria is dependent on certain environmental factors: moisture, temperature (50–140°F), oxygen, and a source of nourishment such as slime or algae.

Not all the ways in which Legionnaires' disease can be spread are known to us at this time; however, we do know that it can be spread through the air. Centers for Disease Control (CDC) states (in its *Questions and Answers on Legionnaires' disease*, CDC No. 28L0343779) that there is no evidence that Legionnaires' disease is spread person to person.

Air-conditioning cooling towers and evaporative condensers have been the source of most outbreaks to date and the bacterium is commonly found in both. Unfortunately, we do not know if this is an important means of spreading of Legionnaires' disease because other outbreaks have occurred in buildings that did not have air-conditioning.

Not all people are at risk of contacting Legionnaires' disease. The people most at risk include persons:

1. With lowered immunological capacity
2. Who smoke cigarettes and abuse alcohol
3. Who are exposed to high concentrations of *Legionella pneumophila*

Most commonly recognized as a form of pneumonia, the symptoms of Legionnaires' disease usually become apparent 2–10 days after known or presumed exposure to airborne Legionnaires' disease bacteria. A sputum-free cough is common, but sputum production is sometimes associated with the disease. Within less than a day, the victim can experience rapidly rising fever and the onset of chills. Mental confusion, chest pain, abdominal pain, impaired kidney function, and diarrhea are associated manifestations of the disease. CDC estimates that 8,000–18,000 people are hospitalized with the disease each year in the United States.

The obvious question becomes: How do we prevent or control Legionnaires' disease? Good question.

The controls presently being used are targeted on cooling towers and air handling units (condensate drain pans).

Cooling tower procedures used to control bacterial growth vary somewhat on the various regions in a cooling tower system. However, control procedures usually include a good maintenance program, including repair/replacement of damaged components, routine cleaning, and sterilization.

In sterilization, a typical protocol calls for the use of chlorine in a residual solution at about 50 ppm combined with a detergent that is compatible to produce the desired sterilization effect. It is important to ensure that even those spaces that are somewhat inaccessible are properly cleaned of slime and algae accumulations.

Control measures for air handling units: condensate drain pans typically involve keeping the pans clean and checking for proper drainage of fluid—this is important to prevent stagnation and the buildup of slime/algae/bacteria. A cleaning and sterilization program is required anytime algae or slime are found in the unit.

Sick Building Syndrome

The second event that got the public's attention regarding microclimates and the possibility of unhealthy environments contained therein was actually spawned by the first Legionnaires' event and other incidents or complaints that followed. What we are referring to here is *sick building syndrome*.

The term sick building syndrome was coined by an international working group under the *World Health Organization* (WHO) in 1982. The WHO working group studied the literature about indoor climate problems and found that these microclimates in buildings are characterized by the same set of frequently appearing complaints and symptoms. WHO came up with five categories of symptoms exemplified by some complaints reported by occupants supposed to suffer from sick building syndrome (SBS). These categories are listed in the following:

1. **Sensory irritation in eyes, nose, and throat:** Pain, sensation of dryness, smarting feeling, stinging, irritation, hoarseness, voice problems.
2. **Neurological or general health symptoms:** Headache, sluggishness, mental fatigue, reduced memory, reduced capability to concentrate, dizziness, intoxication, nausea and vomiting, tiredness.
3. **Skin irritation**: Pain, reddening, smarting or itching sensations, dry skin.
4. **Nonspecific hypersensitivity reactions:** Running nose and eyes, asthma-like symptoms among non-asthmatics, sounds from the respiratory system.
5. **Odor and taste symptoms**: Changed sensitivity of olfactory or gustatory sense, unpleasant olfactory or gustatory perceptions.

In the past similar symptoms had been used to define other syndromes such as the building disease, the building illness syndrome, building-related illness, or the tight-fitting office syndrome, which in many cases appear to be synonyms for the sick building syndrome; thus, the WHO definition of the SBS worked to combine these syndromes into one general definition or summary. A summary compiled by WHO (1982, 1984) and Molhave (1986) of this combined definition includes the five categories of symptoms listed earlier and also:

1. Irritation of mucous membranes in eye, nose, and throat is among the most frequent symptoms
2. Other symptoms, for example, from lower airways or from internal organs, should be infrequent.
3. A large majority of occupants report symptoms.
4. The symptoms appear especially frequent in one building or in part of it.
5. No evident causality can be identified in relation either to exposures or to occupant sensitivity.

The WHO group suggested the possibility that the SBS symptoms have a common causality and mechanism (WHO, 1982). However, the existence of SBS is still a postulate because the descriptions of the symptoms in the literature are anecdotal and unsystematic (Molhave, 1992).

Indoor Air Pollution

Why is indoor air pollution a problem? As indicated above, recognition that the indoor air environment may be a health problem is a relatively recent emergence. The most significant of indoor air quality concerns are the impact of cigarette smoking, stove and oven operation, and emanations from certain types of particleboard, cement, and other building materials (Wadden & Scheff, 1983).

The significance of the indoor air quality problem became apparent not only because of the Legionnaires' Incident of 1976 and the WHO study of 1982 but also because of another factor that came to the forefront in the mid-1970s: The need to conserve energy. In the early 1970s when hundreds of thousands of people were standing in line to obtain gasoline for their automobiles, it was not difficult to drive home the need to conserve energy supplies.

The resulting impact of energy conservation on inside environments has been substantial. This is especially the case in regards to building modifications that were made to decrease ventilation rates and new construction practices that were incorporated to ensure "tight" buildings to minimize infiltration of outdoor air.

There is some irony in this development, of course. While there is a need to ensure proper building design, construction, and ventilation guidelines to avoid the exposure of inhabitants to unhealthy environments, what really resulted in this mad dash to reduce ventilation rates and "tighten" buildings from infiltration was a tradeoff: energy economics versus air quality.

According to Byrd (2003), indoor air quality (IAQ) refers to the effect, good or bad, of the contents of the air inside a structure on its occupants. Stated differently, indoor air quality (IAQ), in this text, refers to the quality of the air inside workplaces as represented by concentrations of pollutants and thermal (temperature and relative humidity) conditions that affect the health, comfort, and performance of employees. Usually, temperature (too hot or too cold), humidity (too dry or too damp), and air velocity (draftiness or motionlessness) are considered "comfort" rather than indoor air quality issues. Unless they are extreme, they may make someone uncomfortable, but they won't make a person ill. Other factors affecting employees, such as light and noise, are important indoor environmental quality considerations, but are not treated as core elements of indoor air quality problems. Nevertheless, most industrial hygienists must take these factors into account in investigating environmental quality situations.

Byrd (2003) further points out that "good IAQ is the quality of air, which has no unwanted gases or particles in it at concentrations, which will adversely affect someone. Poor IAQ occurs when gases or particles are present at an excessive concentration so as to affect the satisfaction of health of occupants."

In the workplace, poor IAQ may only be annoying to one person, however, at the extreme, it could be fatal to all the occupants in the workplace.

The concentration of the contaminant is crucial. Potentially infectious, toxic, allergenic, or irritating substances are always present in the air. Note that there is nearly always a threshold level below which no effect occurs.

Common Indoor Air Pollutants in the Home

This section takes brief source-by-source look at the most common indoor air pollutants, their potential health effects, and ways to reduce their levels in the home.

Radon

Radon is a noble, nontoxic, colorless, odorless gas produced in the decay of radium-226 and is found everywhere at very low levels. Radon is ubiquitously present in the soil and air near to the surface of the earth. As radon undergoes radioactive decay, it releases an alpha particle, gamma ray, and progeny that quickly decay to release alpha and beta particles and gamma rays. Because radon progeny are electrically charged, they readily attach to particles, producing a radioactive aerosol. It is when radon becomes trapped in buildings and concentrations build up in indoor air that exposure to radon becomes of concern. This is the case because aerosol radon-contaminated particles may be inhaled and deposited in the bifurcations of respiratory airways. Irradiation of tissue at these sites poses a significant risk of lung cancer (depending on exposure dose).

How does radon enter a house?

The most common way in which radon enters a house is through the soil or rock upon which the house is built. The most common source of indoor radon is uranium, which is common to many soils and/or rocks. As uranium breaks down, it releases soil or radon gas, and radon gas breaks down into radon decay products or progeny (commonly called *radon daughters*). Radon gas is transported into buildings by pressure-induced convective flows.

There are other sources of radon, for example, from well water and masonry materials.

Radon levels in a house vary in response to temperature-dependent and wind-dependent pressure differentials and to changes in barometric pressures. When the base of a house is under significant negative pressure, radon transport is enhanced.

According to the EPA, 21,000 lung cancer deaths are related to radon and radon is the second leading cause of lung cancer after smoking. In EPA's booklets, *A Citizen's Guide to Radon*, *Radon Reduction Methods: A Homeowner's Guide*, and *Radon Measurement Proficiency Report* (for each state) the following steps are mentioned as ways to reduce exposure to radon in the home:

1. Measure levels of radon in the home.
2. The state radiation protection office can provide you with information on the availability of detection devices or services.
3. Refer to EPA guidelines in deciding whether and how quickly to take action based on test results.
4. Learn about control measures.
5. Take precautions not to draw larger amounts of radon into the house.
6. Select a qualified contractor to draw up and implement a radon mitigation plan.
7. Stop smoking and discourage smoking in your home.
8. Treat radon-contaminated well water by aerating or filtering through granulated activated charcoal.

Environmental Tobacco Smoke

The use of tobacco products by approximately 43.5 million smokers in the United States results in significant indoor contamination from combustion by-products that pose significant exposures to millions of others who do not smoke but who must breathe contaminated indoor air. Composed of side-stream smoke (smoke that comes from the burning end of a cigarette) and smoke that is exhaled by the smoker, it contains a complex mixture of over 4,700 compounds, including both gases and particles.

According to reports issued in 1986 by the Surgeon General and the National Academy of Sciences, environmental tobacco smoke is a cause of disease, including lung cancer, in both smokers and healthy nonsmokers. Environmental tobacco smoke also increase the lung cancer risk associated with exposures to radon.

The following steps can reduce exposure to environmental tobacco smoke in the office and/or home:

1. Give up smoking and discourage smoking in your home and place of work or require smokers to smoke outdoors.
2. A common method of reducing exposure to indoor air pollutants such as environmental tobacco smoke is ventilation, which works to reduce but not eliminate exposure.

Biological Contaminants

A variety of biological contaminants can cause significant illness and health risks. These include mold and mildew, viruses, animal dander and cat saliva, mites, cockroaches, pollen, and infections form airborne exposures to viruses that cause colds and influenza and bacteria that cause Legionnaires' disease and tuberculosis (TB).

The following steps can reduce exposure to biological contaminants in the home and/or office.

1. Install and use exhaust fans that are vented to the outdoors in kitchens and bathrooms, and vent clothes dryers outdoors.
2. Ventilate the attics and crawl spaces to prevent moisture buildup.
3. Keep water trays in cool mist or ultrasonic humidifiers clean and filled with fresh distilled water daily.
4. Water-damaged carpets and buildings materials should be thoroughly dried and cleaned within 24 hours.
5. Maintain good housekeeping practices both in the home and office.

Combustion Byproducts

Combustion byproducts are released into indoor air from a variety of sources. These include unvented kerosene and gas space heaters, woodstoves, fireplaces, gas stoves, and hot water heaters. The major pollutants released from these sources are carbon monoxide, nitrogen dioxide, and particles.

The following steps can reduce exposure to combustion products in the home (and/ or office):

1. Fuel-burning unvented space heaters should only be operated using great care and special safety precautions.
2. Install and use exhaust fans over gas cooking stoves and ranges, and keep the burners properly adjusted.
3. Furnaces, flues, and chimneys should be inspected annually, and any needed repairs should be made promptly.
4. Woodstove emissions should be kept to a minimum.

Household Products

A large variety of organic compounds are widely used in household products because of their useful characteristics, such as the ability to dissolve substances and evaporate quickly. Cleaning, disinfecting, cosmetic, degreasing, hobby products all contain organic solvents, as do paints, varnishes, and waxes. All of these products can release organic compounds while being used, and when they are stored.

The following steps can reduce exposure to household organic compounds:

1. Always follow label instructions carefully.
2. Throw away partially full containers of chemicals safely.
3. Limit the amount you buy.

Pesticides

Pesticides represent a special case of chemical contamination of buildings where the EPA estimates 80–90% of most people's exposure in the air occurs. These products are extremely dangerous if not used properly.

The following steps can reduce exposure to pesticides in the home:

1. Read the label and follow directions.
2. Use pesticides only in well-ventilated areas.
3. Dispose unwanted pesticides safely.

Asbestos

Asbestos became a major indoor air quality concern in the U.S. in the late 1970s. Asbestos is a mineral fiber commonly used in a variety of building materials and has been identified as having the potential (when friable) to cause cancer in humans.

The following steps can reduce exposure to asbestos in the home (and/or office):

1. Do not cut, rip, or sand asbestos-containing materials.
2. When you need to remove or clean up asbestos, use a professional, trained contractor.

Why is IAQ Important to Workplace Owners?

Workplace structures (buildings) exist to protect workers from the elements and to otherwise support worker activity. Workplace buildings should not make workers sick, cause them discomfort, or otherwise inhibit their ability to perform. How effectively a workplace building functions to support its workers and how efficiently the workplace building operates to keep costs manageable is a measure of the workplace building's performance.

The growing proliferation of chemical pollutants in industrial and consumer products, the tendency toward tighter building envelopes and reduced ventilation to save energy, and pressures to defer maintenance and other building services to reduce costs have fostered indoor air quality problems in many workplace buildings. Employee complaints of odors, stale and stuffy air, and symptoms of illness or discomfort breed undesirable conflicts between workplace occupants and workplace managers. Lawsuits sometimes follow.

If indoor air quality is not well managed on a daily basis, remediation of ensuing problems and/or resolution in court can be extremely costly. Moreover, air quality problems in the workplace can lead to reduced worker performance. So it helps to understand the causes and consequences of indoor air quality and to manage your workplace buildings to avoid these problems.

Worker Symptoms Associated with Poor Air Quality

Worker responses to pollutants, climatic factors, and other stressors such as noise and light are generally categorized according to the type and degree of responses and the time frame in which they occur. Workplace managers should be generally familiar with these categories, leaving detailed knowledge to industrial hygienists.

- **Acute effects**—Acute effects are those that occur immediately (e.g., within 24 hours) after exposure. Chemicals released from building materials may cause headaches, or mold spores may result in itchy eyes and runny noses in sensitive individuals shortly after exposure. Generally, these effects are not long lasting and disappear shortly after exposure ends. However, exposure to some biocontaminants (fungi, bacteria, viruses) resulting from moisture problems, poor maintenance, or inadequate ventilation have been known to cause serious, sometimes life-threatening respiratory diseases, which themselves can lead to chronic respiratory conditions.
- **Chronic effects**—Chronic effects are long-lasting responses to long-term or frequently repeated exposures. Long-term exposures to even low concentrations of some chemicals may induce chronic effects. Cancer is the most commonly associated long-term health consequence of exposure to indoor air contaminants. For example, long-term exposures to environmental tobacco smoke, radon, asbestos, and benzene increase cancer risk.
- **Discomfort**—Discomfort is typically associated with climatic conditions but workplace building contaminants may also be implicated. Workers complain of being too hot or too cold or experience eye, nose, or throat irritation because of low humidity. However, reported symptoms can be difficult to interpret. Complaints that the air is "too dry" may result from irritation from particles on the mucous membranes rather than low humidity, or "stuffy air" may mean that the temperature is too warm or there is lack of air movement, or "stale air" may mean that there is a mild but difficult to identify odor. These conditions may be

unpleasant and cause discomfort among workers, but there is usually no serious health implication involved. Absenteeism, work performance, and employee morale, however, can be seriously affected when building managers fail to resolve these complaints.

- **Performance effects**—Significant measurable changes in worker's ability to concentrate or perform mental or physical tasks have been shown to result from modest changes in temperature and relative humidity. In addition, recent studies suggest that the similar effects are associated with indoor pollution due to lack of ventilation or the presence of pollution sources. Estimates of performance losses from poor indoor air quality for all buildings suggest a 2–4% loss on average. Future research should further document and quantify these effects.

Building Factors Affecting Indoor Air Quality

Building factors affecting indoor air quality can be grouped into two factors: Factors affecting indoor climate and factors affecting indoor air pollution.

- **Factors affecting indoor climate**—The thermal environment (temperature, relative humidity, and airflow) are important dimensions of indoor air quality for several reasons. First, many complaints of poor indoor air may be resolved by simply altering the temperature or relative humidity. Second, people who are thermally uncomfortable will have a lower tolerance to other building discomforts. Third, the rate at which chemicals are released from building material is usually higher at higher building temperatures. Thus, if occupants are too warm, it is also likely that they are being exposed to higher pollutant levels.

- **Factors affecting indoor air pollution**—Much of the building fabric, its furnishings and equipment, its occupants and their activities produce pollution. In a well-functioning building, some of these pollutants will be directly exhausted to the outdoors, and some will be removed as outdoor air enters that building and replaces the air inside. The air outside may also contain contaminants that will be brought inside in this process. This air exchange is brought about by the mechanical introduction of outdoor air (outdoor air ventilation rate), the mechanical exhaust of indoor air, and the air exchanged through the building envelope (infiltration and exfiltration).

Pollutants inside can travel through the building as air flows from areas of higher atmospheric pressure to areas of lower atmospheric pressure. Some of these pathways are planned and deliberate so as to draw pollutants away from occupants, but problems arise when unintended flows draw contaminants into occupied areas. In addition, some contaminants may be removed from the air through natural processes, as with the adsorption of chemicals by surfaces or the settling of particles onto surfaces. Removal processes may also be deliberately incorporated into the building systems. Air filtration devices, for example, are commonly incorporated into building ventilation systems.

Thus, the factors most important to understanding indoor pollution are (a) indoor sources of pollution, (b) outdoor sources of pollution, (c) ventilation parameters, (d) airflow patterns and pressure relationships, and (e) air filtration systems.

Types of Workplace Air Pollutants

Common pollutants or pollutant classes of concern in commercial buildings along with common sources of these pollutants are provided in Table 16.3.

Sources of Workplace Air Pollutants

Air quality is affected by the presence of various types of contaminants in the air. Some are in the form of gases. These would be generally classified as toxic chemicals. The types of interest are combustion products (carbon monoxide, nitrogen dioxide), volatile organic compounds (formaldehyde, solvents, perfumes and fragrances, etc.), and semi-volatile organic compounds (pesticides). Other pollutants are in the form of animal dander, etc.); soot; particles from buildings, furnishings and occupants such as fiberglass, gypsum powder, paper dust, lint from clothing, carpet fibers, etc.; dirt (sandy and earthy material), etc.

Burge and Hoyer (1998) point out many specific sources for contaminants that result in adverse health effects in the workplace, including the workers (contagious diseases, carriage of allergens, and other agents on clothing); building compounds (VOCs, particles, fibers); contamination of building components (allergens, microbial agents, pesticides); and outdoor air (microorganisms, allergens, and chemical air pollutants).

When workers complain of IAQ problems, the industrial hygienist is called upon to determine if the problem really is an IAQ problem. If he/she determines that some form of contaminant is present in the workplace, proper remedial action is required. This usually includes removing the source of the contamination.

****

Table 16.3   Indoor Pollutants and Potential Sources

Pollutant or Pollutant Class	Potential Sources
Environmental tobacco smoke	Lighted cigarettes, cigars, pipes
Combustion contaminants	Furnaces, generators, gas or kerosene space heaters, tobacco products, outdoor air, vehicles.
Biological contaminants	Wet or damp materials, cooling towers, humidifiers, cooling coils or drain pans, damp duct insulation or filters, condensation, re-entrained sanitary exhausts, bird droppings, cockroaches or rodents, dust mites on upholstered furniture or carpeting, body odors.
Volatile organic compounds (VOCs)	Paints, stains, varnishes, solvents, pesticides, adhesives, wood preservatives, waxes, polishes, cleansers, lubricants, sealants, dyes, air fresheners, fuels, plastics, copy machines, printers, tobacco products, perfumes, dry cleaned clothing.
Formaldehyde fabrics	Particle board, plywood, cabinetry, furniture.
Soil gases (radon, sewer gas, VOCs)	Soil and rock (radon), sewer drain leak, dry drain methane traps, leaking underground storage tanks, land fill.
Pesticides	Termiticides, insecticides, rodenticides, fungicides, disinfectants, herbicides.

THE CASE OF THE STICKY HEAD

In 1996 (on a daily basis), a supervisor complained about a sticky, perfume-laden, gooey, messy substance that accumulated on his bald head every time he sat at his desk in his second floor office. Wondering what was causing this unusual occurrence, the supervisor finally reported the mysterious daily accumulation of goo to the organizational safety professional.

The organizational safety and environmental manager, a certified industrial hygienist (CIH) not only saw the humor in the supervisor's goo report but also understood the more serious implication: something was afoul (in more than one way) with second floor IAQ. In particular, something was peculiar, strange, and not right about the operation of the second floor's HVAC system.

The safety and environmental manager quickly identified the perfume-laden goo delivery vehicle: the very large-diameter ventilation supply diffuser—located directly above the bald-headed supervisor's desk chair. The safety and environmental manager had attached a plain piece of white copy paper that hung a few inches below the diffuser's pin-holed sized outlets; she waited a few hours.

After a two-hour wait, the safety and environmental manager removed the paper from the diffuser. She discovered that what she was now holding in her hand was no longer a plain, white piece of copy paper but instead a perfect sheet of yellowish, sticky flypaper. Additionally, there was no need to hold the sticky paper close to her breathing zone to notice the heavy, sweet smell of perfumed hairspray oozing, wafting from it.

So, with the problem in hand, so to speak, the safety and environmental manager then had to determine why the women's hairspray, perfume, and whatever else was coming out of the ventilation supply diffuser right above the supervisor's desk chair. The air supply should have been from the building's make up air supply, which comes from outside the building, but obviously it was coming from another source—the wrong source.

The safety and environmental manager began her investigation of the perfumed air source by tracing the overhead ventilation ducting hand-over-hand from room to room. Approximately 100 feet along the overhead ducting from the bald-headed supervisor's office the safety professional noticed that all the ductwork came together into a central, square-boxed metal unit. She noticed that there were two of these central, square-boxed ducts; one was labeled "Supply" and the other "Exhaust."

Standing on an eight-foot stepladder and craning her neck ceiling-ward, the safety and environmental manager was at first confused by the octopi-like ductwork entering and leaving the octopi-like body of supply and exhaust boxes, which acted like distribution plenum areas for moving air into (supply) or out of (exhaust) the system. Trying to follow the maze of ducting to and from these distribution boxes was not easy. However, one thing she noticed right away; the supply ductwork from the bald-headed supervisor's office was not connected to the distribution box that was stenciled Supply but instead was attached to the box stenciled Exhaust. She quickly realized that the supervisor's line was incorrectly attached to the wrong distribution box. To fix the problem seemed simple enough: disconnect the supply duct from the exhaust distribution plenum and connect it to the supply distribution box as it should be.

The fix was not as simple as she first imagined. After thinking about it a few minutes, she realized that there was another problem. Where was all the perfumed, sticky air coming from? To answer this problem, she traced each of the input side ductwork pipes connected to the distribution box back to the point where their exhaust diffusers were located to pick up air that was to be exhausted.

During the ductwork tracing operation, the safety and environmental manager discovered other problems—all the fire/smoke damper lever arms installed at connections where ducts formed junctions with other ducts were in the wrong position. The dampers (located inside the ductwork) direct air flow where wanted or prevent it from moving where not wanted (as fire/smoke dampers, they protect ductwork penetrations in walls or floors that have a fire resistance rating and perform operational smoke control in static or dynamic smoke management systems). With the dampers in the wrong positions, air flow was being directed where it was not supposed to flow or there was no flow at all in some ducts.

Even though she identified more than a dozen dampers that were positioned in the improper position, the safety and environmental manager left the dampers in the position she found them in (she did not want to change anything that would prevent her from understanding the problem). She continued her hand-over-hand duct tracing. This process took a few hours but finally lead to her gaining full understanding of the problem. At the terminal end points of the ducts she found 24-ceiling intake exhaust diffusers positioned above a large second floor area which had been partitioned into 60 separate office spaces. The office spaces were occupied by 60 administrative clerks. All of the clerks were female.

The safety and environmental manager now had a pretty good inclination as to the source of the sticky hair spray and perfume that the bald-headed supervisor had complained about. However, as a professional she wanted to actually observe what she assumed to be the case—to be sure of her assumptions. Thus, she spent the next full workday walking through the partitioned office area at various times and observed the clerks within their partitioned work areas. She noticed that at lunch time, many of the woman used hair spray to groom their hair and several others applied perfume from spray bottles. She also noticed that just a few minutes before completing the workday and leaving the building for home, many of the clerks performed the same routine of hair spray and perfume application.

The safety and environmental manager observed a couple of these hair spray applications and noticed that she could actually see some of the aerosol from spray cans directed at each head of hair. She also noticed that some of the aerosol was pulled upward, toward the nearest exhaust diffuser. Once inside the diffuser, the aerosol-laden air flowed directly to the distribution box area where it should have been pulled into the main exhaust duct to the exterior of the building. Instead, when the aerosol-laden air stream left the distribution box, it was pushed into the wrongly connected air supply duct that ventilated the bald-headed supervisor's office.

In correcting the problem, the safety professional hired a ventilation contractor who specialized in correcting HVAC problems. The contractor was tasked with a threefold project: remove the sticky material (hairspray and perfume residue) coating the interior exhaust ducting; reconnect correctly the proper ductwork to its proper distribution box; and place each smoke/fire damper in the correct position. While placing the dampers in their designed positions, the safety professional had the contractor fasten and lock each mechanical damper (those that had to be manipulated by hand to change position) so that no one could intentionally and incorrectly reposition the damper without unlocking them. The automatic smoke and fire dampers were left unlocked to close in case of fire or smoke.

After the contractor completed the work, the safety and environmental manager assembled a team of observers and placed an observer at an exterior exhaust outlet duct, several observers at various supply and exhaust ducts, and she took up station right beneath the bald-headed supervisor's supply diffuser. Each observer (equipped with walkie-talkie) was directed to direct a spray can of observation tracer smoke spray into exhaust diffusers while the other observers simply watched to ensure that none of the smoke came out of the wrong diffuser. The safety and environmental manager watched to make sure that the bald-headed supervisor's air supply diffuser continued to supply conditioned air with conditioned fresh air without any trace of smoke.

After four separate successful tests in a row, the safety and environmental manager directed each observer to leave their posts and assemble in the conference room. In the conference room, the safety observer listened as each observer gave his or her report of what they had observed. Satisfied with the results of the study, the safety observer thanked and dismissed the observers. Later she reported to the bald-headed supervisor that he no longer had to fear a sticky head while working in his office. A week later, the safety professional conducted an employee meeting and stressed the importance of the building's HVAC system. She also explained that the building's HVAC system was not to be tampered with or adjusted by anyone without her permission.

***

Tables 16.4 and 16.5 identify indoor and outdoor sources (respectively) of contaminants commonly found in the workplace and offer some measures for maintaining control of these contaminants.

Table 16.5 identifies common sources of contaminants that are introduced from outside buildings. These contaminants frequently find their way inside through the building shell, openings, or other pathways to the inside.

Indoor Air Contaminant Transport

Air contaminants reach worker breathing zones by traveling from the source to the worker by various pathways. Normally, the contaminants travel with the flow of air. Air moves from areas of high pressure to areas of low pressure. That is why controlling workplace air pressure is an integral part of controlling pollution and enhancing building IAQ performance.

Air movements should be from occupants, toward a source, and out of the building rather than from the source to the occupants and out the building. Pressure differences will control the direction of air motion and the extent of occupant exposure.

Driving forces change pressure relationships and create airflow. Common driving forces are identified in Table 16.6.

Table 16.4   Indoor Sources of Contaminants

Category/Common Sources	Mitigation and Control
Housekeeping and Maintenance	
• Cleanser	• Use low-emitting products
• Waxes and polishes	• Avoid aerosols and sprays
• Disinfectants	• Dilute to proper strength
• Air fresheners	• Do not overuse; use during unoccupied hours
• Adhesives	• Use proper protocol when diluting and mixing
• Janitor's/storage closets	• Store properly with containers closed and lid tight
• Wet mops	• Use exhaust ventilation for storage spaces (eliminate
• Drain cleaners	return air)
• Vacuuming	• Clean mops, store mop top up to dry
• Paints and coatings	• Avoid "air fresheners"—clean and exhaust instead
• Solvents	• Use high-efficiency vacuum bags/filters
• Pesticides	• Use integrated Pest Management
• Lubricants	
Occupant-Related Sources	
• Tobacco products	• Smoking policy
• Office equipment (printers/copiers)	• Use exhaust ventilation with pressure control for major
• Cooking/microwave	local sources
• Art supplies	• Low-emitting art supplies/marking pens
• Marking pens	• Avoid paper clutter
• Paper products	• Education material for occupants and staff
• Personal products (e.g., perfume)	
• Tracked in dirt/pollen	
Building Uses as Major Sources	
• Print/photocopy shop	• Use exhaust ventilation & pressure control
• Dry cleaning	• Use exhaust hoods where appropriate; check hood
• Science laboratory	airflows
• Medical office	
• Hair/nail salon	
• Cafeteria	
• Pet store	
Building-Related Sources	
• Plywood/compressed wood	• Use low-emitting sources
• Construction adhesives	• Air out in an open/ventilated area before installing
• Asbestos products	• Increase ventilation rates during and after installing
• Insulation	• Keep material dry prior to enclosing
• Wall/floor coverings (vinyl/plastic)	
• Carpets/carpet adhesives	
• Wet building products	
• Transformers	
• Upholstered furniture	
• Renovation/remodeling	
HVAC system	
• Contaminated filters	• Perform HVAC preventive maintenance
• Contaminated duct lining	• Change filter
• Dirty drain pans	• Clean drain pans; proper slope and drainage
• Humidifiers	• Use portable water for humidification

(Continued)

Table 16.4   (Continued)

Category/Common Sources	Mitigation and Control
• Lubricants • Refrigerants • Mechanical room • Maintenance activities • Combustion appliances	• Keep duct lining dry; move lining outside of duct of Boilers/furnaces/stoves/generators if possible • Fix leaks/clean spills • Maintain spotless mechanical room (not a storage area) • Avoid back drafting • Check/maintain flues from boiler to outside • Keep combustion appliances properly tuned • Disallow unvented combustion appliances • Perform polluting activities during unoccupied hours
Moisture	
• Mold	• Keep building dry
Vehicles	
• Underground/attached garage	• Use exhaust ventilation • Maintain garage under negative pressure relative to the building • Check air flow patterns frequently • Monitor CO

Table 16.5   Outdoor Sources of Contaminants

Category/Common Sources	Mitigation and Control
Ambient Outdoor Air	
• Air quality in the general area	• Filtration or air cleaning of intake air
Vehicular Sources	
• Local vehicular traffic • Vehicle idling areas • Loading dock	• Locate air intake away from source • Require engines shut off at loading dock • Pressurize building/zone • Add vestibules/sealed doors near source
Commercial/Manufacturing Sources	
• Laundry or dry cleaning • Paint shop • Restaurant • Photo-processing • Automotive shop/gas station • Electronics manufacture/assembly • Various industrial operations	• locate air intake away from source • pressurize building relative to outdoors • consider air-cleaning options for outdoor air intake • use landscaping to block or redirect flow of contaminants
Utilities/Public Works	
• Utility power plant • Incinerator • Water treatment plant	
Agricultural	
• Pesticide spraying • Processing or packing plants • Ponds	
Construction/Demolitions	
	• Pressurize building • Use walk-off mats

(Continued)

Table 16.5   Outdoor Sources of Contaminants (*Continued*)

Category/Common Sources	Mitigation and Control
Building Exhaust	
• Bathrooms exhaust • Restaurant exhaust • Air handler relief vent • Exhaust from major tenant (e.g., dry cleaner)	• Separate exhaust or relief from air intake • Pressurize building
Water Sources	
• Pools of water on roof • Cooling tower mist	• Proper roof drainage • Separate air intake from source of water • Treat and maintain cooling tower water
Birds and Rodents	
• Fecal contaminants • Bird nesting	• Bird proof intake grills • Consider vertical grills
Building Operations and Maintenance	
• Trash and refuse area • Chemical/fertilizer/grounds keeping storage • Painting/roofing/sanding	• Separate source from air intake • Keep source area clean/lids on tight • Isolate storage area from occupied areas
Ground Sources	
• Soil gas • Sewer gas • Underground fuel storage tanks	• Depressurize soil • Seal foundation and penetrations to foundations • Keep air ducts away from ground sources

Table 16.6   Major Driving Forces

Driving Force	Effect
Wind	Positive pressure is created on the windward side causing infiltration, and negative pressure on the leeward side causing exfiltration, though wind direction can be varied due to surrounding structures.
Stack effect	When the air inside is warmer than outside, it rises, sometimes creating a column—rising up stairwells, elevator shafts, vertical pipe chases etc. This buoyant force of the air results in positive pressure on the higher floors and negative pressure on the lower floors and a neutral pressure plane somewhere between.
HVAC/fans	Fans are designed to push air in a directional flow and create positive pressure in front and negative pressure behind the fan.
Flues and Exhaust	Exhausted air from a building will reduce the building air pressure relative to the outdoors. Air exhausted will be replaced either through infiltration or through planned outdoor air intake vents.
Elevators	The pumping action of a moving elevator can push air out of or draw air into the elevator shaft as it moves.

Indoor Air Dispersion Parameters

Several parameters (some characterize observed patterns of contaminant distribution and others characterize flow features, such as stability) are used to characterize the dispersion of a contaminant inside a room. Table 16.7 lists these parameters and groups them under four headings: contaminant distribution, temperature distribution, stability, and buoyancy of the room air, and supply air conditions. Because of overlap between these subjects, some parameters appear more than once.

Table 16.7 Parameters Used to Characterize Contaminant Dispersion in Rooms

Contaminant Distribution	Temperature Distribution	Stability/Buoyancy Room Air	Supply Air Conditions
Contaminant Concentration	Air Diffusion Performance	Index Reynolds Number	Purging Effectiveness of Inlet
Local Mean Age of Air	Temperature Effectiveness	Rayleigh Number	
Purging Effectiveness of Inlets	Effective Draft Temperature	Grashof Number	Reynolds Number
Local Specific Contaminant-Accumulating Index		Froude Number Richardson Number	Froude Number Archimedes Number
Air Change Efficiency		Flux Richardson Number	
Ventilation Effectiveness Factor		Buoyancy Flux	
Relative Ventilation Efficiency			

Source: ASHARAE, Handbook: HVAC Systems & Equipment (2000), Fundamentals (2001), Refrigeration (2002) and HVAC Applications (2003).

Did You Know?

Flow parameters such as the mean age of air are difficult but not impossible to calculate experimentally. They are used mainly as a tool to help interpret data from numerical simulations of contaminant dispersion (Peng & Davidson, 1999).

Parameters

- **Contaminant concentration**—indicator of contaminant distribution in a room, that is, the mass of contaminant per unit volume of air (measured in kg/m^3).
- **Local mean age of air**—is the average time it takes for air to travel from the inlet to any point P in the room (Di Tommaso et al., 1999).
- **Purging effectiveness of inlets**—is a quantity that can be used to identify the relative performance of each inlet in a room where there are multiple inlets.
- **Local specific contaminant-accumulating index**—a general index capable of reflecting the interaction between the ventilation flow and a specific contaminant source.
- **Air change efficiency (ACE)**—is a measure of how effectively the air present in a room is replaced by fresh air from the ventilation system (Tommaso et al., 1999). It is the ratio of room mean age that would exist if the air in the room was completely missed to the average time of replacement of the room.
- **Ventilation effectiveness factor (VEF)**—is defined as the ratio of two contaminant concentration differentials: the contaminant concentration in the supply air (typically zero) and the contaminant concentration in the room under complete mixing conditions (Zhang et al. 2001).
- **Relative ventilation efficiency**—is the ratio of the local mean age that would exist if the air in the room were completely mixed to the local mean age that is actually measured at a point.

- **Air diffusion performance index (ADPI)**—is primarily a measure of occupant comfort rather than an indicator of contaminant concentrations. It expresses the percentage of locations in an occupied zone that meet air movement and temperature specifications for comfort.
- **Temperature effectiveness**—is similar in concept to ventilation effectiveness and reflects the ability of a ventilation system to remove heat.
- **Effective draft temperature**—indicates the feeling of coolness due to air motion.
- **Reynolds number**—expresses the ratio of the inertial forces to viscous forces.
- **Rayleigh number**—characterizes natural convection flows.
- **Grashof number**—is equivalent to the Rayleigh number divided by the Prandtl number (a dimensionless number approximating the ratio of viscosity and thermal diffusivity).
- **Froude number**—a dimensionless number used to characterize flow through corridors and doorways and in combined displacement and wind ventilation cases.
- **Richardson number**—characterizes the importance of buoyancy.
- **Flux Richardson number**—is used to characterize the stabilizing effect of stratification on turbulence.
- **Buoyancy flux**—used to characterize buoyancy driven flows.
- **Archimedes number**—conditions of the supplied air are often characterized by the discharge Archimedes number, which expresses the ratio of the buoyancy forces to momentum forces or the strength of natural convection to forced convection.

Common Airflow Pathways

Contaminants travel along pathways—sometimes over great distances. Pathways may lead from an indoor source to an indoor location or from an outdoor source to an indoor location.

The location experiencing a pollution problem may be close by, in the same or an adjacent area, but it may be a great distance from, and/or on a different floor from a contaminant source.

Knowledge of common pathways helps to track down the source and/or prevent contaminants from reaching building occupants (see Table 16.8).

Major IAQ Contaminants

Safety and environmental managers spend a large portion of their time working with and mitigating air contaminant problems in the workplace. The list of potential contaminants workers might be exposed to while working is extensive. There are, however, a few major chemical-/material-derived air contaminants (other than those poisonous gases and materials that are automatically top priorities for the industrial hygienist to investigate and mitigate) that are considered extremely hazardous. These too garner the safety and environmental manager's immediate attention and remedial action(s). In this section, we focus on asbestos, silica, formaldehyde, and lead as those hazardous contaminants (keeping in mind that there are others) requiring the immediate attention of the industrial hygienist.

Table 16.8   Common Airflow Pathways for Contaminants

Common Pathway	Comment
Indoors	
Stairwell/Elevator shaft	The stack effect brings about air flow by drawing air toward these chases on the lower floors and elevator shaft away from these chases on the higher floors, affecting the flow of contaminants
Vertical electrical or plumbing chases	
Receptacles, outlets, openings	Contaminants can easily enter and exit building cavities and thereby move from space to space.
Duct or plenum	Contaminants are commonly carried by the HVAC system throughout the occupied spaces.
Duct or plenum leakage	Duct leakage accounts for significant unplanned air flow and energy loss in buildings.
Flue or exhaust leakage	Leaks from sanitary exhausts or combustion flues can cause serious health problems.
Room spaces	Air and contaminants move within a room or through doors and corridors to adjoining spaces.
Outdoors to Indoors	
Indoor air intake	Polluted outdoor air or exhaust air can enter the building through the air intake.
Windows/doors	A negatively pressurized building will draw air and outside pollutants into the building through any available opening.
Substructures/slab penetrations	Radon and other soil gases and moisture laden air or microbial contaminated air often travel through crawlspaces and other substructures into the building.

Asbestos Exposure (OSHA, 2002)

Asbestos is the name given to a group of naturally occurring minerals widely used in certain products, such as building materials and vehicle brakes, to resist heat and corrosion. Asbestos includes chrysotile, amosite, crocidolite, tremolite, anthophyllite, actinolite, and any of these materials that have been chemically treated and/or altered. Typically, asbestos appears as a whitish, fibrous material, which may release fibers that range in texture from coarse to silky; however, airborne fibers that can cause health damage may be too small to seen with the naked eye.

An estimated 1.3 million employees in construction and general industry face significant asbestos exposure on the job. Heaviest exposures occur in the construction industry, particularly during the removal of asbestos during renovation or demolition (abatement). Employees are also likely to be exposed during the manufacture of asbestos products (such as textiles, friction products, insulation, and other building materials) and automotive brake and clutch repair work.

The inhalation of asbestos fibers by workers can cause serious diseases of the lungs and other organs that may not appear until years after the exposure has occurred. For instance, asbestosis can cause a buildup of scar-like tissue in the lungs and result in loss of lung function that often progresses to disability and death. As mentioned, asbestos fibers associated with these health risks are too small to be seen with the naked eye, and smokers are at higher risk of developing some asbestos-related diseases. For example, exposure to asbestos can cause asbestosis (scarring of the lungs resulting in loss of lung function that often progresses to disability and to death); mesothelioma (cancer affecting the membranes lining the lungs and abdomen); lung cancer; and cancers of the esophagus, stomach, colon, and rectum.

OSHA has issued the following three standards to assist safety and environmental managers with compliance and to protect workers from exposure to asbestos in the workplace:

- 29 CFR 1926.1101 covers construction work, including alteration, repair, renovation, and demolition of structures containing asbestos.
- 29 CFR 1915.1001 covers asbestos exposure during work in shipyards.
- 29 CFR 1910.1001 applies to asbestos exposure in general industry, such as exposure during brake and clutch repair, custodial work, and manufacture of asbestos-containing products.

The standards for the construction and shipyard industries classify the hazards of asbestos work activities and prescribe particular requirements for each classification:

- Class I—is the most potentially hazardous class of asbestos jobs and involves the removal of thermal system insulation and sprayed on or troweled-on surfacing of asbestos-containing materials or presumed asbestos-containing materials.
- Class II—includes the removal of other types of asbestos-containing materials that are not thermal systems insulation, such as resilient flooring and roofing materials containing asbestos.
- Class III—focuses on repair and maintenance operations where asbestos-containing materials are disturbed.
- Class IV—pertains to custodial activities where employees clean up asbestos-containing waste and debris.

There are equivalent regulations in states with OSHA-approved state plans.

Permissible Exposure Limits

Employee exposure to asbestos must not exceed 0.1 fibers per cubic centimeter (f/cc) of air, averaged over an 8-hour work shift. Short-term exposure must also be limited to not more than 1 f/cc, averaged over 30 minutes. Rotation of employees to achieve compliance with either permissible exposure limit (PEL) is prohibited.

Exposure Monitoring

In construction and shipyard work, unless the industrial hygienist is able to demonstrate that employee exposures will be below the PELs (a "negative exposure assessment"), it is generally a requirement that monitoring for workers in Class I and II regulated areas be conducted. For workers in other operations where exposures are expected to exceed one of the PELs, periodic monitoring must be conducted. In general industry, for workers who may be exposed above a PEL or above the excursion limit, initial monitoring must be conducted. Subsequent monitoring at reasonable intervals must be conducted, and in no case at intervals greater than 6 months for employees exposed above a PEL.

Competent Person

In all operations involving asbestos removal (abatement), employers must name a "competent person" qualified and authorized to ensure worker safety and health, as required by Subpart C., "General Safety and Health Provisions for Construction" (29 CFR 1926.20). Under the requirements for safety and health prevention programs, the competent person must frequently inspect job sites, materials, and equipment. A fully trained and licensed industrial hygienist often fills this role.

In addition, for Class I jobs, the competent person must inspect onsite at least once during each work shift and upon employee request. For Class II and III jobs, the competent person must inspect often enough to assess changing conditions and upon employee request.

Regulated Areas

In general industry and construction, regulated areas must be established where the 8-hour TWA or 30-minute excursions values for airborne asbestos exceed the PELs. Only authorized persons wearing appropriate respirators can enter a regulated area. In regulated areas, eating, smoking, drinking, chewing tobacco or gum, and applying cosmetics are prohibited. Warning signs must be displayed at each regulated area and must be posted at all approaches to regulated areas.

Methods of Compliance

In both general industry and construction, employers must control exposures using engineering controls, to the extent feasible. Where engineering controls are not feasible to meet the exposure limit, they must be used to reduce employee exposures to the lowest levels attainable and must be supplemented by the use of respiratory protection.

Respirators

In general industry and construction, the level of exposure determines what type of respirator is required; the standards specify the respirator to be used. Keep in mind that respirators must be used during all Class I asbestos jobs. Refer to 29 CFR 1926.103 for further guidance on when respirators must be worn.

Labels

Caution labels must be placed on all raw materials, mixtures, scrap, waste, debris, and other products containing asbestos fibers.

Protective Clothing

For any employee exposed to airborne concentrations of asbestos that exceed the PEL, the employer must provide and require the use of protective clothing such as coveralls or similar full-body clothing, head coverings, gloves, and foot covering. Wherever the possibility of eye irritation exists, face shields, vented goggles, or other appropriate protective equipment must be provided and worn.

Training

For employees involved in each identified work classification, training must be provided. The specific training requirements depend upon the particular class of work being performed. In general industry, training must be provided to all employees exposed above a PEL. Asbestos awareness training must also be provided to employees who perform housekeeping operations covered by the standard. Warning labels must be placed on all asbestos products, containers, and installed construction materials when feasible.

Recordkeeping

The employer must keep an accurate record of all measurements taken to monitor employee exposure to asbestos. This record is to include: the date of measurement, operation involving exposure, sampling and analytical methods used, and evidence of their accuracy; number, duration, and results of samples taken; type of respiratory protective devices worn; name, social security number, and the results of all employee exposure measurements. This record must be kept for 30 years.

Hygiene Facilities and Practices

Clean change rooms must be furnished by employers for employees who work in areas where exposure is above the TWA and/or excursion limit. Two lockers or storage facilities must be furnished and separated to prevent contamination of the employee's street clothes from protective work clothing and equipment. Showers must be furnished so that employees may shower at the end of the work shift. Employees must enter and exit the regulated area through the decontamination area.

The equipment room must be supplied with impermeable, labeled bags and containers for the containment and disposal of contaminated protective clothing and equipment.

Lunchroom facilities for those employees must have a positive pressure, filtered air supply and be readily accessible to employees. Employees must wash their hands and faces prior to eating, drinking or smoking. The employer must ensure that employees do not enter lunchroom facilities with protective work clothing or equipment unless surface fibers have been removed from the clothing or equipment.

Employees may not smoke in work areas where they are occupationally exposed to asbestos.

Medical Exams

In general industry, exposed employees must have a preplacement physical examination before being assigned to an occupation exposed to airborne concentrations of asbestos at or above the action level or the excursion level. The physical examination must include chest X-ray, medical and work history, and pulmonary function tests. Subsequent exams must be given annually and upon termination of employment, though chest X-rays are required annually only for older workers whose first asbestos exposure occurred more than 10 years ago.

In construction, examinations must be made available annually for workers exposed above the action level or excursion limit for 30 or more days per year or who are required to wear negative pressure respirators; chest X-rays are at the discretion of the physician.

Silica Exposure

Crystalline silica (SiO_2) is a major component of the earth's crust. In pure, natural form, SiO_2 crystals are minute, very hard, translucent, and colorless. Most mined minerals contain some SiO_2. "Crystalline" refers to the orientation of SiO_2 molecules in a fixed pattern as opposed to a nonperiodic, random molecular arrangement defined as amorphous (e.g., diatomaceous earth). Therefore, silica exposure occurs in a wide variety of settings, such as mining, quarrying, and stone cutting operations; ceramics and vitreous enameling; and in use of filters for paints and rubber. The wide use and multiple applications of silica in industrial applications combine to make silica a major occupational health hazard (silicosis), which can lead to death.

Silicosis is a disabling, nonreversible, and sometimes fatal lung disease caused by overexposure to respirable crystalline silica. More than one million U.S. workers are exposed to crystalline silica, and each year more than 250 die from silicosis (see Table 16.9-A and 16.9-B). There is no cure for the disease, but it is 100 percent preventable if employers, workers, and health professionals work together to reduce exposures.

Table 16.9-A and 16.9-B—Deaths from Silica in the Workplace. The first column is the occupation title. The second column, Proportional Mortality Ratio (PMR) is the observed number of deaths from silicosis per occupation divided by the expected number of deaths. Therefore, a value of one indicates no additional risk. A value of ten would indicate a risk ten times greater than normal risk of silicosis. The first table provides risk by occupation and the second provides risk by industry.

Table 16.9-A Deaths From Silica in the Workplace

Occupation	PMR
Miscellaneous metal and plastic machine operators	168.44
Hand molders and shapers, except jewelers	64.12
Crushing and grinding machine operators	50.97
Hand molding, casting, and forming occupations	35.70
Molding and casting machine operators	30.60
Mining machine operators	19.61
Mining occupations (not elsewhere classified)	15.33
Construction trades (not elsewhere classified)	14.77
Grinding, abrading, buffing, and polishing machine operators	8.47
Heavy equipment mechanics	7.72
Miscellaneous material moving equipment operators	6.92
Millwrights	6.56
Crane and tower operators	6.02
Brickmasons and stonemasons	4.71
Painters, construction, and maintenance	4.50
Furnace, kiln, oven operators, except food	4.10
Laborers, except construction	3.79
Operating engineers	3.56
Welders and cutters	3.01
Machine operators, not specified	2.86
Not specified mechanics and repairers	2.84
Supervisors, production occupations	2.73
Construction laborers	2.14
Machinists	1.79
Janitors and cleaners	1.78

Table 16.9-B Deaths From Silica in the Workplace

Industry	PMR
Metal mining	69.51
Miscellaneous nonmetallic mineral and stone products	55.31
Nonmetallic mining and quarrying, except fuel	49.77
Iron and steel foundries	31.15
Pottery and related products	30.73
Structural clay products	27.82
Coal mining	9.26
Blast furnaces, steelworks, rolling and finishing mills	6.49
Miscellaneous fabricated metal products	5.87
Miscellaneous retail stores	4.63
Machinery, except electrical (not elsewhere classified)	3.96
Other primary metal industries	3.63
Industrial and miscellaneous chemicals	2.72
Not specified manufacturing industries	2.67
Construction	1.82

Source: *Work-Related Lung Disease Surveillance Report* (2002). National Institute for Occupational Safety and Health, U.S. Department of Health and Human Services, Public Health Service, Centers for Disease Control and Prevention, Tables 3–8; DHHA (NIOSH) Publication No. 96–134. Publications Dissemination, EID, National Institute for Occupational Safety and Health, 4676 Columbia Parkway, Cincinnati, OH.

Guidelines for Control of Occupational Exposure to Silica

In accordance with the Occupational Safety and Health Administration's (OSHA) standard for air contaminants (29 CFR 1910.1000), employee exposure to airborne crystalline silica shall not exceed an 8-hour time-weighted average limit (variable) as stated in 29 CFR 1910.1000, Table Z-3, or a limit set by a state agency whenever a state-administered Occupational Safety and Health Plan is in effect.

As mandated by OSHA, the first mandatory requirement is that employee exposure be eliminated through the implementation of feasible engineering controls (e.g., dust suppression and ventilation). After all such controls are implemented and they do not control to the permissible exposure, each employer must rotate its employees to the extent possible in order to reduce exposure. Only when all engineering or administrative controls have been implemented, and the level of respirable silica still exceeds permissible exposure limits, may an employer rely on a respirator program pursuant to the mandatory requirements of 29 CFR1910.134. Generally where working conditions or other practices constitute recognized hazards likely to cause death or serious physical harm, they must be corrected.

Formaldehyde (HCHO) Exposure

Formaldehyde (HCHO) is a colorless, flammable gas with a pungent suffocating odor. Formaldehyde is common to the chemical industry. It is the most important aldehyde produced commercially, and is used in the preparation of urea-formaldehyde and phenol-formaldehyde resins. It is also produced during the combustion of organic materials and is a component of smoke.

The major sources in workplace settings are in manufacturing processes (used in the paper, photographic, and clothing industries) and building materials. Building materials may contain phenol, urea, thiourea, or melamine resins that contain

HCHO. Degradation of HCHO resins can occur when these materials become damp from exposure to high relative humidity, or if the HCHO materials are saturated with water during flooding, or when leaks occur. The release of HCHO occurs when the acid catalysts involved in the resin formulation are reactivated. When temperatures and relative humidity increase, out-gassing increases (DOH, Wash., 2003).

Formaldehyde exposure is most common through gas-phase inhalation. However, it can also occur through liquid-phase skin absorption. Workers can be exposed during direct production, treatment of materials, and production of resins. Health care professionals; pathology and histology technicians; and teachers and students who handle preserved specimens are potentially at high risk.

Studies indicate that formaldehyde is a potential human carcinogen. Airborne concentrations above 0.1 ppm can cause irritation of the eyes, nose, and throat. The severity of irritation increase as concentrations increase; at 100 ppm it is immediately dangerous to life and health. Dermal contact causes various skin reactions including sensitization, which might force persons thus sensitized to find other work.

OSHA requires that the employer conduct initial monitoring to identify all employees who are exposed to formaldehyde at or above the action level or STEL and to accurately determine the exposure of each employee so identified. If the exposure level is maintained below the STEL and the action level, employers may discontinue exposure monitoring, until there is a change which could affect exposure levels. The employer must also monitor employee exposure promptly, upon receiving reports of formaldehyde-related signs and symptoms.

In regard to exposure control, the best prevention is provided by source control (if possible). The selection of HCHO-free or low-emitting products, such as exterior-grade plywood that uses phenol HCHO resins, for indoor use is the best starting point.

Secondary controls include filtration, sealants, and fumigation treatments. Filtration can be achieved using selected adsorbents. Sealants involve coating the materials in question with two or three coats of nitrocellulose varnish, or water-based polyurethane. Three coats of these materials can reduce outgassing by as much as 90 percent.

Training is required at least annually for all employees exposed to formaldehyde concentrations of 0.1 ppm or greater. The training will increase employees' awareness of specific hazards in their workplace and of the control measures employed. The training also will assist successful medical surveillance and medical removal programs. These provisions will only be effective if employees know what signs or symptoms are related to the health effects of formaldehyde and if they are periodically encouraged to do so.

Lead Exposure

Lead has been poisoning workers for thousands of years. Most occupational over exposures to lead have been found in the construction trades, such as plumbing, welding, and painting. In plumbing, soft solder (banned for many uses in the U.S.), used chiefly for soldering tinplate and copper pipe joints, is an alloy of lead and tin. Although the use of lead-based paint in residential applications has been banned, since lead-based paint inhibits the rusting and corrosion of iron and steel, it is still used on construction projects. Significant lead exposures can also arise from removing paint from surfaces previously coated with lead-based paint. According to OSHA 93-47 (2003), the operations that generate lead dust and fume include the following:

- Flame-torch cutting, welding, the use of heat guns, sanding, scraping, and grinding of lead-painted surfaces in repair, reconstruction, dismantling, and demolition work
- Abrasive blasting of structures containing lead-based paints
- Use of torches and heat guns, and sanding, scraping, and grinding lead-based paint surfaces during remodeling or abating lead-based paint
- Maintaining process equipment or exhaust duct work.

Health Effects of Lead

There are several routes of entry in which lead enters the body. When absorbed into the body in certain doses, lead is a toxic substance. Lead can be absorbed into the body by inhalation and ingestion. Except for certain organic lead compounds not covered by OSHA's Lead Standard (29 CFR 1926.62), such as tetraethyl lead, lead, when scattered in the air as a dust, fume, or mist, can be absorbed into the body by inhalation.

A significant portion of the lead that can be inhaled or ingested gets into the blood stream. Once in the blood stream, lead is circulated throughout the body and stored in various organs and tissues. Some of this lead is quickly filtered out of the body and excreted, but some remains in the blood and other tissues. As exposure to lead continues, the amount stored in the body will increase if more lead is being absorbed that is being excreted. Cumulative exposure to lead, which is typical in construction settings, may result in damage to the blood, nervous system, kidneys, bones, heart, and reproductive system and contributes to high blood pressure. Some of the symptoms of lead poisoning include the following:

- Poor appetite
- Dizziness
- Pallor
- Headache
- Irritability/anxiety
- Constipation
- Sleeplessness
- Weakness
- Insomnia
- "Lead line" in gums
- Fine tremors
- Hyperactivity
- "Wrist drop" (weakness of extensor muscles)
- Excessive tiredness
- Numbness
- Muscle and joint pain or soreness
- Nausea
- Reproductive difficulties

Lead Standard Definitions

According to OSHA's Lead Standard, the terms listed below have the following meanings:

- **Action level**—means employee exposure, without regard to the use of respirators, to an airborne concentration of lead of 30 micrograms per cubic meter of air (30 µg/m³), averaged over an eight-hour period.
- **Permissible exposure limit (PEL)**—means the concentration of airborne lead to which an average person may be exposed without harmful effects. OSHA has established a PEL of fifty micrograms per cubic meter of air (50 µg/m³) averaged over an eight-hour period. If an employee is exposed to lead for more than eight hours in any work day, the permissible exposure limit, a time-weighted average (TWA) for that day, shall be reduced according to the following formula:

Maximum permissible limit (in µg/m³) = 400 × hours worked in the day.

When respirators are used to supplement engineering and administrative controls to comply with the PEL and all the requirements of the lead standard's respiratory protection rules have been met, employee exposure, for the purpose of determining whether the employer has complied with the PEL, may be considered to be at the level provided by the protection factor of the respirator for those periods the respirator is worn. Those periods may be averaged with exposure levels during periods when respirators are not worn to determine the employee's daily TWA exposure.
- **µg/m³**—means micrograms per cubic meter of air. A microgram is one millionth of a gram. There are 454 grams in a pound.

Worker Lead Protection Program

The employer is responsible for the development and implementation of a worker lead protection program. This program is essential in minimizing worker risk of lead exposure.

The most effective way to protect workers is to minimize exposure through the use of engineering controls and good work practices.

At the minimum, the following elements should be included in the employer's worker protection program for employees exposed to lead:

- Hazard determination, including exposure assessment
- Engineering and work practice controls
- Respiratory protection
- PPE (protective work clothing and equipment)
- Housekeeping
- Hygiene facilities and practices
- Medical surveillance and provisions for medical removal
- Employee information and training
- Signs
- Recordkeeping

Mold Control

Molds can be found almost anywhere; they can grow on virtually any organic substance as long as moisture and oxygen are present. The earliest known writings that appear to discuss mold infestation and remediation (removal, cleaning up) are found in Leviticus, Chapter 14, of the Bible's Old Testament.

Where are molds "typically" found?

As mentioned, name any spot or place; they have been found growing in office buildings, schools, automobiles, in private homes, and other locations where water and organic matter are left unattended. Mold is not a new issue—just one which, until recently, has received little attention by regulators in the United States. That is, there are no state or federal statutes or regulations regarding molds and IAQ.

Molds reproduce by making spores that usually cannot be seen without magnification. Mold spores waft through the indoor and outdoor air continually. When mold spores land on a damp spot indoors, they may begin growing and digesting whatever they are growing on in order to survive. Molds generally destroy the things they grow on (USEPA, 2001).

The key to limiting mold exposure is to prevent the germination and growth of mold. Since mold requires water to grow, it is important to prevent moisture problems in buildings. Moisture problems can have many causes, including uncontrolled humidity. Some moisture problems in workplace buildings have been linked to changes in building construction practices during the 1970s, '80s, and '90s. Some of these changes have resulted in buildings that are tightly sealed but may lack adequate ventilation, potentially leading to moisture buildup. Building materials, such as drywall, may not allow moisture to escape easily. Moisture problems may include roof leaks, landscaping or gutters that direct water into or under the building, and unvented combustion appliances. Delayed maintenance or insufficient maintenance are also associated with moisture problems in buildings. Moisture problems in temporary structures have frequently been associated with mold problems.

Building maintenance personnel, architects, and builders need to know effective means of avoiding mold growth that might arise from maintenance and construction practices. Locating and cleaning existing growths are also paramount to decreasing the health effects of mold contamination. Using proper cleaning techniques is important because molds are incredibly resilient and adaptable (Davis, 2001).

Molds can elicit a variety of health responses in humans. The extent to which an individual may be affected depends upon his or her state of health, susceptibility to disease, the organisms with which he or she came in contact, and the duration and severity of exposure (Ammann, 2000). Some people experience temporary effects that disappear when they vacate infested areas (Burge, 1997). In others, the effects of exposure may be long term or permanent (Yang, 2001).

It should be noted that systemic infections caused by molds are not common. Normal, healthy individuals can resist systemic infection from airborne molds.

Those at risk for system fungal infection are severely immunocompromised individuals such as those with HIV/AIDs, individuals who have had organ or bone marrow transplants, and persons undergoing chemotherapy.

In 1994, an outbreak of Stachybotrys chartarum in Cleveland, Ohio was believed by some to have caused pulmonary hemorrhage in infants. Sixteen of the infants died. CDC sponsored a review of the cases and concluded that the scientific evidence provided did not warrant the conclusion that inhaled mold was the cause of the illnesses in the infants. However, the panel also stated that further research was warranted, as the study design for the original research appeared to be flawed (CDC, 1999).

Below is a list of mold components known to elicit a response in humans:

- *Volatile organic compounds (VOCs)*—"Molds produce a large number of volatile organic compounds. These chemicals are responsible for the musty odors produced by growing molds" (McNeel and Kreutzer, 1996). VOCs also provide the odor in cheeses, and the "off" taste of mold infested foods. Exposure to high levels of volatile organic compounds, affect the central nervous system (CNS), producing such symptoms as headaches, attention deficit, inability to concentrate, and dizziness (Ammann, 2000). According to McNeel, at present the specific contribution of mold volatile organic compounds in building-related health problems has not been studied. Also, mold volatile organic compounds are likely responsible for only a small fraction of total VOCs indoors (Davis, 2001).
- *Allergens*—all molds, because of the presence of allergens on spores, have the potential to cause an allergic reaction in susceptible humans (Rose, 1999). Allergic reactions are believed to be the most common exposure reaction to molds. These reactions can range from mild, transitory responses, like runny eyes, runny nose, throat irritation, coughing, and sneezing; to severe, chronic illnesses such as sinusitis and asthma (Ammann, 2000).
- *Mycotoxins*—are natural organic compounds that are capable of initiating a toxic response in vertebrates (McNeel and Kreutzer, 1996). Some molds are capable of producing mycotoxins. Molds known to potentially produce mycotoxins that have been isolated in infestations causing adverse health effects include certain species of *Acremonium, Alternaria, Aspergillus, Chaetomium, Cladosporium, Fusarium, Paecilomyces, Penicillium, Stachybotrys,* and *Trichoderma* (Yang, 2001).

While a certain type of mold or mold strain may have the genetic potential for producing mycotoxins, specific environmental conditions are believed to be needed for the mycotoxins to be produced. In other words, although a given mold might have the potential to produce mycotoxins, it will not produce them if the appropriate environmental conditions are not present (USEPA, 2001).

Currently, the specific conditions that cause mycotoxin production are not fully understood. The USEPA recognizes that mycotoxins have a tendency to concentrate in fungal spores and that there is limited information currently available regarding the process involved in fungal spore release. As a result, USEPA is currently conducting research in an effort to determine "the environmental conditions required for sporulation, emission, aerosolization, dissemination, and transport of [Stachybotrys] into the air" (USEPA, 2001).

Mold Prevention

As mentioned, the key to mold control is moisture control. Solve moisture problems before they become mold problems. Several mold prevention tips are listed below.

- Fix leaky plumbing and leaks in the building envelope as soon as possible.
- Watch for condensation and wet spots. Fix sources(s) of moisture problem(s) as soon as possible.
- Prevent moisture due to condensation by increasing surface temperature or reducing the moisture level in air (humidity). To increase surface temperature, insulate or increase air circulation. To reduce the moisture level in air, repair leaks, increase ventilation (if outside air is cold and dry), or dehumidify (if outdoor air is warm and humid).
- Keep heating, ventilation, and air-conditioning (HVAC) drip pans clean, flowing properly, and unobstructed.
- Vent moisture-generating appliances, such as dryers, to the outside where possible.
- Perform regular build/HVAC inspections and maintenance as scheduled.
- Maintain low indoor humidity, below 60% relative humidity (RH), ideally 30–50%, if possible.
- Clean and dry wet or damp spots within 48 hours.
- Don't let foundations stay wet. Provide drainage and slope the ground away from the foundation.

Mold Remediation

At the present time, there are no standardized recommendations for mold remediation; however, USEPA is working on guidelines. There are certain aspects of mold cleanup, however, which are agreed upon by many practitioners in the field.

- A common sense approach should be taken when assessing mold growth. For example, it is generally believed those small amounts of growth, like those commonly found on shower walls pose no immediate health risk to most individuals.
- Persons with respiratory problems, a compromised immune system, or fragile health should not participate in cleanup operations.
- Cleanup crews should be properly attired. Mold should not be allowed to touch bare skin. Eyes and lungs should be protected from aerosol exposure.
- Adequate ventilation should be provided while, at the same time, containing the infestation in an effort to avoid spreading mold to other areas.
- The source of moisture must be stopped and all areas infested with mold thoroughly cleaned. If thorough cleaning is not possible due to the nature of the material (porous versus semi- and nonporous), all contaminated areas should be removed.

Safety tips that should be followed when remediating moisture and mold problems include:

- Do not touch mold or moldy items with bare hands.
- Do not get mold or mold spores in your eyes.
- Do not breathe in mold or mold spores.
- Consider using PPE when disturbing mold. The minimum PPE is a respirator, gloves, and eye protection.

Mold Cleanup Methods

A variety of mold cleanup methods are available for remediating damage to building materials and furnishings caused by moisture control problems and mold growth.

These include wet vacuum, damp wipe, HEPA vacuum, and the removal of damaged materials and sealing off in plastic bags. The specific method or group of methods used will depend on the type of material affected.

A COTTAGE INDUSTRY IS BORN

As a result of the Legionnaires' Incident in Philadelphia, the WHO study and significant amounts of media attention, Legionnaires' disease, sick building syndrome, and indoor air pollution became common terms used by common working people in the workplace. An offshoot of this media attention was a new cottage industry consisting of so-called "experts" in indoor air pollution who commenced selling their services to conduct sick building surveys. From about 1985 until the early 1990s, this new industry was booming. Since then, many of these new enterprises folded their operations because of a lack of business; the initial scare wore off.

During its heyday, the movement to solve the sick building syndrome problem resulted in some air pollutants being identified as the culprits or potential culprits in causing the SBS. One such category of air pollutants in nonindustrial environments was identified as volatile organic compounds (VOCs). In 1989 WHO classified (according to boiling-point ranges) the entire range of organic pollutants into four groups: (1) Very volatile (gaseous compounds), (2) volatile organic compounds, (3) semivolatile organic compounds, and (4) organic compounds associated with particulate matter.

FALL PROTECTION

Fall protection is the series of steps taken to cause reasonable elimination or control of the injurious effects of an unintentional fall while accessing or working.

. . . Fall hazard distance begins and is measured from the level of a workstation on which a worker must initially step and where a fall hazard exists. It ends with the greatest distance of possible continuous fall, including steps, openings, projections, roofs, and direction of fall (interior or exterior). Protection is required to keep workers from striking objects and to avoid pendulum swing, crushing, and foreseeable impact with any part of the body to which injury could occur.

The object of elevated fall protection is to convert the hazard to a slip or minor fall at the very worst—a fall from which hopefully no injury occurs (J. L. Ellis, 1988).

Because injuries received from falls in the workplace are such a common occurrence—in a typical year more than 10,000 workers will lose their lives in falls—safety and environmental managers not only need to be aware of fall hazards, but also of the need to institute a Fall Protection Safety Program that includes the components shown in Figure 16.2 (Kohr, 1989). In addition to the preceding information provided by Kohr, we might ask just how frequent and serious are accidents/incidents related to falls? Let's look at a few telling facts about falls in the workplace. The National Safety Council's annual report typically "predicts" 1,400 or more deaths, and more than 400,000 disabling injuries each year from falls. Falls are the leading cause of disabling injuries in the United States, accounting for close to 18% of all workers'

Figure 16.2   Elements of a Fall Protection Safety Program

compensation claims. According to the Bureau of Labor Statistics (BLS), in 2013 there were 699 deaths from falls, slips, or trips.

According to R.L. Kohr, the primary causes of falls are:

1. A foreign object on the walking surface
2. A design flaw in the walking surface
3. Slippery surfaces
4. An individual's impaired physical condition

Historically, which industries have the most injuries as a result of falls? Note, as you might imagine, that the construction industry (42% of all injuries resulting from falls) has the largest percentage of injuries. A NSC Accident Facts publication (1984 & 1985) reported that 70% of reported falls were from scaffolds, 14% from roofs, and another 14% were from barrels, boxes, equipment, or furniture. Eisma (1990) reports that 85% of falls from elevation resulted in lost workdays, and 20% resulted in death. In a much more recent report provided by the BLS, falls caused 681 or 14.5 percent of workplace deaths in 2011.

Obviously, as the above data clearly indicates, falls are a problem that the safety and environmental manager must continually face. In this chapter, we discuss fall protection and the steps the safety and environmental manager should take to lessen their impact.

Fall Protection: Defining the Problem

When attempting to install a Fall Protection Safety Program into any organization, safety and environmental managers must first define their needs (what the organization requires). The actual needs of any type of fall protection program are going to be driven mainly by the type of work the organization does. Obviously, if the company is involved in construction, the needs are rather straightforward, because much of the work conducted will include the necessity of doing elevated work. However, this might also be the case for various trades, such as carpentry, for example. Public utility and transportation work might also require elevated work. The factor that may surprise you is the large percentage of falls from elevation that occur in the manufacturing industry.

To define the problem associated with all types of falls, let's examine what falls are all about. None of us has a problem understanding what a fall from a high-rise construction project involves. It is simply a fall from elevation. However, in the workplace, worker injuries result from types of falls other than falling from elevations. Falls in the workplace also include slips, trips, and stair falls, as well as elevated falls. *Slips* and *trips* are falls on the same level. *Stair falls* are falls on one or more levels. *Elevated falls* are from one level to another. In the following sections, each of these types of falls is discussed in greater detail, but first we discuss the physical factors at work in causing a fall. Note: remember that the safety and environmental manager must address and work to reduce or eliminate all types of falls.

Physical Factors at Work in a Fall

The safety and environmental manager is a student and practitioner of science. In a moment of humor, workers sometimes say, "The bigger they are, the harder they fall," and "It's not the fall that's so bad, it's the sudden stop when you hit the ground." Though this is often the common view, many would-be practitioners in the safety and environmental health field are often surprised to find out that science not only plays a role in falls, but that slips, trips, and falls actually involve three well-known laws of science: friction, momentum, and gravity.

Friction is the resistance between things, such as between work shoes and the workplace walking surface. Without friction, workers are likely to slip and fall. Probably the best example of this phenomenon is a slip on ice. On icy surfaces, shoes can't grip the surface normally, causing a loss of traction and a fall.

Momentum (in physics) is the product of the mass of a body and its linear velocity. Simply put, momentum is affected by speed and size of the moving object. Momentum is best understood if we translate the humorous sayings above to: The more you weigh and the faster you move, the harder you fall if you slip or trip.

Gravity is (on earth) the force of attraction between any object in the earth's gravitational field and the earth itself. Simply put, gravity is the force that pulls you to the ground once a fall is in process. If someone loses balance and begins to fall, they are going to hit the ground. The human body is equipped with mechanisms that work to prevent falls (loss of balance or center of gravity). These mechanisms include the eyes, ears, and muscles, which all work to keep the human body close to its natural center of balance. However, if this center of balance shifts too far, a fall will occur if balance can't be restored to normal.

Because gravity obviously has the same effect on all of us here on earth, it has always been surprising to us to discover how such a well-known (but often ignored) basic law of science is so often and conveniently ignored by various industries. For example, we commonly (even in this day and age) come face-to-face with company owners or workplace foremen who ignore the laws of gravity and require their workers to perform "daring" (and extremely dangerous) feats in the workplace. The worker (who needs the job and the security it provides) is led to believe that gravity is something that is not important to them, but only important in movies about space travel, perhaps. Obviously, this is a dangerous mind-set and practice that the company safety and environmental manager must not tolerate.

Slips: Falls on the Same Level

In its simplest form, a slip is a loss of balance caused by too little friction between the feet and surfaces walked or worked on. The more technical explanation refers to a slip resulting in a sliding motion, when the friction between the feet (shoe sole surface) and the surface is too little. This slip (loss of traction), in turn, often leads to a loss of balance. The result is a fall.

Slips can be caused by a number of design factors and work practices, individually or in combination. Design factors include footwear, floor surfaces, personal characteristics, and the work task.

Footwear is an important consideration in the prevention of a slip-fall. Not only is the condition of the footwear important in fall prevention, but also the composition, shape, and style. For industrial applications, the organizational safety professional should ensure that only approved safety shoes are worn. Safety shoes should not only be designed to include toe protection; they should also include slip-resistant soles.

For floor surfaces, design, installation, composition and condition, gradient, modifications by protective coatings and cleaning/waxing agents, and illumination are all important elements that must be taken into consideration in providing safe floor surfaces in the workplace. Common solutions used to make floor surfaces slip-resistant include grooving, gritting, matting, and grating.

Personal characteristics (physical make-up or disabilities, age, physical health, emotional state, agility, and attentiveness) are also factors important to consider in making walking and working surfaces slip-resistant for workers.

Work task design also plays an important role in causing and/or preventing slip-falls.

Slips can also be the result of work practices that cause walking surfaces to be constantly wet—wet from spills, or wet or slippery from weather hazards like snow and ice. Workplace supervisors and workers (and the safety and environmental manager) must follow safe work practices and exercise vigilance to ensure such conditions do not occur or are remediated as quickly as possible when they do. This type of problem is much more common than we might realize. How often have workers spilled oil or some other slippery chemical on the workplace floor then just walked away from the spill—leaving this common slip hazard for another worker to step on, slip on, and fall? The common workplace safe work practice and housekeeping rule should be to clean up spills right away. Another unsafe work practice that commonly leads to slip-falls is when the worker is in a hurry, rushing to finish whatever he or she is attempting to accomplish.

Trips: Falls on the Same Level

Have you ever considered what happens when a worker trips? If you are a safety and environmental manager, you should. Trips normally occur whenever a worker's foot contacts an object that causes him or her to lose balance. However, you do not always have to come into contact with an object to trip. Too much friction between the foot and walking surface may cause trips.

Like slips, trips are commonly caused when the worker is rushing, hurrying to complete whatever he or she is doing. The problem with hurrying is, of course, that the victim's attention is usually focused on anything but possible trip hazards.

Another common factor that leads to a trip is the practice of carrying objects that are too large for the worker to adequately see the walking surface in front of him or her.

Lighting also plays a critical role in preventing trips. Inadequate lighting fixtures, burned-out light bulbs, and lights that are turned off all increase the opportunity for trips to occur.

Again, as in the prevention of trips, housekeeping plays an important role in prevention. Good workplace housekeeping practices include keeping passageways clean and uncluttered; arranging equipment so that it doesn't interfere with walkways or pedestrian traffic; keeping working areas clear of extension or power tool cords; eliminating loose footing on stairs, steps, and floors; and properly storing gangplanks and ramps.

Stair Falls: Falls on One or More Levels

One of the first things any conscientious safety and environmental manager should do when first hired (and should continue throughout his or her tenure) is to become completely familiar with the applicable literature that describes workplace hazards, their frequency of occurrence, and the recommended hazard control methods.

For falls from stairs, the best publication we know is the one provided by the Department of Labor, Bureau of Labor Statistics titled *Injuries Resulting From Falls on Stairs (Bulletin 2214)*, August 1984. This particular booklet is excellent because it not only provides statistical data, but also is an eye-opener on the way many of the injuries occur (causal factors). For example, it is widely known and accepted that stairs are a high-risk area. It is also accepted that a loss of balance can occur from a slip or trip while a worker (or any person) is traveling up or down a stairway. However, for the safety engineer the question becomes why—why are stairs so hazardous? What are the causal factors? Bulletin 2214 comes in handy in trying to answer questions like these. For example, Bulletin 2214 points out that the vast majority of falls on stairs occur when traveling down the stairs not holding the handrail. This is an important point for two reasons: (1) The safety and environmental manager can focus training on this important point; and (2) the safety and environmental manager can ensure that handrails are not only in place in all stairways, but are also in good repair.

Loss of traction is the common cause of the highest number of stairway slipping and falling accidents. Again, this is where good housekeeping practices come into play. Many of the stairway slipping and falling accidents happen because of water or other liquid on steps. Along with improper housekeeping practices, stairs can also become hazardous whenever they are improperly designed, installed, and/or neglected. Safe work practices should also be considered. A work practice that allows the worker to carry or reach for objects while climbing stairs is not a good work practice.

Elevated Falls: Falls From One Level to Another

When workers are working from elevated scaffolds, ladders, platforms, and other surfaces, the risk of serious injury from an elevated fall is increased exponentially whenever the worker has a loss of balance resulting from a slip or trip. Unfortunately, our experience has shown that often the practice of various supervisors and companies requires workers to perform work from elevated areas to use some type of device (handrail or handline), which they are supposed to grab onto break their fall. In our judgment (and the judgment of most experienced safety and environmental managers), this is not fall protection. These types of jerry-rigged devices are not acceptable substitutes for guardrails, appropriate mid-rails, and toeboards. OSHA requires guardrails to be 42" nominal, mid-rails 21", and toeboards 4".

J.E. Ellis (1988) makes a good point in that "unlike many workplace hazards, few, if any, 'near-miss' incidents help people learn to appreciate the seriousness of elevated falls" (p 28). When you consider that losing one's balance from an elevation of 10 to 200 feet or more usually leaves little chance to avoid serious or fatal injury, Ellis' statement makes a lot of sense.

Fall Protection Measures

Just about anyone can talk about the hazards and dangers inherent in slips, trips, and falls from elevation. Under *29 CFR Subpart M, Fall Protection, 1926.501*, employers must assess the workplace to determine if the walking or working surfaces on which employees are to work have the strength and structural integrity to safely support workers. Accordingly, the real goal should be on how to prevent slips, trips, and falls from elevation from occurring in the first place. This is the safety and environmental manager's goal.

But how does he or she accomplish this?

Good question.

Recall that in Figure 16.2, we highlighted the major components that make up a good fall protection program. This is where the safety and environmental manager comes in. After determining that workers may be required to perform elevated work, then obviously the safety engineer responsible for the safety and health of such workers needs to develop a fall protection program, which includes the components shown in Figure 16.2. Even if a particular company does not require workers to work from elevated locations, remember that every workplace still has slip and trip hazards that must be guarded against. The need for a company fall protection program may still be necessary.

Along with installing a company fall protection program that includes the components displayed in Figure 16.2, we have three other important recommendations: (1) preplanning before beginning any elevated work (e.g., a Scaffold Safety Program); (2) establishing a written policy and developing rules; and (3) a written safe work practice designed to prevent falls.

Preplanning is all about thinking through the job at hand. For example, if exterior refurbishing work is to be accomplished on a chemical storage tank that is 80 feet in height, erecting scaffolding is probably required. Properly erecting scaffolding takes both preplanning and a great deal of skill. If scaffolding is to be used, the organization responsible for erecting the scaffolding should have a Scaffold Safety Program. To assist you in this requirement, we have provided a sample Scaffold Safety Program (one that has been used for years with great success) below, along with a sample written fall protection policy and a sample fall protection safe work practice.

Scaffold Safety Program

***

Scaffold Safety Program
(A Sample)

I. INTRODUCTION

The Occupational Safety and Health Administration (OSHA) regulates scaffolding used in both general industry and construction. The company has a responsibility to instruct each employee who may use scaffolding in the recognition and avoidance of unsafe conditions, proper construction, use, placement, and care. This program was designed to reduce falls from elevations.

The Scaffold Safety Program covers general safe work practices and requirements. Because each work center has a different type of scaffolding system or may rent a different type of scaffolding, hands-on training detailing the proper erection and care of scaffolding must be conducted in-house.

This program also specifies the inspection of company-owned or rented scaffolding.

II. GENERAL SAFETY REQUIREMENTS
FOR ERECTING SCAFFOLDING

Competent persons shall oversee the erection and dismantling of scaffolding systems. A competent person is a person who has read the manufacturer's directions and who has had training on the company's Scaffolding Safety Program. A competent person is defined as a person who is capable of identifying existing and predictable hazards in the surroundings, or work conditions that are hazardous, and who has authorization to take prompt corrective measures to eliminate the hazard.

A. FOOTING/ANCHORAGE—The footing or anchorage for scaffolds must be sound, rigid, and capable of carrying the maximum intended load without settling or displacement. Unstable objects such as barrels, boxes, loose brick, or concrete blocks cannot be used to support scaffolds or planks.

B. GUARDRAILS/TOE-BOARDS—Scaffolds erected greater than 10 feet in height above the ground floor shall have guardrails and toeboards on all open sides and ends. Scaffolds that are 4 to 10 feet in height and have a minimum horizontal dimension, in either direction, of less than 45 inches, must have guardrails and toeboards on all open sides and ends. The standard size for guardrails is 42 inches in height with a mid-rail at 21 inches. The standard size for toeboards is four inches. If guardrails and toeboards are not used, employees working on the scaffold must wear fall protection equipment. Fall protection equipment includes a full body harness tied off to an anchor point (eyebolt, pipe, etc.) that will support the employee's weight. Body harnesses and lanyards must meet OSHA regulations and the American National Standards Institute (ANSI) standards. While erecting and dismantling scaffolding, employees may have to wear fall protection. For example, if employees are in the process of erecting or dismantling scaffolding and the handrails and toeboards are removed and the scaffolding is greater than 10 feet above the ground floor, fall protection is necessary.

C. SCREENS—Where persons are required to work or pass under an in-use scaffold, a screen between the toeboard and the guardrail extending along the entire opening must be installed. This screen must be #18 inch gauge U.S. Standard wire 1/2" mesh, or the equivalent.

D. TIMBER SCAFFOLD FRAMING—All load-carrying timber members of scaffold framing shall be a minimum of 1500 fiber (stress grade) construction grade lumber.

E. PLANKING—When prefabricated, manufactured planking (flooring) is not used, all wood planking must be Scaffold Grade or equivalent. Southern Pine classification D165 and D172 should be used. The scaffold plank must be scaffold grade solid-sawn lumber with a nominal size of 2 x 10 inches or larger. All planking of platforms must be overlapped at least 12 inches or secured from movement. Scaffold planks must extend over their end supports not less than 6 inches and not more than 12 inches. Platforms need to be tightly planked for the full width of the scaffold except for necessary entrance opening.

F. LOADING—Never allow more than three employees on the same scaffold board at a time. The maximum permissible span for 1.12 x 9 inch or wider plank of full thickness is 4 feet with an average loading of 50 pounds per square foot (PSF).

G. ACCESS LADDER—An access ladder or equivalent safe access must be provided for scaffolding. Always check your scaffolding. Some types of scaffolding have the access ladder built in, however, with other types, you must provide a ladder for safe access. The ladder must be located so that when it is in use, it will not have a tendency to tip the scaffold. A landing platform must be provided at intervals not to exceed 35 feet.

H. SWAYING/DISPLACEMENT—The poles, legs, or uprights of scaffolds must be secured and rigidly braced to prevent swaying and displacement. All pole scaffolds need to be securely guyed or tied to the building or structure. If the height and length of a scaffolding system exceeds 25 feet, the scaffold needs to be secured at intervals not greater than 25 feet vertically and horizontally.

I. OVERHEAD PROTECTION—Employees working on a scaffold with overhead hazards must be provided with protection. For example, employees working on scaffolding under low overhead pipes or obstructions must wear hard hats. Hard hats must also be worn while workers are erecting and dismantling scaffolding.

J. SLIPPERY CONDITIONS—Slippery conditions such as chemical spills (oil, polymer, etc.) need to be quickly eliminated.

K. CROSS BRACING—Cross bracing must be installed across the width of the scaffold at least every third set of posts horizontally, and every fourth runner vertically. The bracing shall extend diagonally from the inner and outer runners upward to the next outer and inner runners.

L. ROLLING PLATFORMS—All tools and materials need to be secured or removed from the platform before the platform is moved. Employees must not be on rolling platforms when they are moved.

M. TAG LINES/HOIST ARMS—A tag line should be used to keep the material that is hoisted from damaging the scaffold assembly. Hoist arms can be used to lift scaffold components or light materials to the work platform. However, hoist arms will often introduce an overturning force to the scaffold assembly, one that must be compensated for by tying the scaffolding system to the structure.

N. OUTRIGGER SCAFFOLDS—Outrigger beams must not extend more than 6 feet beyond the face of the building. The inboard end of outrigger beams (measured from the fulcrum point to anchorage point) cannot be less than 1.5 times the outboard end in length. The beams must rest on edge, the sides must be plumb, and the edges must be horizontal. The fulcrum point of the beam must rest on a secure bearing at least 6 inches in each horizontal dimension. The beam must be secured in place against movement, and must be securely braced at the fulcrum point against tipping. The inboard ends of outrigger beams must be securely anchored, either by means of struts bearing against sills in contact with the overhead beams or ceiling, by means of tension members secured to the floor joists underfoot, or by both if necessary. The inboard ends of outrigger beams need to be secured against tripping, and the entire supporting structure must be securely braced in both directions to prevent any horizontal movement. If a registered professional engineer designs outrigger scaffolds, the scaffold must be constructed and erected in accordance with such design. If outrigger scaffolds are not designed by a professional engineer, they need to be constructed and erected in accordance with Table 16.10.

O. DESIGN—Stationary and rolling scaffolding should be erected and dismantled following the manufacturer's instructions. If stationary scaffolding is to exceed 125 feet in height, a registered professional engineer must design it. If rolling scaffolding is to exceed 60 feet in height, a professional engineer must design it. When tube

Table 16.10 Outrigger Scaffold Requirements

Maximum Scaffold Load	Light Duty (25 psf)	Medium Duty (50 psf)
Outrigger size	2 × 10″	3 × 10″
Maximum Outrigger	10″	6′
Spacing	0″	0″
Planking	2 × 10″	2 × 10″
Guardrail	2 × 4″	2 × 4″
Guardrail	2 × 4″	2 × 4″
Toeboards	4″ (minimum)	4″ (minimum)

and coupler scaffolding is used, no dissimilar metals can be used together. The use of shore or lean-to scaffolds is strictly prohibited.

P. WEATHER—Working outside on scaffolding during periods of high winds or storms is prohibited. Working outside on scaffolds covered with ice or snow is prohibited.

III. GUIDELINES FOR PURCHASING AND RENTING SCAFFOLDING

1. Manufactured scaffolding that is purchased or rented must meet OSHA's Scaffolding Standard 1926.451 and 1910.28.
2. Manufactured scaffolding that is purchased or rented must be capable of supporting at least four times the designed working load. If wire or fiber rope is used in scaffold supports, it must be capable of supporting at least six times the designed working load.
3. Scaffolding systems that are purchased or rented must not be broken, bent, excessively rusted, or altered, and must be free from all other structural defects.

IV. INSPECTION OF SCAFFOLDING

1. Rented scaffolding systems need to be inspected by a competent person before use. If the rented system is found to be broken, bent, excessively rusted, altered, or has other structural defects, it should not be used.
2. A competent person must inspect company scaffolding before use. If the scaffolding system is found to be broken, bent, excessively rusted, altered, or has other structural defects, it should not be used until repaired. A qualified vender must repair scaffolds. Extra parts (such as clamps, coupling pins, cross braces, etc.) that are purchased for a scaffolding system must be manufacturer-approved parts for that scaffolding system.
3. Chemicals will often destroy or damage scaffolding parts. If strong acids or bases are spilled on a scaffolding system, they may severely damage or destroy it. Scaffolding that has had chemical damage shall be evaluated. If parts are found to be damaged or destroyed, the scaffolding system will not be used until replacement parts are procured.

4. If scaffolds are erected for more than one day, they will be inspected daily by a competent person.
5. Competent persons must document all scaffold inspections. This documentation can be noted in a logbook-type fashion.
6. The inspection logbook must be at the work center or work site, available for inspection by the safety and environmental manager, work center supervisor, or regulatory agency.
7. Company work center supervisors are responsible for ensuring that the inspection logbook is kept and that all necessary scaffolding inspections are performed.

V. TRAINING

1. The safety and environmental manager will train employees on the company's Scaffolding Safety Program and general scaffolding safe work practices.
2. Company work center supervisors are responsible for training their employees on how to properly erect and dismantle scaffolding used at their work center.
3. Company work center supervisors are responsible for designating competent persons within their work center. Competent persons are designated on the basis of training and experience.

***

Fall Protection Policy

As with any organizational safety and health program (and the overall safety program in particular) the safety and environmental manager should ensure that a Fall Protection Policy is established and applicable rules are developed. Keep in mind that written safety policies are not only intended to protect employees but also to protect contractors. Note that establishing and developing any organizational safety and environmental health policy is only part of the job. Making sure that every employee has been trained on the policy is the other, more difficult task. This may sound like a simplistic statement, but you might be surprised at the number of organizations we have audited with excellent written polices, but when employees are asked about these policies, they shake their heads. "The organization's Fall Protection Policy? Gee, I don't think we have one of those," they say. This kind of misinformation or lack of information is the safety and environmental manager's worst nightmare. You must guard against it.

We have provided a sample written policy for elevated work operations. From this sample (which is an adaptation of a similar policy used by Kaiser Aluminum & Chemical Corporation, 1986), you should be able to gain an appreciation for the type of written policy you should have at your company.

***

Elevated Work Policy
(A Sample)

I. PURPOSE

To establish a standard policy to ensure elevated work at heights of eight (8) feet or greater conducted by company and contractor personnel are done so in the safest manner possible.

II. OBJECTIVES

A. To prevent employee death and/or minimize employee injury resulting from falls from elevated work locations.
B. To identify and label elevated work sites requiring fall protection equipment.
C. To establish minimum standards for all protection equipment and systems and their applications.

III. SCOPE

All free fall hazards will be guarded against by the use of permanent or semipermanent guardrail assemblies. When a Free Fall Hazard cannot be prevented through such measures, personal fall protection equipment must be used.

This policy applies to working at elevated work locations that are eight (8) feet or more above floor or grade level. It covers activities such as (but not limited to): work in pipe racks, on sloped roofs, on unguarded scaffolding, ship cargo holds, when working from suspended scaffolds, floats or boatswains chairs; working on tank tops; and when working inside or outside any process structure not equipped with appropriate guarded work platforms.

IV. PROCEDURE

A. Supervisor Planning
  1. The supervisor will be responsible for evaluating the need for a personal fall protection system as an integral part of preplanning a job.
  2. When work must be performed at recognized unguarded elevated heights, the supervisor may select either option below:
     a. Option 1—Fall Prevention
        Eliminate the free fall hazard during all phases of the job (traveling to and from elevated work areas as well as during the performance of the task at the elevated work area) by means of temporary scaffolding, platforms, railings, manlifts, ladders, etc.

Note: Every effort must be taken to minimize the potential of the free fall hazards to individuals installing temporary or permanent fall protection system.

b. Option 2—Fall Protection System

By selecting and installing a personal fall protection system, eliminating the free fall hazard (greater than 6 feet) when traveling to and from the elevated work area, as well as during performance of the task at the elevated work areas.

Note: Every effort must be taken to minimize the potential of the free fall hazards to individuals installing temporary or permanent fall protection system.

B. User Responsibility

1. Each employee assigned to work at elevated heights has the responsibility of thoroughly inspecting the personal fall protection system's anchor points, connecting means (i.e., lanyard or device) and body holding devices (i.e., harnesses) prior to using the system. Any problems noted with any of the above must be brought to the attention of the supervisor.

2. Any questions concerning the type of personal fall protection systems best suited for a particular job, as well as systems installation (e.g., anchor point type/strength) should be directed to the safety and environmental manager.

***

THERMAL STRESS

Exposure to heat or cold can lead to serious illness. Factors such as physical activity, clothing, wind, humidity, working and living conditions, age, and health influence if a person will get ill. There are several ways to lessen the chances of succumbing to exposure. Battling the elements safely includes protecting skin from excessive exposure to subfreezing temperatures and protecting skin from excessive exposure to the sun (Cyr and Johnson, 2002).

Appropriately controlling the temperature, humidity, and air distribution in work areas is an important part of providing a safe and healthy workplace. A work environment in which the temperature is not properly controlled can be uncomfortable. Extremes of either heat or cold can be more than uncomfortable—they can be dangerous. Heat stress and cold stress are major concerns of modern health and safety and environmental managers. This section provides the information they need to know in order to overcome the hazards associated with extreme temperatures.

Thermal Comfort

Thermal comfort in the workplace is a function of a number of different factors. Temperature, humidity, air distribution, personal preference, and acclimatization are all determinants of comfort in the workplace. However, determining optimum conditions is not a simple process (Alpaugh, 1988).

To fully understand the hazards posed by temperature extremes, safety and environmental managers must be familiar with several basic concepts related to thermal energy. The most important of these are summarized here:

- **Conduction**—is the transfer of heat between two bodies that are touching or from one location to another within a body. For example, if an employee touches a workpiece that has just been welded and is still hot, heat will be conducted from the workpiece to the hand. Of course, the result of this heat transfer is a burn.
- **Convection**—is the transfer of heat from one location to another by way of a moving medium (a gas or a liquid). Convection ovens use this principle to transfer heat from an electrode by way of gases in the air to whatever is being baked.
- **Metabolic heat**—is produced within a body as a result of activity that burns energy. All humans produce metabolic heat. This is why a room that is comfortable when occupied by just a few people may become uncomfortable when it is crowded. Unless the thermostat is lowered to compensate, the metabolic heat of a crowd will cause the temperature of a room to rise to an uncomfortable level.
- **Environmental heat**—is produced by external sources. Gas or electric heating systems produce environmental heat, as do sources of electricity and a number of industrial processes.
- **Radiant heat**—is the result of electromagnetic nonionizing energy that is transmitted through space without the movement of matter within that space.

The Body's Response to Heat

Operations involving high air temperatures, radiant heat sources, high humidity, direct physical contact with hot objects, or strenuous physical activities have a high potential for inducing heat stress in employees engaged in such operations. Such places include: Iron and steel foundries, nonferrous foundries, brick-firing and ceramic plants, glass products facilities, rubber products factories, electrical utilities (particularly boiler rooms), bakeries, confectioneries, commercial kitchens, laundries, food canneries, chemical plants, mining sites, smelters, and steam tunnels.

Outdoor operations conducted in hot weather, such as construction, refining, asbestos removal, and hazardous waste site activities, especially those that require workers to wear semipermeable or impermeable protective clothing, are also likely to cause heat stress among exposed workers (OSHA, 2003).

The human body is equipped to maintain an appropriate balance between the metabolic heat it produces and the environmental heat to which it is exposed. Sweating and the subsequent evaporation of the sweat are the body's way of trying to maintain an acceptable temperature balance.

According to Alpaugh (1988), this balance can be expressed as a function of the various factors in the following equation.

$$H = M \pm R \pm C - E \qquad\qquad (16.4)$$

where
    H = body heat
    M = internal heat gain (metabolic)
    R = radiant heat gain
    C = convection heat gain
    E = evaporation (cooling)

The ideal balance when applying the equation is no new heat gain. As long as heat gained from radiation, convection, and a metabolic process does not exceed that lost through the evaporation induced by sweating, the body experiences no stress or hazard. However, when heat gain from any source of sources is more than the body can compensate for by sweating, the result is **heat stress**.

There are several causal factors involved in heat stress. These include

1. Age, weight, degree of physical fitness, degree of acclimatization, metabolism, use of alcohol or drugs, and a variety of medical conditions such as hypertension all affect a person's sensitivity to heat. However, even the type of clothing worn must be considered. Prior heat injury predisposes an individual to additional injury.
2. It is difficult to predict just who will be affected and when, because individual susceptibility varies. In addition, environmental factors include more than the ambient air temperature. Radiant heat, air movement, conduction, and relative humidity all affect an individual's response to heat (OSHA, 2003).

The American Conference of Governmental Industrial Hygienists (1992) states that workers should not be permitted to work when their deep body temperature exceeds 38°C (100.4°F).

1. **Heat** is a measure of energy in terms of quantity.
2. A **calorie** is the amount of energy in terms of quantity.
3. **Evaporative cooling** takes place when sweat evaporates from the skin. High humidity reduces the rate of evaporation and thus reduces the effectiveness of the body's primary cooling mechanism.
4. **Metabolic heat** is a by-product of the body's activity.

Heat Disorders and Health Effects

According to OSHA (2003), heat stress can manifest itself in a number of ways depending on the level of stress. The most common types of heat stress are heat stroke, heat exhaustion, heat cramps, heat rash, transient heat fatigue, and chronic heat fatigue. These various types of heat stress can cause a number of undesirable bodily reactions including prickly heat, inadequate venous return to the heart, inadequate blood flow to vital body parts, circulatory shock, cramps, thirst, and fatigue.

1. Heat Stroke occurs when the body's system of temperature regulation fails and body temperature rises to critical levels. This condition is caused by a combination of highly variable factors, and its occurrence is difficult to predict. Heat stroke is very dangerous and should be dealt with immediately, because it can be fatal. The primary signs and symptoms of heat stroke are confusion; irrational behavior; loss of consciousness; convulsions; a lack of sweating (usually); hot, dry skin; and an abnormally high body temperature, for example, a victim of heat stroke will have a rectal temperature of 104.5°F or higher that will typically continue to climb.

- If a worker shows signs of possible heat stroke, professional medical treatment should be obtained immediately. The worker should be placed in a shady area and the outer clothing should be removed. The worker's skin should be wetted and air movement around the worker should be increased to improve evaporative cooling until professional methods of cooling are initiated and the seriousness of the condition can be assessed. Fluids should be replaced as soon as possible. The medical outcome of an episode of heat stroke depends on the victim's physical fitness and the timing and effectiveness of first aid treatment.

2. **Heat exhaustion** is a type of heat stress that occurs as a result of water and/or salt depletion. Employees working in the heat should have such fluids readily available and drink them frequently. Electrolyte imbalance is a problem with heat exhaustion and heat cramps. When people sweat in response to exertion and environmental heat, they lose more than just water. They also lost salt and electrolytes. **Electrolytes** are minerals that are needed in order for the body to maintain the proper metabolism and in order for cells to produce energy. Loss of electrolytes causes these functions to break down. For this reason it is important to use commercially produced drinks that contain water, salt, sugar, potassium, or electrolytes to replace those lost through sweating.

   - The signs and symptoms of heat exhaustion are headache, nausea, vertigo, weakness, thirst, and giddiness. Fortunately, this condition responds readily to prompt treatment. Heat exhaustion should not be dismissed lightly, however, for several reasons. One of the principal is the fainting associated with heat exhaustion can be dangerous because the victim may be operating machinery or controlling an operation that should not be left unattended.

   - A victim of heat exhaustion should be moved to a cool but not cold environment and allowed to rest lying down. Fluids should be taken slowly but steadily by mouth until the urine volume indicates that the body's fluid level is once again in balance.

3. Performing hard physical labor in a hot environment usually causes **heat cramps**.

   - This is a type of heat stress that occurs as a result of salt and potassium depletion. Observable symptoms are primarily muscle spasms that are typically felt in the arms, legs, and abdomen.

   - To prevent heat cramps, acclimatize workers to the hot environment gradually over a period of at least a week. Then ensure that fluid replacement is accomplished with a commercially available carbohydrate-electrolyte replacement product that contains the appropriate amount of salt, potassium, and electrolytes.

4. **Heat rashes** are the most common problem in hot work environments. This is a type of heat that manifests itself as small raised bumps or blisters that cover a portion of the body and give off a prickly sensation that can cause discomfort. It is caused by prolonged exposure to hot and humid conditions in which the body is continuously covered with sweat that does not evaporate because of the high humidity. In most cases, heat rashes will disappear when the affected individual returns to a cool environment.

5. **Heat fatigue** is a type of heat stress that manifests itself primarily because of the victim's lack of acclimatization. Well-conditioned employees who are properly acclimatized will suffer this form of heat stress less frequently and less severely

than poorly conditioned employees will. Consequently, preventing heat fatigue involves physical conditioning and acclimatization, because there is no treatment for heat fatigue except to remove the heat stress before a more serious heat-related condition develops.

Cold Hazards

Temperature hazards are generally thought of as relating to extremes of heat. This is natural because most workplace temperature hazards do relate to heat. However, temperature extremes at the other end of the spectrum—cold—can also be hazardous. Employees who work outdoors in colder climates and employees who work indoors in such jobs as meatpacking are subjected to cold hazards.

There are four factors that contribute to cold stress: cold temperature, high or cold wind, dampness, and cold water. These factors, alone or in combination, draw heat away from the body (Greaney, 2000).

OSHA (1998) expresses cold stress through its cold stress equation. That is,

Low Temperature + Wind Speed + Wetness = Injuries & Illness

The major injuries associated with extremes of cold can be classified as being either generalized or localized. A generalized injury from extremes of cold is hypothermia. Localized injuries include frostbite, frostnip, and trenchfoot.

- **Hypothermia**—results when the body is unable to produce enough heat to replace the heat loss to the environment. It may occur at air temperatures up to 65°F and it is when the body uses its defense mechanisms to help maintain its core temperature.
- **Frostbite**—is an irreversible condition in which the skin freezes, causing ice crystals to form between cells. The toes, fingers, nose, ears, and cheeks, are the most common sites of frostbite.
- **Frostnip**—is less severe than frostbite. It causes the skin to turn white and typically occurs on the face and other exposed parts of the body. There is no tissue damage; however, if the exposed area is not either covered or removed from exposure to the cold, frostnip can become frostbite.
- **Trenchfoot**—is caused by continuous exposure to cold water. It may occur in wet, cold environments or through actual immersion in water.

Windchill Factor

The *windchill factor* increases the level of hazard posed by extremes of cold. Safety and environmental managers need to understand this concept and how to make it part of their deliberations when developing strategies to prevent cold stress injuries.

OFFICE SAFETY

We commonly think of the standard office workplace as a haven from safety and health hazards. Even experienced safety and environmental managers can hold this misperception. This is the case, of course, because office areas are often thought of as plush environments, with even plusher accommodations and accoutrements: thick pile carpet, fancy desks, a cafeteria, television, spit-and-polished restroom facilities, and air-conditioning . . . the Internet, Email, iPhone stations, etc. Not until a catastrophic event occurs (a building collapse during an earthquake, a devastating fire, or an electrocution) does the average person (and/or office worker) realize that office environments can be hazardous—and in fact, deadly.

The nonbeliever might, along with a very large population of workers, feel that catastrophic events such as those we mentioned are rare and therefore not that big of a deal. If you are a victim of such an event, your view, of course, is quite different. Historical records indicate that the highest severity and significant losses (because of the large population of office workers) are usually associated with such events.

Safety and environmental managers handling office safety and health issues must look at the big picture. Yes, without a doubt, precautions must be taken to ensure certain procedures are in place to lessen the devastating effects of catastrophic risk. The safety and environmental manager should also deal with the risks associated with other office-related events —not catastrophic ones, but injury threatening, just the same—slips, trips, falls, cuts, indoor air quality, VDT effects, and strains.

The safety and environmental manager who overlooks the need for an Office Safety Program is falling into a trap—into an untenable situation—whereby he or she has failed to recognize that hazards are omnipresent. They must be guarded against in all workplace areas, including office spaces.

Implementation of an Office Safety Program

In implementing an office safety program, the safety and environmental manager must be cognizant of the fact that office areas are typically safer than most other workplaces. At the same time, however, the safety and environmental managers must be aware of the fact that hazards that can potentially cause illness and injury do exist. Situations, materials, and equipment that can lead to illness or injury can include the following:

- tripping hazards
- back injury
- falling
- file cabinets
- shock
- chemical exposure
- computers

The safety and environmental manager should ensure that the above elements are included in the Office Safety Program. One of the first elements the safety and environmental manager should include in the Office Safety Program (after having carefully evaluated all office spaces) is to write an *Emergency Action Plan.*

Emergency Action Plan

The Emergency Action Plan usually focuses on evacuation. Many offices are located in high-rise buildings, where a fire is most certainly a life-threatening situation and where evacuation is difficult. The larger the high-rise office complex, the larger numbers of people are potentially at risk, because many of the offices may be well beyond the normal reach of ground firefighting equipment, and relatively inaccessible—except from inside the building. Even many of the modern high-rise buildings constructed of fire-resistant materials are dangerous because of their contents and decor. These contents typically include many decorative elements made from materials that present fire hazards and produce dangerous smoke. Remember, fire's smoke and gases kill more people than does the fire itself.

One of the primary duties of the safety and environmental manager is to inspect high-rise office buildings, to evaluate and control combustible fire loads prior to the onset of fire. In the control effort, considerations must be made for both the vertical and horizontal components. One control method that has been tried, tested, and proved effective is sprinkler installations, which can reduce the losses dramatically. Along with sprinkler systems, ventilation systems should also be carefully evaluated. Properly designed ventilation systems can work to dramatically control smoke in high-rise fires.

Lighting and Eye Hazards

The need for artificial illumination is nowhere more pronounced than in office areas. For this reason, consideration of *lighting and eye hazards* must be taken into account when implementing the Office Safety Program. The safety and environmental manager must recognize that the design of artificial illumination systems has an impact on the performance, responses, and comfort of those using the environment in office spaces. Illumination design is both an art and a science. The scientific aspects include the measurement of various lighting parameters and the design of energy-efficient lighting systems. The artistic side comes into play in combining light sources to create, for example, a particular mood in a medical office, to highlight a display in a department store, or to complement a particular color scheme (Sanders & McCormick, 1993).

The question that often confronts the designer and safety and environmental manager is how much illumination is enough? This question has occupied the attention of various illumination experts for many years. In an office environment, where detail work is often performed, where computers have replaced typewriters, and where writing and reading are common work practices, careful consideration to exactly how much illumination is enough, how much is too much, and how much is too little must be given. To design an energy-efficient lighting system to supply the proper amount of illumination without creating glare is a significant achievement. The safety engineer should be acquainted with lighting requirements and the effects of lighting on workers, and must possess enough knowledge about lighting systems and their possible effects on workers to make an intelligent decision on the type of illuminating engineering required for each particular office situation.

Hazard Communication

Another important element in any Office Safety Program is Hazard Communication. In Chapter 5, we discussed OSHA's Hazard Communication Program in detail, and you probably had little difficulty recognizing that hazardous materials are common in most industrial workplaces. However, in office areas? Surprisingly, though chemicals are not as prevalent in the office as in the industrial workplace setting, they do exist.

The problem for the safety and environmental manager is that often common office chemicals and the hazards they present are either overlooked or underestimated. Though offices are not typically production process areas, they still require cleaning. Many of these cleaning materials are hazardous to workers. Most offices use copiers and printers. These essential machines commonly use chemical toners and inks, some of which fall into the hazardous category. Standard office practices allow workers to add toner or change cartridges as needed. Office workers are also commonly exposed to paper correction fluids, white board cleaners, and markers used on various types of charts. Though many of these chemicals may or may not be hazardous, the safety and environmental manager holds the responsibility of making this decision. Usually consulting the MSDS or reading the label easily makes this determination. If the chemical is found to be hazardous, the safety and environmental manager must take steps to ensure that the office workforce is aware of the hazards.

Office Equipment

Another important element that must be considered in an Office Safety Program is office equipment. Office equipment is like air-conditioning and interior illumination; we take them for granted until they malfunction: until the air-conditioning fails, lights go out, facsimile machines, copiers, computers, and printers don't work. Otherwise, office machines, ranging from the very simple to the complex, from the electrically powered to the handheld and mechanically powered, from the desktop computer to the stapler, are just part of the office and are often overlooked.

Nishiyama (1987) points out that automation has introduced a new set of safety and health problems into the office environment. Morrok & Yamoda (1987) identified the following problems associated with office automation: eye fatigue, seeing double images and complementary colors, headache, yawny feelings, unwillingness to talk, shoulder fatigue, neck fatigue, dryness in the throat, sleepy feelings, and whole-body tiredness.

Most legislation dealing with office automation concerns standards for video display terminals (VDT) interaction. For example, in some regions of the United States VDT users are required to (1) have an eye examination every year; (2) have an adjustable chair with adjustable back-rest height and tension; and (3) take a fifteen-minute break from the VDT every hour. Legislation introduced in other regions is similar to this proposal (Amick, 1987; Spellman& Whitting, 2006).

The safety and environmental manager must not ignore office equipment; it has the potential for harm any time it is used incorrectly, hastily, or without care and caution. While it is true that office equipment presents relatively minor hazards (such as cuts or punctures or bruises), some could present a much greater hazard—electrocution, for example. Obviously, the best way to avoid such hazards is to ensure that they are operated correctly by workers, workers who think about what they are doing and how they are doing it.

However, because most employees give office machines little or no thought (unless they malfunction), they usually give little thought to what they are doing and how they are doing it while operating office machines. Inattention to the task at hand is often responsible, not just for short-term, minor injury, but also for long-term repetitive stress injury—carpal tunnel syndrome, for example. These two areas are what the safety engineer must guard against. How? No doubt this can be and often is a perplexing question.

General Operating Guidelines

Hazard control in the office, designed to protect employees from injury by office machines, begins with devising general operating guidelines, informing employees of these guidelines, posting them, and following up to ensure that the guidelines are actually being followed.

What should be included in these guidelines? The following sample guidelines on usage of office equipment are generally applicable to most office equipment.

1. Employees are not authorized to use office machines they are not qualified to operate. Before operating any office machine, first read the equipment directions or seek assistance from a qualified employee.
2. Ensure that guards are in place on equipment that requires them before using that equipment.
3. Do not use any office machine that appears defective in any way.
4. When office machines are mobile, do not place them too near the edge of a table or desk.
5. If the office machine tends to move during operation, make sure it is secure.

Along with employee awareness of electrical safety hazards and mechanical office equipment that can cut or puncture (e.g., hand-operated paper cutters), employees must also be made aware of the hazards caused by file cabinets (e.g., open drawer trip hazards), shelves (e.g., placing heavy objects on top shelves versus on lower shelves), and desks (e.g., damaged desks that could tear, rip, cut, or puncture body parts). As a safety and environmental manager, spend some time walking through office areas on normal workdays. Ask questions—who is responsible for changing lamps in lighting fixtures? Who hangs pictures? How do you access those files? As we said, office workers often aren't thinking "safety," so you must.

OFFICE SAFETY VERSUS WORKPLACE SAFETY

"Familiarity breeds contempt" is cliché, and like all clichés, it holds a great deal of truth. Workers think of their offices as safe places and don't stop to think about working safely. Because workers often use the easiest or quickest method to accomplish a task, and because offices frequently do not have common tools easily available for use, workers frequently jerry-rig techniques for taking care of something they need to do—from using the heel of their shoe to hammer in a nail to using the rolling desk chair as a step stool to reach a lighting fixture or high shelf.

Of course, the worker involved in such a process isn't thinking about safety at all. Office workers are notoriously unaware that their jobs involve hazards at all. They are, in fact, often using the same questionable technique they would use at home for the same sort of task. That does not mean, however, that they aren't risking serious injury. A concerned safety engineer would blanch and have to bite his or her tongue to keep from completely alienating the worker discovered standing on a castered office chair with a contoured seat, attempting to lift a 25 lb. file box from a shelf above his or her head, with the only possible landing for a potential fall onto the computer desk and terminal.

We did manage to contain our initial reaction, however, and gently pointed out the immediate potential dangers. We then ordered the equipment needed to accomplish this common task safely and worked on a storage reconfiguration that stored often-used files in a more accessible location.

Bloodborne Pathogens

A fairly recent peril has been added to the list of possible office-related hazards: exposure to *bloodborne pathogens*. When accidents, injuries, or illnesses occur in the office, employees need to know how to respond safely and correctly. This is particularly the case for those office employees who render first aid. The safety engineer must look at this potential life-threatening area with particular attention. Specifically, the safety and environmental manager should ensure that a proper accident response scenario is in writing and in place to detail exact response procedures. An office procedure entailing the procedures to be followed when rendering first aid services must be in place.

Another important aspect cannot be overlooked: accident reporting. Employees must be trained to report all on-the-job accidents, no matter what their level of severity. Accidents that involve release of body fluids that other employees come into contact with must be reported and proper medical response effected (the victim and the employees who came in contact should be medically evaluated and offered a Hepatitis B vaccination). Employees must be thoroughly trained on avoiding bloodborne pathogens.

Security

Unfortunately, the next element that should be included in an office safety program has to do with employee *security*. Employees must be informed of the dangers involved not only with sexual harassment in the workplace, but also with avoiding workplace violence, preventing theft, elevator security, and parking lot security.

One of the key steps in attempting to provide employee security is employee awareness. Employees must be constantly reminded that their security is an issue and must be trained on how to deal with potential problems. For example, if an angry or hostile coworker or customer confronts an employee, he or she should be trained to follow certain precautionary guidelines which include:

1. Stay calm and listen.
2. Maintain eye contact.
3. Be patient and courteous.
4. Try to keep the situation in your control—when you talk, keep talking.

A simple procedure such as the one just listed is rather easy to list—but never easy, obviously, to actually put into play when required. However, employees should have some type of training in what to do when confronted by a potentially violent person. Employees also need to be informed on how to prevent theft, on elevator security, and on how to deal with sexual harassment.

The safety and environmental manager should pay particular attention to parking lot security. That a well-lighted parking lot has the tendency to increase employee security is a well-known fact. Good parking lot lighting is not foolproof, however—nothing really is. As the safety engineer, you should carefully inspect and study employee parking areas, not only to ensure that adequate lighting is available, but also to evaluate such areas to see if anything else can be done to increase security (e.g., install video cameras, etc.).

Ergonomics

Ergonomics is another element the safety and environmental manager must consider when developing an office safety program. Ergonomics, as explained earlier, is a term more prevalently used in Europe and the rest of the world than in the United States. It simply means arranging the environment to fit the person in it. Applied to an office environment, it is probably best defined as, "Working to stay comfortable, and staying comfortable to work" (J. J. Keller, 1997). Ergonomic hazards in the office setting are normally associated with repetitive work—much of it done on computer terminals. Much of the industrialized workforce now uses video display terminals (VDTs) regularly. For the safety engineer, VDT hazard control considerations include workplace surveys and redesigns, equipment and chair adjustments allowing proper body alignment, foot and wrist rests, document holds, exercise techniques, and work-rest regimens (CoVan, 1995).

Safety Outside the Office

To this point we have discussed safety primarily inside the office, but the safety and environmental manager should include another element in the office safety program: safety outside the office. This element specifically relates office employee protection whenever the employee leaves the office and walks into a company production area, for example. Common office practice allows office workers to use company vehicles to run errands and so forth. Make sure workers are aware of safe highway driving techniques. Another factor that has to be considered is weather. Eventually all office workers leave the office and enter the outdoor world. When this occurs during severe weather events, the employee's safety could be at risk. Injury or death on the way to or from work is simply not worth the risk.

THOUGHT-PROVOKING QUESTIONS

- Discuss the similarities and differences between the job functions of safety and environmental managers and industrial hygienists.
- Why were workplace injuries and occupational diseases regarded as separate problems prior to the OSH Act?

- What are the three parts that make up the industrial hygiene paradigm?
- What two factors affect an industrial hygienist's function within an organization?
- What is the chief difference between safety engineering and industrial hygiene? What does this mean in terms of potential professional growth and job responsibility?
- What control measures might an industrial hygienist recommend as correction measures?
- Define stressors.
- How much detailed knowledge should an industrial hygienist have concerning workplace processes and operations?
- What stressors are of greatest concern to industrial hygienists?
- What principle areas should the industrial hygienist be concerned with?
- Define toxicology.
- What's the difference between toxic and hazard?
- Discuss dose-response relationship and threshold level. What effect does time have on these?
- What are air contaminant values? How are they used?
- What are the common routes of entry for toxicants into the body? Which is the most common for the worksite?
- What is the difference between acute and chronic exposure?
- What information should you have about a substance to determine whether or not it is hazardous?
- What are the categories of airborne contaminants? Define them.
- How does a safety engineer or industrial hygienist test for airborne contaminants?
- Define engineering controls.
- Describe how ventilation can control airborne hazards.
- Three elements ensure ventilation systems are safe and effective. What are they, and why is each important?
- What is industrial ventilation's purpose?
- How does natural ventilation occur?
- What functions do installed ventilation systems perform?
- What four components make up a ventilation system?
- What is the difference between exhaust and supply ventilation?
- How are pressures in a ventilation system measured?
- What is a water gauge and how does it work?
- What three ventilation pressures should safety and environmental managers and industrial hygienists be familiar with? Why? Define the pressures. How are they expressed?
- How are pressure measurements taken? What tools are needed? How do they work?
- Discuss local exhaust ventilation. How does it work?
- When is local exhaust ventilation the proper method of control?
- How can safety and environmental managers ensure that the ventilation system is and remains effective?
- What are ventilation system limitations?
- Define general ventilation systems. Define dilution ventilation systems. What are their applications and purposes? How do you determine which to install?
- What is the key operation in dilution ventilation?

- Define noise. How is it measured?
- What are the steps in a noise level survey, and how is a survey used to meet OSHA criteria?
- Define maximum noise levels and minimum noise levels. How are they determined, and why are they important?
- At what level does OSHA become concerned about workplace noise levels?
- Discuss time-weighted average.
- What are the differences between continuous and intermittent or impact-type noise makers?
- Discuss workplace noise and PEL.
- How do you calculate daily dose?
- A person has been exposed to the below noise levels during one work day. Has this person been exposed to too much noise in an 8-hour period?
  - 85 dBA for 3 hrs
  - 90 dBA for 2.75 hrs
  - 95 dBA for 1.5 hrs
  - 100 dBA for .75 hrs
- What dBA is the upper limit?
- How can a safety and environmental manager eliminate noise in the workplace, and why should it happen?
- How does a safety and environmental manager determine an "acceptable level of noise?"
- How does a safety and environmental manager accomplish creating an "acceptable level?"
- What three engineering control components can a safety and environmental manager use to control and reduce noise?
- What health and safety threats are posed by vibration?
- Define whole-body and segmented vibration. How are they different?
- How does whole-body vibration affect workers?
- How does segmented vibration affect workers?
- What control measures are effective against vibration?
- Why should safety and environmental managers be included in design or renovation projects at the planning level, as well as at the installation and construction stages of a project?
- What are administrative controls and what is their purpose?
- Define PELs and TLVs. Why are they important? What are their individual strengths?
- What is the chief problem in limiting exposure duration?
- What are the three time periods of average exposure?
- What role does good housekeeping practice play in administrative control?
- Why are materials handling and transferring procedures important?
- How can leak detection programs protect workers?
- What role does training play in administrative control?
- What's the safety engineer's biggest concern with PPE as a method of control? What's PPE's biggest drawback?
- What are the three principle types of PPE?

- Discuss PPE and worker sense of security.
- Define fall protection
- What industries have the highest rates of death and injury from falls? Why?
- What's the difference between a slip and a trip? Why is the distinction important?
- What laws of science work to cause falls? How does each affect how falls occur?
- What design factors are involved in causing and preventing slips?
- What are frequent causes of slips? Why?
- What are some of the causes of trips? How can many of these be eliminated?
- Why are stairs so hazardous? What simple things can workers do to prevent stair falls?
- Are handrails useful for stopping falls? Why or why not?
- What should guardrails consist of?
- What three components make up a safety program for fall protection?
- What special considerations should be made for scaffold work? Why does scaffolding merit special concern?
- Define "competent person" as the term applies to scaffolding safety regulations.
- What does fall protection PPE consist of?
- What are the loading rules for scaffolding? Why are they important?
- What PPE should be provided for protection against overhead hazards?
- What special attention and concern do outriggers merit? Why?
- What schedule should the safety and environmental manager follow for inspection of scaffolding?
- Why is a fall protection policy important?
- Define free fall.
- What are some of the risks and hazards common to office spaces?
- What's an emergency action plan, and why is it important?
- What problems with lighting are common in offices?
- How does HazCom apply to offices and office personnel?
- Are bloodborne pathogens a consideration? What should a program covering this entail?
- What considerations should the safety and environmental manager take for employee security? What areas need attention for safety?
- What should the safety and environmental manager be concerned with in areas outside the office?

REFERENCES AND RECOMMENDED READING

A Citizen's Guide to Radon. USEPA, 1986.

Alpaugh, E.L. *Fundamentals of Industrial Hygiene*, 3rd ed. Revised by T.J. Hogan. Chicago: National Safety Council: 259–260.

American Conference of Governmental Industrial Hygienists. *Guidelines for Assessment and Control of Aerosols.* Cincinnati, OH: ACGIH, 1997.

American Industrial Hygiene Association. *Field Guide for the Determination of Biological Contaminants in Environmental Samples.* Fairfax, VA: AIHA, 1996.

American Industrial Hygiene Association. *Practitioners Approach to Indoor Air Quality Investigations.* Fairfax, VA: AIHA, 1989.

Amick, B., III. "The Impacts of Office Automation on the Quality of Worklife: Considerations for United States Policy." In *Occupational Health and Safety in Automation and Robotics*, edited by K. Noro, 232. Chicago: National Safety Council, 1987.

Ammann, H. "Is Indoor Mold Contamination a Threat?" Washington State Department of Health. Accessed August 9, 2003. http://www.doh.wa.gov/ehp/ocha/mold.html.

American Conference of Governmental Industrial Hygienists. *1992–1993 Threshold Limit Values for Chemical Substances and Physical Agents and Biological Exposure Indices.* Cincinnati, OH: American Conference of Governmental Industrial Hygienists, 1992.

Burge, H.A. "The Fungi: How They Grow and Their Effects on Human Health." *Heating/ Piping/Air Conditioning* July 1997: 69–75.

Burge, H.A., and M.E. Hoyer. "Indoor Air Quality." In *The Occupational Environment—Its Evaluation and Control*, edited by S.R. DiNardi. Fairfax, VA: American Industrial Hygiene Association, 1998.

Byrd, R.R. *IAQ FAG Part 1.* Glendale, CA: Machado Environmental Corporation, 2003.

Centers for Disease Control. *Reports of Members of the CDC External Expert Panel on Acute Idiopathic Pulmonary Hemorrhage in Infants: A Synthesis.* Washington, DC: Centers for Disease Control, 1999.

Code of Federal Regulations. Lead Exposure in Construction, title 29, sec. 1926.

Code of Federal Regulations. Occupational Safety and Health Standards, title 29, sec. 1910.

CoVan, J. *Safety Engineering.* New York: Wiley, 1995.

Cyr, D.L., and S.B. Johnson. *Battling the Elements Safely.* Accessed May 15, 2015. http:// umaine.edu/publications/2312e/

Davis, P.J. *Molds, Toxic Molds, and Indoor Air Quality.* Sacramento, CA: California Research Bureau, California State Library, 2001.

Di Tommaso, R.M., E. Nino, and G.V. Fracastro. "Influence of the Boundary Thermal Conditions on the Air Change Efficiency Indexes." *Indoor Air* 9 (1999): 63–69.

Dockery, D.W., and J.D. Spengler. "Indoor-outdoor Relationships of Respirable Sulfates and Particles." *Atmospheric Environment* 15 (1981): 335–343.

Eisma, T.L. "Rules Change: Worker Training Helps Simplify Fall Protection." *Occupational Health & Safety* March 1990: 52–54.

Ellis, J.E. *Introduction to Fall Protection.* Des Plaines, IL: American Society of Safety Engineers, 1988.

Ferry, T. *Safety and Health Management Planning.* New York: Van Nostrand Reinhold, 1990.

Grimaldi, J.V., and R.H. Simonds. *Safety Management*, 5th ed. Homewood, IL: Irwin, 1989.

Hammer, W. *Occupational Safety Management and Engineering*, 4th ed. Englewood Cliffs, NJ: Prentice Hall, 1989.

Hazard, W.G. "Industrial Ventilation." In *Fundamentals of Industrial Hygiene*, 3rd ed., edited by B. Plog. Chicago: National Safety Council, 1988.

Hollowell, C.D., et al. "Impact of Infiltration and Ventilation on Indoor Air Quality." *ASHRAE Journal* July 1979: 49–53.

Hollowell, C.D. et al. "Impact of Energy Conservation in Buildings on Health." In *Changing Energy Use Futures*, edited by R.A. Razzolare and C.B. Smith. New York: Pergamon, 1979.

"Indoor Air Facts No. 1." *EPA and Indoor Air Quality.* USEPA, 1987.

Keller, J.J. *Office Safety Handbook.* Neenah, WI: J. J. Keller & Associates, Inc., 1997.

Kohr, R.L. "Slip Slidin' Away," *Safety & Health* 140, no. 5 (1989): 52.

Leviticus 14.

McNeel, S., and Kreutzer, R. (1996). "Fungi & Indoor Air Quality" *Health & Environment Digest* Vol 10, No. 2 (May/June).

Molhave, L., 1986. Indoor Air Quality in Relation to Sensory Irritation Due to Volatile Organic Compounds. *ASHRAE Trans.* 92(1): Publication #2954, p. 306–316.

Morooka, K., and S. Yamoda. "Multivariate Analysis of Fatigue on VDT Work." In *Occupational Health and Safety in Automation and Robotics*, edited by K. Noro. Chicago: National Safety Council, 1987: 236.

National Safety Council. *Accident Facts 1984 & 1985*. Chicago: National Safety Council.

National Safety Council, *Accident Facts*, current edition.

NIOSH. *The Industrial Environment, Its Evaluation and Control*. Cincinnati, OH: NIOSH, 1973.

Nishiyama, K. "Introduction and Spread of VDT Work and Its Occupational Health Problems in Japan." In *Occupational Health and Safety in Automation and Robotics*, edited by K. Noro. Chicago: National Safety Council, 1987: 251.

Occupational Safety and Health Administration. *The Cold Stress Equation* (OSHA3156). Washington, DC: U.S. Department of Labor, 1998.

Occupational Safety and Health Administration. *OSHA Technical Manual 4: Heat stress*. Washington, DC: U.S. Department of Labor, 2003.

Olishifski, J.B., and B.A. Plog. "Overview of Industrial Hygiene." *Fundamentals of Industrial Hygiene*, 3rd ed. Chicago: National Safety Council, 1988.

Passon, et al. "Sick-Building Syndrome and Building-Related Illnesses." Medical Laboratory Observer 28, no. 7 (1996): 84–95.

Pater, R. "Fallsafe: Reducing Injuries from Slips & Falls." *Professional Safety* October 1985: 15–18.

Peng, S.H., and I. Davidson. "Performance Evaluation of a Displacement Ventilation System for Improving Indoor Air Quality: A Numerical Study." In *8th International Conference on Indoor Air Quality and Climate*. Edinburgh, Scotland, 1999.

"Residential Air Cleaners Indoor Air Facts No. 7," *Air and Radiation*. USEPA, 1990.

Rose, C.F. "Antigens." *ACGIH Bioaerosols Assessment and Control*. Cincinnati, OH: American Conference of Governmental Industrial Hygienists, 1999: 25–1 to 25–11.

Sanders, M. S., and E.J. McCormick. *Human Factors in Engineering and Design*, 7th ed. New York: McGraw-Hill, 1993.

Spellman, F.R., and N. Whiting. *Environmental Science and Technology: Concepts and Applications*, 2nd ed. Rockville, MD: Government Institutes, 2006.

Spengler, J.D., et al. "Sulfur Dioxide and Nitrogen Dioxide Levels Inside and Outside Homes and the Implications on Health Effects Research." *Environmental Science & Technology* 13 (1979): 1276–1280.

The Inside Story—A Guide to Indoor Air Quality. USEPA, 1988.

U.S. Department of Labor, Bureau of Labor Statistics. *Injuries Resulting From Fall From Elevation*, Bulletin 2195. Washington, DC: U.S. Government Printing Office, 1984.

USEPA, Indoor Environmental Management Branch. "Children's Health Initiative: Toxic Mold." Accessed January 9, 2008. http://epa.gov.appedwww/crb/iemb/child.htm.

USEPA. *Indoor Air Facts No. 4 (revised) Sick Building Syndrome*. Accessed August 9, 2007. http://www.epa.gov/iaq/pubs/sbs.html.

U.S. Environmental Protection Agency. *Building Air Quality: a Guide for Building Owners and Facility Managers*. Washington, DC: EPA, 1991.

Wadden, R.A., and P.A. Scheff. *Indoor Air Pollution: Characteristics, Predications, and Control*. New York: John Wiley & Sons, 1983.

Washington State Department of Health, Office of Environmental Health and Safety. *Formaldehyde*. Accessed August 2003. http://www.doh.wa.gov/ehp/ts/IAQ/Formaldehyde.htm.

Woods, J.E. "Environmental Implications of Conservation and Solar Space Heating." Engineering Research Institute, Iowa State University, Ames, Iowa, BEUL 80-3. Meeting of the New York Academy of Sciences, New York. January 16, 1980.

World Health Organization (WHO). *Indoor Air Pollutants, Exposure and Health Effects Assessment. Euro-Reports and Studies No. 78: Working Group Report.* Copenhagen: WHO Regional Office, 1982.

World Health Organization (WHO). *Indoor Air Quality Research Euro-Reports and Studies No. 103.* Copenhagen: WHO Regional Office for Europe, 1984.

Yang, C.S. "Toxic Effects of Some Common Indoor Fungi." Accessed August 6, 2006. http://www.envirovillage.com/Newsletters/Enviros/No4_09. htm.

Zhang, Y., X. Wang, G.L. Riskowski, and L.L. Christianson. "Quantifying Ventilation Effectiveness for Air Quality Control." Transactions of the American Society of Agricultural and Biological engineers (ASABE) 44, no. 2 (2001): 385–390.

Chapter 17

Safety and Health Training

Throughout this text, one of the major points we make and remake is to emphasize the importance of employee safety and health training. This emphasis has been for good reason. Without a doubt, providing routine safety and health training for workers is one of the most important job duties of the safety and environmental manager or other safety professional in any industry. Indeed, most managers know the importance of safety training, but what is not well known is that specific training requirements are detailed in OSHA, DOT, and EPA regulations. Certain OSHA regulations, for example, state or imply that the employer is responsible for providing training and knowledge to the worker. Employees must be apprised of all hazards to which they are exposed, along with relevant symptoms, appropriate emergency treatment, and proper conditions and precautions of safe use or exposure.

To this point in the text, several OSHA safety and health standards or programs have been featured. Employers must comply with these standards and must also require workers to comply. More than 100 OSHA, DOT, and EPA safety and health regulations contain training requirements.

Note that although OSHA requires training (OSHA 2056, 1991—Revised), it does not always specify what is required of the employer or entity that provides the training. Safety and environmental managers must create their own training programs. Information and instruction on safety and health issues in the workplace are foundational to building a viable organizational safety program. Workers cannot be expected to perform their assigned tasks safely unless they are aware of the hazards or the potential hazards involved with each job assignment (F. R. Spellman, 1996).

INTRODUCTION

Hoover et al (1989) points out that in discussions with safety professionals over the years, when asked, "If you could only keep one part of your activities, which would you keep?" the answer most often given is "training." Certainly training is important, but even more important is a well-thought-out program, one that is well balanced, aggressive, has continuity, and gives management "the best return for the dollar" (p. 19).

Safety and environmental managers and other health and safety professionals play a key role in ensuring that all employees at all levels receive the appropriate types and amounts of safety training. What this really means is that safety professionals must also play an active role in preparing, presenting, arranging for the application of, and evaluating safety and health training.

In this chapter, we cover those elements required to make up an organizational safety and health training program. Specifically, we discuss the requirements for a written training program, the need to conduct training, recordkeeping requirements, and the need to evaluate the organization training program to ensure that is both current and effective.

Written Training Program

Throughout this text we have emphasized the need for the safety and environmental manager to ensure that just about anything and everything he or she does or says should be in writing. The organization's training program should be a formal written program.

So, the logical question is: "What should the written safety and health training program consist of?" Safety organizations such as the National Safety Council and practicing safety and health professionals generally cite six basic steps for developing a training program. These are:

1. Identifying training needs
2. Formulating training objectives
3. Gathering materials and developing course outlines
4. Selecting training methods and techniques
5. Conducting training session
6. Evaluating training programs

Identifying Training Needs

In the chapter opening, we pointed out that more than 100 OSHA (and other regulatory agencies) standards and regulations have training requirements. Let's take a closer look at the OSHA training requirements.

The Occupational Safety and Health Act (or OSH Act) of 1970 mandates that employers provide health and safety training. The OSH Act requires the following:

• Education and training programs for employees
• Establishment and maintenance of proper working conditions and precautions
• Provision of information about all hazards employees will be exposed to on the job
• Provision of information about the symptoms of exposure to toxic chemicals and other substances that may be present in the workplace
• Provision of information about emergency treatment procedures

Simply put, the legal reasons for training are quite clear. The safety and environmental manager should utilize OSHA's guidance on training—OSHA's training requirements are a good place to start in formulating the organization's safety program. Specifically, if the organization comes under OSHA or some other regulatory agency, you should review the regulator's training requirements and make a determination if their standards or regulations apply to the organization. For example, if the organization's workers are required to wear respirators in the normal performance of their duties, then OSHA's 29 CFR 1910.134 *Respiratory Protection Standard* certainly comes into play. In reviewing 1910.134, in particular .134(b)(3), it is quite clear that training is required: "The user shall be instructed and trained in the proper use of respirators and their limitations." Anyone who has worked with OSHA standards would have little difficulty recognizing this respirator training requirement as being a requirement in just about all of OSHA's standards.

On the other hand, the safety and environmental manager not only needs to determine what type of training is required for his or her organization, but also needs to determine what training is not required. For example, OSHA, under its 29 CFR 1910.138 *Powered Industrial Trucks Standard* requires forklift operators to be trained. Specifically, the standard states, "Only trained and authorized operators shall be permitted to operate a powered industrial truck (forklift and other type powered trucks). Methods shall be devised to train operators in the safe operation of powered industrial trucks." Obviously, if the safety and environmental manager determines that his or her organization has forklifts, rents forklifts, borrows forklifts, and for any reason requires workers to operate such machines, then training is required. However, if the organization does not require its workers to operate forklifts or other powered trucks, training is not required.

Thus, if a potential hazard is present in the facility for which OSHA specifically requires training, developing and conducting training for the particular hazard should be a top priority. Basically, what it comes down to is that the safety and environmental manager needs to tailor his or her organization's safety and health training program to suit the needs of the organization and to ensure compliance with the regulations.

OSHA's general industry standards (including their training requirements) are numerous and often quite complex. With this in mind, an attempt is made in Table 17.1 to simplify the process by identifying typical industry training requirements. Note, however, that Table 17.1 is not "all-inclusive" and that the frequency of training is not listed.

In addition to the training listed in Table 17.1, the safety and environmental manager should consider integrating several other topics into the safety and health training program. These include heat and cold stress, environmental rules and regulations, ergonomics, first aid/CPR, and other general occupational safety and health topics.

Safety and health training should also be focused on any new equipment or new process introduced into the workplace. It is difficult (if not impossible), obviously, to expect employees to safely operate any new equipment or process unless they have been properly trained to do so.

Any time any regulation or company standard operating procedure is modified, changed or revised for any reason, training or refresher training should be conducted.

Table 17.1 OSHA Standards that Require Training (A Representative Sample) Standard

1. Hazard Communication Standard (1910.1200)
2. Employee Emergency Plans and Fire Prevention Plans (1910.38)
3. Powered Platforms for Building Maintenance (1910.66)
4. Confined Space Entry (1910.146)
5. Lockout/Tagout (1910.147)
6. Respiratory Protection (1910.134)
7. Hearing Conservation (1910.95)
8. Flammable and Combustible Liquids (1910.106)
9. Ventilation (1910.94)
10. Explosives and Blasting Agents (1910.109)
11. Ionizing Radiation (1910.96)
12. Hazardous Waste Operations and Emergency Response—HAZWOPER (1910.120)
13. Specifications for Accident Prevention Signs and Tags (1910.145)
14. Welding, Cutting, and Brazing (1910.252)
15. Machine Guarding (1910.217-218)
16. Powered Industrial Trucks (1910.178)
17. Fire Brigades (1910.156)
18. Portable Fire Extinguishers (1910.157)
19. Bloodborne Pathogens (1910.1030)
20. Fire Extinguishing Systems (1910.160)
21. Personal Protective Equipment (1910.132-138)
22. Electrical Safety (1910.301-339)
23. Excavation Safety (29 CFR 1926.650-652)
24. Process Safety Management (29 CFR 1910.119)
25. Storage and Handling of Liquefied Petroleum Gases (1910.110)

The safety and environmental manager should also look at past performance to determine if a need for additional training exists. For example, if employees have been trained on how to perform a job function in a safe manner, but injury reports and/or equipment failures caused by employee errors increase, then obviously there is a need to improve employee performance. Administering refresher training best brings this about.

The safety and environmental manager should also look at the individual processes and individual job requirements. If a certain process is highly advanced and very difficult to operate correctly (and safely), and if the workers must remember very difficult information, refresher training on an ongoing basis is called for.

Formulating Program Goals and Objectives

Once the training needs have been identified, it is necessary to quantify through goals and objectives what the workers will gain by completing the training. Effective, clearly stated goals and objectives specifically describe what workers are expected to do, to do better, or to discontinue doing after completion of the training. They are used to determine if workers have attained the desired level of proficiency. Training goals and objectives also help the safety professional to determine or measure the cost effectiveness of the program. Goals and objectives, like the training program, should be in writing and presented to each trainee so he or she clearly understands what the training entails and what he or she is required to learn.

Course Description

Communicating to the trainees the importance of any formal training program is important. This is best accomplished by using a syllabus. Each formal safety and health training program to be presented to workers should have a course syllabus containing the title, course objectives, course outline, evaluation, teaching methodology, and references. The syllabus conveys to the learner the expectation, the knowledge to be acquired, the work to be accomplished, and evaluation criteria. Obviously, the syllabus should be developed before the training program begins and handed out before the class begins.

Selecting Training Methods and Techniques

Selecting or developing the training methods, materials, and techniques are important. During this process the safety and environmental manager should keep in mind that the training methods and techniques used should simulate the actual job function as closely as possible. The actual training methods employed depend on the material, the size of the class, and the instructor's style.

Conducting Training Sessions

In determining what is to be covered in each training session, instructors should also decide what teaching method to employ, approximately how much time it will take, what visual aids or demonstrations to use, and what materials must be ready to hand out to workers. In our experience, when training sessions include the use of demonstrations and other visual methods, the training is more acceptable to the workers. They buy in on the presentation, which means they are likely to learn much more than they otherwise would.

Evaluating the Training Program

The primary way the effectiveness of the training program is determined is by evaluating it. In fact, the most critical part of the training process (any training process) is that of evaluation. Training programs are typically evaluated based on results, and asking questions, both in formal testing and in surveying the trainees, can give the presenter valuable information on how to make training most effective. Did the training correct the deficiency. Did the training supply the information the workers need to perform their assigned duties in a safe, productive fashion? Evaluation is all about feedback. Feedback is obtained by asking the questions above and by asking whether or not the training was helpful. Did the workers get what they wanted? Was there anything that they would have liked done differently? It should be apparent that the answers to these questions can best be obtained by asking workers who have been trained. Using a formal, preprinted course evaluation form is a practical, proven technique. Those parts of the program receiving poor evaluations should be revised promptly.

Recordkeeping

We have stated previously that no matter how much training an organization provides, no matter how detailed it is, no matter its quality and content, and no matter how effective it is, if the training is not properly documented, the regulators and/or courts of law will consider that the training has not been done.

Beyond the regulatory and legal requirements of documenting training, documentation is also required to provide a record for the manager to review to determine who has and who has not been trained. Obviously, this is important. Additionally, many training sessions must be routinely repeated (refresher training) on a periodic basis. Keeping up-to-date and accurate records helps in this process.

THOUGHT-PROVOKING QUESTIONS

- What elements of training should the safety and environmental manager be involved in?
- What are the basic steps for developing a safety program?
- How does the safety and environmental manager determine training needs? What training should go beyond OSHA regulation requirements?
- How should the safety and environmental manager determine when training is not needed?
- What should accident and injury records tell the safety and environmental manager about training?
- What should a job's degree of difficulty trigger for training requirements?
- How should training direction be determined?
- What level of documentation should training include?
- What information should be included in the syllabus? Who should receive one?
- What training techniques are most effective?
- How should training effectiveness be determined?
- Aside from regulation requirements, what other functions can recordkeeping on training provide?

REFERENCES AND RECOMMENDED READING

Hoover, R.L., R.L. Hancock, K.L. Hylton, O.B. Dickerson, and G.E. Harris. *Health, Safety, and Environmental Control.* New York: Van Nostrand Reinhold, 1989.

LaBar, G. "Worker Training: An Investment in Safety," *Occupational Hazards* August 1991: 25.

Laing, P.M., ed. *Supervisor's Training Manual,* 7th ed. Chicago, 1991: 35.

Occupational Safety and Health Administration. *All About OSHA.* U.S. Department of Labor, (OSHA 2056).

Parachin, V. "10 Tips for Powerful Presentations." *Training* July/August 1990: 71–83.

Spellman, F. R. *Safe Work Practices for Wastewater Treatment Plants.* Lancaster, PA: Technomic Publishing Company, 1996.

Index

accident investigation:
  process for investigating accidents, 55;
  realities of investigations, 60;
  references and recommended reading,
    62;
  safety professionals and investigations,
    60;
  supervisor's first report of accident, 57;
  thought-provoking questions, 61;
  what an accident investigation is, 53;
  what an accident is, 53
administrative and engineering controls:
  hearing safety, 218
air pollution:
  workplace environmental concerns, 283,
    288
air quality:
  workplace environmental concerns, 278
arc welding:
  fire and hot work safety, 121
asbestos:
  workplace environmental concerns, 285
assigning managers:
  certified professionals, 13;
  guidelines, 10;
  hiring managers, 11
atmospheric testing:
  confined space safety, 172
audiometric testing:
  hearing safety, 215
audit items:

hazard communication and hazardous
    waste, 77

barriers and shields:
  confined space safety, 149
basic and applied sciences:
  qualifications of managers, 15
behavior-based models:
  managing aspects, 28
benchmarking:
  managing aspects, 31
biological contaminants:
  workplace environmental concerns, 284
bloodborne pathogens:
  workplace environmental concerns, 330

CERCLA:
  hazard communication and hazardous
    waste, 86
certified safety professionals:
  educational qualifications, 13
changes in the workplace:
  legal ramifications, 8;
  regulatory influence, 8
chemical engineering exposure:
  setting the stage, 7
chemicals:
  EPA's lists of hazardous wastes, 83;
  hazardous chemicals, 81;
  OSHA standards, 86;
  toxic chemicals, 81;

what a hazardous waste is, 83;
where hazardous wastes come
    from, 84;
why we are concerned about hazardous
    wastes, 84;
workplace environmental concerns, 259
civil engineering exposure:
setting the stage, 6
classification of hazards:
hazard communication and hazardous
    waste, 68
clothing and jewelry:
machine guarding, 249
cold hazards:
workplace environmental concerns, 325
combustion by-products:
workplace environmental concerns, 284
common indoor air pollutants in the home:
workplace environmental concerns, 283
communication:
hazard communication and hazardous
    waste, 63
computer systems:
data protection, 102
confined space safety:
alternative protection methods, 169;
atmospheric testing, 172;
barriers and shields, 149;
certification exam, 156;
definitions, 141;
duties of attendants, 166;
duties of authorized entrants, 165;
duties of entry supervisors, 167;
entry program, 140;
evaluating the workplace, 144;
ingress and egress equipment, 149;
ladders, 149;
lighting, 148;
miscellaneous equipment, 151;
occupational safety and health
    administration, 140;
permits, 146, 153, 170;
personal protective equipment, 147;
pre-entry requirements, 151;
references and recommended reading,
    176;
rescue equipment, 150;
rescue operations, 167, 169;
system for permits, 153;
testing and monitoring equipment, 147;

thought-provoking questions, 175;
tragedy in 1993, 137;
training, 154;
ventilating equipment, 147;
written program for permits, 146
contractors:
lockout/tagout, 135
controlling:
managing aspects, 27
cutting safety:
fire and hot work safety, 123

data protection:
emergency response and workplace
    security, 102
definitions:
confined space safety, 141;
fire and hot work safety, 113;
hazard communication and hazardous
    waste, 75, 80, 81;
lockout/tagout, 129;
machine guarding, 239;
respiratory protection, 189, 193;
terminology, 38
directing:
managing aspects, 27
duties of managers:
accident investigation, 51, 60;
examples of duties, 19;
setting the stage, 19

economics and management:
setting the stage, 21
education:
certified professionals, 13;
recommended programs for managers, 9;
safety and environmental managers, 4;
three Es paradigm, 18
electrical engineering exposure:
setting the stage, 7
electrical safety:
control of hazards, 223;
design of electrical systems, 222;
grounding and bonding, 227;
hazards involved, 221;
references and recommended reading,
    229;
thought-provoking questions, 228
elevated work policies:
workplace environmental concerns, 320

emergency response and workplace security:
  analysis of workplace security, 99;
  buildings, workstations, and areas, 100;
  considerations for workplace safety, 97;
  control and prevention, 99;
  data protection, 102
  equipment protection, 102;
  familiarity with people you work with, 101;
  hazard assessment, 99;
  judging by appearances, 98;
  OSHA and emergency response, 93;
  plans for emergency response, 93;
  prevent random acts of violence, 101;
  protection from harm, 103;
  references and recommended reading, 104;
  safety for all concerned, 103;
  security equipment, 100;
  survey of workplace, 99;
  theft protection, 101;
  thought-provoking questions, 104;
  threat assessment teams, 98;
  work practice controls and procedures, 100
employing managers:
  certified professionals, 13;
  guidelines, 10;
  hiring managers, 11
enforcement:
  three Es paradigm, 19
engineering:
  three Es paradigm, 17
engineering controls:
  workplace environmental concerns, 263
environmental aspects:
  qualifications of managers, 16
environmental concerns:
  workplace industrial hygiene, 254
environmental controls:
  workplace environmental concerns, 262
environmental engineering exposure:
  setting the stage, 7
Environmental Protection Agency (EPA):
  lists of hazardous wastes, 83
equipment protection:
  emergency response and workplace security, 102
ergonomics:
  elements of ergonomics program, 233;

employee participation, 234;
hazard prevention and control, 234;
management commitment, 234;
medical management, 234;
occupational safety and health administration, 232;
program evaluation, 234;
references and recommended reading, 235;
thought-provoking questions, 23, 5
training, 235;
workplace environmental concerns, 331
exams:
  confined space safety, 156
explosions:
  emergency response plans, 93
extremely hazardous substances:
  hazard communication and hazardous waste, 81

facilities and equipment design:
  qualifications of managers, 16
fall protection:
  workplace environmental concerns, 309, 319
fines:
  payment of fines, 9
fire and hot work safety:
  arc welding, 121;
  cutting safety, 123;
  definitions, 113;
  fire extinguishers, 109;
  fire prevention and control, 108;
  fire watch requirements, 114;
  industrial facilities, 106;
  miscellaneous measures, 110;
  occupational safety and health administration, 106;
  permit programs, 112;
  permits, 110, 113;
  references and recommended reading, 124;
  safe work practices, 113;
  terminology, 113;
  thought-provoking questions, 123;
  welding safety programs, 114
fire prevention and protection:
  emergency response plans, 93;
  qualifications of managers, 15
floods and flooding:

emergency response plans, 93
formaldehyde exposure:
  workplace environmental concerns, 302

generalist nature of managers:
  setting the stage, 4
Globally Harmonized System:
  benefits of HazCom with GHS, 65
grounding and bonding:
  electrical safety, 227

hazard communication and hazardous waste:
  audit items, 77;
  benefits of HazCom with GHS, 65;
  better communication for worker safety
    and health, 65;
  CERCLA, 86;
  classification of hazards, 68;
  definitions, 75, 80, 81;
  EPA's lists of hazardous wastes, 83;
  extremely hazardous substances, 81;
  failure to communicate, 63;
  handling of hazardous waste, 78;
  hazardous chemicals, 81;
  hazardous materials, 80;
  hazardous substances, 81;
  hazardous waste legislation, 85;
  hazardous wastes, 81;
  label changes under the revised HCS, 68;
  major changes to the HazCom standard,
    67;
  occupational safety and health
    professionals, 72;
  OSHA standards, 86;
  phase-in period for the HazCom standard,
    66;
  pictograms, 69;
  program management and evaluation, 88;
  RCRA's definition of a hazardous
    substance, 81;
  references and recommended reading, 90;
  resource conservation and recovery act,
    85;
  SDS changes under the revised HCS, 70;
  terminology, 75;
  thought-provoking questions, 89;
  throwaway nature of society, 79;
  toxic chemicals, 81;
  what a hazardous waste is, 83;

where hazardous wastes come from, 84;
  why we are concerned about hazardous
    wastes, 84
hearing safety:
  administrative and engineering controls,
    218;
  audiometric testing, 215;
  designation of audiometric evaluation
    procedures, 215;
  designation of hearing protection devices,
    216;
  hearing conservation program, 210, 213;
  hearing protection, 216;
  occupational noise exposure, 209;
  occupational safety and health
    administration, 208;
  prevention of noise-induced hearing loss,
    207;
  record keeping, 217;
  references and recommended reading,
    219;
  safe work practices, 217;
  sound level survey, 214;
  thought-provoking questions, 218;
  training, 216
heat stress:
  workplace environmental concerns, 321
hiring managers:
  setting the stage, 11
household products:
  workplace environmental concerns, 285

indoor air quality:
  workplace environmental concerns, 278
industrial engineering exposure:
  setting the stage, 6
industrial facilities:
  fire and hot work safety, 106
industrial hygiene:
  workplace environmental concerns, 254
industrial relations:
  responsibilities of managers, 21
ingress and egress equipment:
  confined space safety, 149
inspections:
  lockout/tagout, 134;
  respiratory protection, 196
investigations:
  accident investigation, 51

labels:
  changes in labeling under the revised
      HCS, 68;
  pictograms, 69;
  SDS changes under the revised HCS, 70;
  workplace environmental concerns, 299
ladders:
  confined space safety, 149
lead exposure:
  workplace environmental concerns, 303
legal ramifications:
  changes in the workplace, 8
Legionnaire's disease:
  workplace environmental concerns, 279
legislation:
  CERCLA, 86;
  hazardous waste legislation, 85;
  resource conservation and recovery act, 85
lighting:
  confined space safety, 148
lockout/tagout:
  application and compliance, 131;
  authorized employees, 131;
  contractors, 135;
  definitions, 129;
  energy analysis, 132;
  machine guarding, 250;
  methods of informing outside contractors
      of procedures, 135;
  periodic inspections, 134;
  procedures, 133;
  references and recommended reading, 136;
  removal of lockout/tagout in absence of
      employee, 135;
  sample program, 131;
  special conditions, 135;
  thought-provoking questions, 136;
  training, 134;
  use of locks, 132;
  use of tags, 131

machine guarding:
  checklist for machine guarding, 247;
  clothing and jewelry, 249;
  common safeguarding methods, 240;
  definitions, 239;
  devices used to safeguard machinery,
      241;

  feeding and ejection methods, 244;
  location/distance safeguards, 244;
  lockout/tagout, 250;
  mechanical hazards, 237;
  miscellaneous safeguarding accessories,
      244;
  motions and actions, 240;
  references and recommended reading,
      251;
  safeguarding defined, 239;
  safe work practices, 245;
  terminology, 239;
  thought-provoking questions, 250;
  training, 246;
  types of machine safeguards, 239;
  warnings, 249
major accidents and disasters:
  accident investigation, 51;
  advances in technology, 6;
  hazardous waste legislation, 85
managing aspects:
  behavior-based models, 28;
  benchmarking, 31;
  controlling, 27;
  directing, 27;
  organizing, 26;
  planning, 26;
  references and recommended reading, 35;
  right way to manage, 26;
  thought-provoking questions, 34;
  total quality management paradigm, 33
mechanical engineering exposure:
  setting the stage, 7
medical emergencies:
  emergency response plans, 93
medical exams:
  workplace environmental concerns, 300
medical surveillance:
  respiratory protection, 198
mold control:
  workplace environmental concerns, 306,
      308

noise exposure:
  hearing safety, 207;
  workplace environmental concerns, 269

occupational safety and health
  administration (OSHA):

confined space safety, 140;
emergency response and workplace
    security, 93;
ergonomics, 232;
fire and hot work safety, 106;
hearing safety, 208;
personal protective equipment, 181, 182;
respiratory protection, 189;
standards for hazardous wastes, 86;
workplace environmental concerns, 261
occupational safety and health professionals:
hazard communication and hazardous
    waste, 72
office safety:
workplace environmental concerns, 325,
    329
organizing:
managing aspects, 26

payment of fines:
setting the stage, 9
permits:
confined space safety, 146, 153, 170;
fire and hot work safety, 110, 113
personal protective equipment:
case study 10.1, 180;
confined space safety, 147;
hazard assessment, 183;
occupational safety and health
    administration, 181, 182;
references and recommended reading,
    188;
sample training guide, 184;
thought-provoking questions, 187;
training, 183;
workplace environmental concerns, 277
pesticides:
workplace environmental concerns, 285
pictograms:
hazard communication and hazardous
    waste, 69
planning:
emergency response plans, 93
managing aspects, 26
precepts of management:
setting the stage, 1
preventing accidents and disasters:
primary responsibility of managers, 7
profiling:
judging by appearances, 98

program management and evaluation:
hazard communication and hazardous
    waste, 88;
qualifications of managers, 15
protective clothing:
workplace environmental concerns, 299

qualifications of managers:
educational qualifications, 13;
setting the stage, 3;
thought-provoking questions, 22

radon:
workplace environmental concerns, 283
recycling:
throwaway nature of society, 79
regulations:
hazardous waste legislation, 85
regulatory influence:
changes in the workplace, 8;
setting the stage, 8
reports:
supervisor's first report of accident, 57
rescue equipment:
confined space safety, 150
rescue operations:
confined space safety, 167, 169
resource conservation and recovery act:
hazard communication and hazardous
    waste, 85
respiratory protection:
definitions, 189, 193;
documentation procedures, 200;
fit-testing of respirators, 197;
inspection of respirators, 196;
maintenance, cleaning, and storage of
    respirators, 196;
medical surveillance, 198;
occupational safety and health
    administration, 189;
program administration, 202;
program evaluation, 201;
program operation, 202;
references and recommended reading,
    205;
responsibilities of department directors,
    192;
safe use of SCBA/supplied air respirators,
    200;
sample evaluation checklist, 202;

sample written protection program, 191;
selection and distribution of respirators,
    196;
terminology, 189, 193;
thought-provoking questions, 204;
training, 199;
types of respirators, 195;
workplace environmental concerns, 299
right way to manage:
managing aspects, 26

safety and health training:
identifying training needs, 340;
importance of training, 339;
recordkeeping, 343;
references and recommended reading,
    344;
thought-provoking questions, 344;
written training programs, 340
safety data sheets:
changes under the revised HCS, 70
scaffold safety:
workplace environmental concerns, 315
security equipment:
emergency response and workplace
    security, 100
setting the stage:
assigning safety and health persons, 10;
certified safety professionals, 13;
chemical engineering exposure, 7;
civil engineering exposure, 6;
a closer look at management, 3;
duties of managers, 19;
economics and management, 21;
education and the manager, 4;
electrical engineering exposure, 7;
environmental engineering exposure, 7;
hiring managers, 11;
industrial engineering exposure, 6;
legal ramifications, 8;
major accidents and disasters, 6;
mechanical engineering exposure, 7;
payment of fines, 9;
qualifications of managers, 3;
references and recommended reading, 22;
regulatory influence, 8;
safety and environmental management
    precepts, 1;
the "S" in safety, 20;
thought-provoking questions, 22;

three Es paradigm, 17;
training programs, 9;
what a safety and environmental manage
    is, 2;
why a manager should be a generalist, 4
sick building syndrome:
workplace environmental concerns, 281
silica exposure:
workplace environmental concerns, 300
sound level survey:
hearing safety, 214
spills:
emergency response plans, 93
statistics:
responsibilities of managers, 21
stressors:
workplace environmental concerns, 257
system safety and product safety:
qualifications of managers, 16

terminology:
confined space safety, 141;
fire and hot work safety, 113;
hazard communication and hazardous
    waste, 75, 80, 81;
lockout/tagout, 129;
machine guarding, 239;
references and recommended reading, 50;
respiratory protection, 189, 193;
technical terms, 38;
thought-provoking questions, 50
testing and monitoring equipment:
confined space safety, 147
tests:
confined space safety, 156
theft protection:
emergency response and workplace
    security, 101
thermal stress:
workplace environmental concerns, 321
threat assessment teams:
emergency response and workplace
    security, 98
three Es paradigm:
education, 18;
enforcement, 19;
engineering, 17
throwaway nature of society:
hazard communication and hazardous
    waste, 79

tobacco smoke:
  workplace environmental concerns, 284
Total Quality Management (TQM) paradigm:
  managing aspects, 33
toxic chemicals:
  EPA's lists of hazardous wastes, 83;
  hazard communication and hazardous
      waste, 81;
  OSHA standards, 86;
  what a hazardous waste is, 83;
  where hazardous wastes come from, 84;
  why we are concerned about hazardous
      wastes, 84;
  workplace environmental concerns, 259;
training:
  confined space safety, 154;
  ergonomics, 235;
  hearing safety, 216;
  lockout/tagout, 134;
  machine guarding, 246;
  personal protective equipment, 183;
  recommended programs for managers, 9;
  respiratory protection, 199;
  responsibilities of managers, 21;
  safety and health training, 339;
  scaffold safety, 319;
  workplace environmental concerns, 299

ventilating equipment:
  confined space safety, 147
ventilation:
  workplace environmental concerns, 263

warnings:
  machine guarding, 249
waste:
  EPA's lists of hazardous wastes, 83;
  extremely hazardous substances, 81;
  handling of hazardous waste, 78;
  hazard communication and hazardous
      waste, 63;
  hazardous chemicals, 81;
  hazardous materials, 80;
  hazardous substances, 81;
  hazardous wastes, 81;
  OSHA standards, 86;
  toxic chemicals, 81;
  what a hazardous waste is, 83;
  where hazardous wastes come from, 84;

  why we are concerned about hazardous
      wastes, 84;
welding safety programs:
  fire and hot work safety, 114
what a manager is:
  setting the stage, 2
wind chill hazards:
  workplace environmental concerns,
      325
workplace environmental concerns:
  administrative controls, 274;
  areas of concern, 258;
  asbestos, 285;
  biological contaminants, 284;
  bloodborne pathogens, 330;
  building factors affecting indoor air
      quality, 287;
  chemicals, 259;
  cold hazards, 325;
  combustion by-products, 284;
  common airflow pathways, 296;
  common indoor air pollutants in the
      home, 283;
  competent person designation, 299;
  elevated work policies, 320;
  engineering controls, 263;
  environmental controls, 262;
  ergonomics, 331;
  fall protection, 309, 319;
  formaldehyde exposure, 302;
  health hazards, 261;
  heat stress, 321;
  household products, 285;
  hygiene facilities and practices, 300;
  importance of IAQ, 285;
  indoor air contaminant transport, 291;
  indoor air pollution, 282;
  indoor air quality, 278;
  industrial hygiene, 254;
  industrial toxicology, 259;
  labels, 299;
  lead exposure, 303;
  Legionnaire's disease, 279;
  major IAQ contaminants, 296;
  medical exams, 300;
  methods of compliance, 299;
  mold control, 306, 308;
  monitoring exposure, 298;
  noise exposure, 269;

occupational safety and health administration, 261;
office safety, 325, 329;
permissible exposure limits, 298;
personal protective equipment, 277;
pesticides, 285;
protective clothing, 299;
radon, 283;
recordkeeping, 300;
references and recommended reading, 334;
regulated areas, 299;
respirators, 299;
safety outside the office, 331;
scaffold safety, 315;
sick building syndrome, 281;
silica exposure, 300;
sources of workplace air pollutants, 288;
stressors, 257;
symptoms associated with poor air quality, 286;
thermal stress, 321;
thought-provoking questions, 331;
tobacco smoke, 284;
toxic chemicals, 259;
training, 299;

types of workplace air pollutants, 288;
ventilation, 263;
vibration control, 274;
wind chill hazards, 325;
workplace security:
analysis of workplace security, 99;
buildings, workstations, and areas, 100;
considerations for workplace safety, 97;
control and prevention, 99;
emergency response and workplace security, 93;
equipment and data protection, 102;
familiarity with people you work with, 101;
hazard control and prevention, 99;
prevent random acts of violence, 101;
protection from harm, 103;
references and recommended reading, 104;
safety for all concerned, 103;
security equipment, 100;
survey of workplace, 99;
theft protection, 101;
thought-provoking questions, 104;
work practice controls and procedures, 100